From Continental Shelf to Slope: Mapping the Oceanic Realm

IUGS/GSL publishing agreement

This volume is published under an agreement between the International Union of Geological Sciences and the Geological Society of London and arises from the IUGS Resourcing Future Generations Programme.

GSL is the publisher of choice for books related to IUGS activities, and the IUGS receives a fee for all books published under this agreement.

Books published under this agreement are subject to the Society's standard rigorous proposal and manuscript review procedures.

Submitting a book proposal
More information about submitting a proposal and producing a book for the Society can be found at https://www.geolsoc.org.uk/proposals.

It is recommended that reference to all or part of this book should be made in one of the following ways:

Asch, K., Kitazato, H. and Vallius, H. (eds) 2022. *From Continental Shelf to Slope: Mapping the Oceanic Realm*. Geological Society, London, Special Publications, **505**, https://doi.org/10.1144/SP505

Battaglini, L., D'Angelo, S. and Fiorentino, A. 2022. Collating European data on geological events in submerged areas: examples of correlation and interpretation from Italian seas. *Geological Society, London, Special Publications*, **505**, 155–167, https://doi.org/10.1144/SP505-2019-96

Geological Society Special Publication No. 505

From Continental Shelf to Slope: Mapping the Oceanic Realm

Edited by

K. Asch

Bundesanstalt für Geowissenschaften und Rohstoffe, Germany

H. Kitazato

Tokyo University of Marine Science and Technology, Japan

and

H. Vallius

Geologian Tutkimuskeskus, Finland

2022
Published by
The Geological Society
London

The Geological Society of London

The Geological Society of London is a not-for-profit organization, and a registered charity (no. 210161). Our aims are to improve knowledge and understanding of the Earth, to promote Earth science education and awareness, and to promote professional excellence and ethical standards in the work of Earth scientists, for the public good. Founded in 1807, we are the oldest geological society in the world. Today, we are a world-leading communicator of Earth science – through scholarly publishing, library and information services, cutting-edge scientific conferences, education activities and outreach to the general public. We also provide impartial scientific information and evidence to support policy-making and public debate about the challenges facing humanity. For more about the Society, please go to https://www.geolsoc.org.uk/

The Geological Society Publishing House (Bath, UK) produces the Society's international journals and books, and acts as European distributor for selected publications of the American Association of Petroleum Geologists (AAPG), the Geological Society of America (GSA), the Society for Sedimentary Geology (SEPM) and the Geologists' Association (GA). GSL Fellows may purchase these societies' publications at a discount. The Society's online bookshop is at https://www.geolsoc.org.uk/bookshop

To find out about joining the Society and benefiting from substantial discounts on publications of GSL and other Societies go to https://www.geolsoc.org.uk/membership or contact the Fellowship Department at: The Geological Society, Burlington House, Piccadilly, London W1J 0BG: Tel. +44 (0)20 7434 9944; Fax +44 (0)20 7439 8975; E-mail: enquiries@geolsoc.org.uk

For information about the Society's meetings, go to https://www.geolsoc.org.uk/events. To find out more about the Society's Corporate Patrons Scheme visit https://www.geolsoc.org.uk/patrons

Proposing a book
If you are interested in proposing a book then please visit: https://www.geolsoc.org.uk/proposals

Published by The Geological Society from:
The Geological Society Publishing House, Unit 7, Brassmill Enterprise Centre, Brassmill Lane, Bath BA1 3JN, UK

The Lyell Collection: www.lyellcollection.org
Online bookshop: www.geolsoc.org.uk/bookshop
Orders: Tel. +44 (0)1225 445046, Fax +44 (0)1225 442836

The publishers make no representation, express or implied, with regard to the accuracy of the information contained in this book and cannot accept any legal responsibility for any errors or omissions that may be made.

Full information on the Society's permissions policy can be found at https://www.geolsoc.org.uk/permissions

British Library Cataloguing in Publication Data

A catalogue record for this book is available from the British Library.
ISBN 978-1-78620-495-0
ISSN 0305-8719

Distributors
For details of international agents and distributors see:
https://www.geolsoc.org.uk/agentsdistributors

Typeset by Nova Techset Private Limited, Bengaluru & Chennai, India
Printed and bound by CPI Group (UK) Ltd, Croydon CR0 4YY, UK

Acknowledgements

This volume was supported by the EMODnet Geology community and project (EASME/EMFF/2018/1.3.1.8/Lot1/SI2.811048). We are indebted to GSJ-AIST and their marine projects for their contributions to this special publication.

Contents

Contents

From Continental Shelf to Slope: Mapping the Oceanic Realm: Introduction

Kristine Asch[1]*, Hiroshi Kitazato[2] and Henry Vallius[3]

[1]Bundesanstalt für Geowissenschaften und Rohstoffe (BGR), Stilleweg 2, 30655 Hannover, Germany

[2]School of Marine Resources and Environment, Tokyo University of Marine Science and Technology, 4-5-7 Konan, Minato-ku, Tokyo 108-8477, Japan

[3]Geologian Tutkimuskeskus, Vuorimiehentie 5, 02150 Espoo, Finland

KA, 0000-0002-7423-7305

*Correspondence: Kristine.Asch@bgr.de;Keca_Asch@yahoo.de

Abstract: This volume covers multidisciplinary Research and Development contributions on geology, geophysics, bathymetric and biologic aspects, towards data sampling and acquisition, data analysis and its results, and innovative ways of data access. It also presents the development of processes to map, harmonize and integrate marine data across EEZ boundaries, an impressive example of which is the European EMODnet (European Marine Observation and Data network) initiative. EMODnet assembles scattered and partially hidden marine data into continentally harmonized geospatial data products for public benefit and increasingly within overseas collaboration. The volume also aims to shed light on an evaluation of biological and mineral resources and environmental assessments at continental shelf to slope depths. Western Pacific examples provide excellent case studies for this topic.

During recent decades, the ocean floor and the marine environment in general have been discovered to contain a wealth of mineral, biological and energy resources. After all, the oceans cover over 70% of Earth and provide about 99% of the living space on the planet.

The ocean basins contain hidden mineral resources, such as manganese nodules, but also industrial minerals such as sand and gravel, and energy resources, particularly oil and gas. In addition, the seas provide a major contribution to renewable resources, since seafloors provide the foundation for the siting and construction of offshore windfarms. Furthermore, the seas represent the habitats for numerous species essential for food sources, biological components that are essential to the environment and biomedical molecules for human health. Like the continents, the oceans encompass fascinating geological and geomorphological features (see Fig. 1, general map of the global offshore bedrock, Asch et al. 2022), which include a wide range of environments for human activities, including fishing, tourism (in particular in coastal areas), ship transportation, construction of submarine infrastructure (including pipelines and cables), and CO_2 sequestration. On the other hand, off-shore environments are a source of natural hazards (e.g. storms, tsunamis, submarine volcanism, submarine landslides and others). The usage opportunities are therefore manifold, from many interest groups, and conflicts between human activities and protection of vulnerable life in the oceanic ecosystems, in particular ecological niches and habitats, are foreseeable and a challenge for the future.

In 2021 the United Nations launched the 'Decade of Ocean Science for Sustainable Development (2021–30)' which emphasized that 'scientific understanding of the ocean's responses to pressures and management action is fundamental for sustainable development. Ocean observations and research are also essential to predict the consequences of change, design mitigation and guide adaptation' (United Nations 2021a). The UN points out that 'the ocean holds the keys to an equitable and sustainable planet.' and that innovative ocean-science solutions need to be unlocked (United Nations 2021b).

However, in this context, it must be stated that 'even with all the technology that we have today – satellites, buoys, underwater vehicles and ship tracks – we have better maps of the surface of Mars and the Moon than we do the bottom of the ocean. We know very, very little about most of the ocean' (Feldmann 2009). So, the ocean floor is still one of the few remaining unexplored regions of the Earth and as such should be a major focus for geoscientific research and mapping.

To understand the sea floor and its complex interdependencies and relationships to the geosphere, biosphere, hydrosphere and atmosphere, it is crucial to observe and map the ocean basins, and to analyse and interpret the mapping results. For the successful achievement of this work, it is critical for scientists to

From: Asch, K., Kitazato, H. and Vallius, H. (eds) 2022. *From Continental Shelf to Slope: Mapping the Oceanic Realm.* Geological Society, London, Special Publications, **505**, 1–5.
First published online July 5, 2022, https://doi.org/10.1144/SP505-2022-96

General Global Map of Seafloor Bedrock Geology

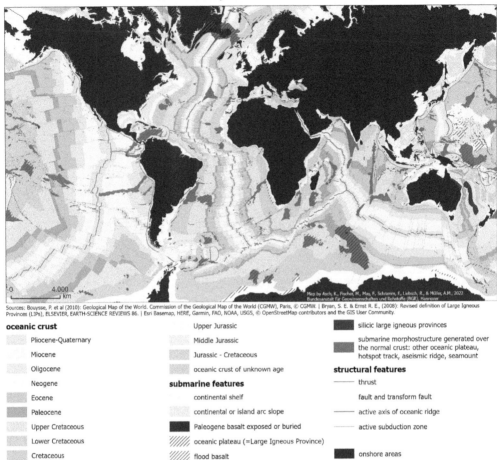

Sources: Bouysse, P. et al (2010): Geological Map of the World. Commission of the Geological Map of the World (CGMW), Paris, © CGMW. | Bryan, S. E. & Ernst R. E., (2008): Revised definition of Large Igneous Provinces (LIPs), ELSEVIER, EARTH-SCIENCE REVIEWS 86. | Esri Basemap, HERE, Garmin, FAO, NOAA, USGS, © OpenStreetMap contributors and the GIS User Community.

oceanic crust

- Pliocene-Quaternary
- Miocene
- Oligocene
- Neogene
- Eocene
- Paleocene
- Upper Cretaceous
- Lower Cretaceous
- Cretaceous

Upper Jurassic

Middle Jurassic

Jurassic - Cretaceous

oceanic crust of unknown age

submarine features

- continental shelf
- continental or island arc slope
- Paleogene basalt exposed or buried
- oceanic plateau (=Large Igneous Province)
- flood basalt

silicic large igneous provinces

submarine morphostructure generated over the normal crust: other oceanic plateau, hotspot track, aseismic ridge, seamount

structural features

- ——— thrust
- ——— fault and transform fault
- ——— active axis of oceanic ridge
- ——— active subduction zone
- onshore areas

Fig. 1. General global map of seafloor bedrock geology. The red lines show the axis of the seafloor spreading. The coloured polygons, coloured in yellow, orange, green, grey and blue hues, indicate the age of the seafloor geological units. The lighter blue, grey and green hues show the off-shore slope, margin and shelf regions, the stronger red hues and hatches Paleogene basalts, flood basalts, large igneous provinces and submarine morphostructures. Generally, almost the entire seafloor is covered by young sediments; these are not presented on the map. Map by Asch, K., Fischer, M., May, F., Schramm, F., Liebsch, R. and Müller, A.M. 2022. Bundesanstalt für Geowissenschaften und Rohstoffe (BGR), Hannover, based on extracts from Bouysse *et al.* (2010) © CGMW, and Bryan and Ernst (2008) using the Esri Basemap, HERE, Garmin, FAO, NOAA, USGS, © OpenStreetMap contributors and the GIS User Community.

co-operate across scientific disciplines engaged in ocean-floor research. This should include geologists, geophysicists, chemists, biologists, oceanographers, and others, in order to attempt to solve the challenges of the conflicting applications. To this end, this special issue is mainly focused on the results of geological, geophysical and geobiological mapping campaigns and projects that have collated and harmonised their results.

Sea-floor mapping confronts special challenges. Since marine mapping campaigns are expensive;

the costs of operating a research vessel at sea, technical equipment, and scientific and technical staff are considerable. Thus, the data acquisition follows the principle: 'collect once, use many times'. The results not only represent the oceans' current state, but also provide the basis for future investigations and applications, as well as evidence of past events.

Most of the marine domain is investigated in much less detail than land areas and there is often only geophysical data to provide information on the marine geological units and structures.

Knowledge is thus heterogeneous: it can be comparatively detailed in small, well-investigated areas while surrounding zones are only patchily investigated. In addition, portrayal rules and classification systems show similar variations to those used for mapping onshore geology (**Breuer and Asch 2021**).

This volume includes multidisciplinary Research and Development contributions from European, Asian and North American seas on the mapping of the seafloor in the disciplines of geology, geography, geophysics, geochemistry, and biology. It includes aspects of data sampling and acquisition, data analysis and their results, and innovative means of data representation. In addition, it also presents the development of mapping, harmonisation and integration of marine information across boundaries of the Exclusive Economic Zones (EEZ). An impressive example of this is the European EMODnet (European Marine Observation and Data network) initiative that was established in 2007 by the European Commission (EC). EMODnet Geology (see contribution by **Vallius** et al. **2020**) assembles scattered and partially hidden marine data into continentally harmonised geospatial data products for public benefit. Since its third phase, overseas collaboration has increasingly become part of the project. Mapping results from the Baltic Sea, North Sea, Irish Sea, Norwegian Sea, Black Sea, Italian seas, and the Portuguese coastal areas have revealed secrets of the marine geological features.

Greene and Barrie (2020) show remarkable mapping of upper plate deformation of the Cascadia subduction zone by geophysical methods and piston coring in the waters around the St. Juan Archipelago near Vancouver. The papers from the EMODnet Geology project outline the collection, harmonisation and public release of available marine geological information. In this book, several regional studies on submarine geological themes are presented: the post-glacial development and Holocene sedimentation processes in the Gulf of Finland (**Ryabchuk** et al. **2020**), the first geomorphological synthesis of the German coastal areas (**Breuer and Asch 2021**), Pliocene deposits of the Black Sea (**Gozhik and Rokitsky 2020**), geological events in the Italian seas (**Battaglini** et al. **2021**), together with economic aspects, including the exploitation of sand deposits off-Iberia (**Mil-Homens** et al. **2020**). Larger analyses of the seafloor off Ireland, Iberia and Norway (**Bøe** et al. **2020**; **O'Toole** et al. **2020**; **Terrinha** et al. **2020**) show the usefulness of a large set of scalable maps available for all to use. A result of the EMODnet project is that these data are no longer hidden in different databases of research institutes or other data holders, but instead become available to all parties interested in the world of marine geology.

The volume also aims to shed light on an evaluation of biological/mineral resources and environmental disaster risk assessments at continental shelf to slope depths of the western Pacific. Precise topographic mapping of the sea floor subsurface using seismic reflection systems provides evidence of geological structures and the superficial distribution of sediments. These data provide fundamental evidence for understanding the characteristics of submarine geology and sedimentary dynamics of coastal areas for evaluating the potential of coastal to slope subsurface geology as mineral/biological resources. The Marine Geology Research Group of the Geological Survey of Japan show clear examples for these data in coastal ocean zones (**Arai 2020**; **Furuyama** et al. **2020**; **Ikehara** et al. **2020**; **Ohta** et al. **2020**; **Sato** et al. **2020**). One such example includes geohazards in the coastal regions, including mega-scale earthquakes and tsunamis that occurred in NE Japan on 11 March, 2011. This inflicted strong damage to coastal inhabitants and economic activities, such as the fishing industry. This leads to the question of how organizations can transfer information to people who have suffered damage of their property or health through disastrous natural events. **Oki** et al. **(2021)** publish a vision paper here focused on how scientists should work together with the general population and local governments. Multidisciplinary approaches provide us with wide varieties of data. The integration of various information captured in geographic information systems (GIS), such as 'habitat mapping', result in composite datasets that are able to aid scientists, the public, government officers and commercial company staff. These datasets can be used as strategic maps to offer a better understanding, together with a better use of sea-floor characteristics of coastal and continental slope seas. The approaches described here certainly facilitate the transfer scientific results from scientists to other stakeholders. However, these procedures should be seen as a first step on the road of 'Science for Society'. In this context, Oki et al. (2021) show two examples of best practice in geological communities.

This volume is complemented by the Quarterly Journal of Engineering Geology and Hydrogeology (QJEG), 54, and here the article by Moses and Vallius (2021) presents various examples of marine geology and topography mapping in the European Seas within the EMODnet Geology project.

Finally, we once again stress how ocean realms both within and beyond national jurisdictions are still one of the last frontiers on the Earth. Global communities should increase working together for sustainable use and protection of the treasures. For this purpose, we need strategic discussion platforms among multiple stakeholder communities to be built and supported.

Acknowledgement The continuous support by the team of the Geological Society of London Publishing House, in particular Bethan Phillips, Rachael Kriefman, Tamzin Anderson, Phil Leat, Caroline Astley and Jo Armstrong, had made it a pleasure to edit this special issue.

We would like to thank the authors of this volume for their motivation to take part in the endeavour to assemble this volume and for their excellent contributions.

Last but not least, we would like to thank the reviewers for their effort, patience and thorough reviews to improve the quality of this volume.

Competing interests The authors declare that they have no known competing financial interests or personal relationships that could have appeared to influence the work reported in this paper.

Author contributions KA: conceptualization (equal), formal analysis (equal), funding acquisition (supporting), project administration (lead), resources (lead), supervision (lead), validation (lead), visualization (lead), writing – original draft (lead), writing – review & editing (lead); **HK:** conceptualization (equal), data curation (supporting), project administration (supporting), writing – review & editing (supporting); **HV:** conceptualization (supporting), funding acquisition (lead), project administration (supporting), supervision (supporting), writing – review & editing (supporting).

Funding This research received no specific grant from any funding agency in the public, commercial, or not-for-profit sectors.

Data availability The datasets generated during and/or analysed during the current study are available at, https://services.bgr.de/geologie/seafloorgeol.

References

Arai, K. 2020. Geological mapping of coastal and offshore Japan (by GSJ-AIST): collecting and utilizing the geological information. *Geological Society, London, Special Publications*, **505**, https://doi.org/10.1144/SP505-2019-95

Asch, K., Fischer, M., May, F., Schramm, F., Liebsch, R. and Müller, A.M. 2022. *General Global Map of Seafloor Bedrock Geology.* Bundesanstalt für Geowissenschaften und Rohstoffe (BGR), Hannover; https://services.bgr.de/geologie/seafloorgeol

Battaglini, L., D'Angelo, S. and Fiorentino, A. 2021. Collating European data on geological events in submerged areas: examples of correlation and interpretation from Italian seas. *Geological Society, London, Special Publications*, **505**, https://doi.org/10.1144/SP505-2019-96

Bøe, R., Bjarnadóttir, L.R. *et al.* 2020. Revealing the secrets of Norway's seafloor – geological mapping within the MAREANO programme and in coastal areas. *Geological Society, London, Special Publications*, **505**, https://doi.org/10.1144/SP505-2019-82

Bouysse, P. *et al.* 2010. *Geological Map of the World.* Commission of the Geological Map of the World (CGMW), Paris, © CGMW.

Breuer, S. and Asch, K.C. 2021. A first approach to a Quaternary geomorphological map of the German seas. *Geological Society, London, Special Publications*, **505**, https://doi.org/10.1144/SP505-2021-24

Bryan, S.E. and Ernst, R.E. 2008. Revised definition of Large Igneous Provinces (LIPs), Elsevier, Earth Science Review 86.

Esri Basemap. HERE, Garmin, FAO, NOAA, USGS, © OpenStreetMap contributors and the GIS User Community.

Feldmann, G. 2009. Oceans: the Great unknown. Interview by *Dan Stillman, Institute for Global Environmental Strategies*, https://www.nasa.gov/audience/forstudents/5-8/features/oceans-the-great-unknown-58.html

Furuyama, S., Sato, T., Arai, K. and Ozaki, M. 2020. Tectonic evolution in the early to Middle Pleistocene off the east coast of the Boso Peninsula, Japan. *Geological Society, London, Special Publications*, **505**, https://doi.org/10.1144/SP505-2019-116

Gozhik, P.F. and Rokitsky, V.E. 2020. The Pliocene deposits of the Black Sea Shelf east of the Danube River Delta. *Geological Society, London, Special Publications*, **505**, https://doi.org/10.1144/SP505-2019-102

Greene, H.G. and Barrie, J.V. 2020. Faulting within the San Juan–southern Gulf Islands Archipelagos, upper plate deformation of the Cascadia subduction complex. *Geological Society, London, Special Publications*, **505**, https://doi.org/10.1144/SP505-2019-125

Ikehara, K., Katayama, H., Sagayama, T. and Irino, T. 2020. Geological controls on dispersal and deposition of river flood sediments on the Hidaka shelf, Northern Japan. *Geological Society, London, Special Publications*, **505**, https://doi.org/10.1144/SP505-2019-114

Mil-Homens, M., Brito, P. *et al.* 2020. Integrated geophysical and sedimentological datasets for assessment of offshore borrow areas: the CHIMERA project (western Portuguese Coast). *Geological Society, London, Special Publications*, **505**, https://doi.org/10.1144/SP505-2019-100

Moses, C. and Vallius, H. 2021. Mapping the geology and topography of the European Seas (European Marine Observation and Data Network, EMODnet). *Quarterly Journal of Engineering Geology and Hydrogeology*, **54**, https://doi.org/10.1144/qjegh2020-131

Ohta, A., Imai, N., Tachibana, Y. and Ikehara, K. 2020. Application of spatial distribution patterns of multi-elements in geochemical maps for provenance and transfer process of marine sediments in Kyushu, western Japan. *Geological Society, London, Special Publications*, **505**, https://doi.org/10.1144/SP505-2019-87

Oki, Y., Kitazato, H., Fujii, T. and Yasukawa, S. 2021. Habitat mapping for human well-being: a tool for reducing risk in disaster-prone coastal environments and human communities. *Geological Society, London, Special Publications*, **505**, https://doi.org/10.1144/SP505-2021-26

O'Toole, R., Judge, M. *et al.* 2020. Mapping Ireland's coastal, shelf and deep-water environments using illustrative case studies to highlight the impact of seabed mapping on the generation of blue knowledge. *Geological Society, London, Special Publications*, **505**, https://doi.org/10.1144/SP505-2019-207

Ryabchuk, D., Sergeev, A. *et al.* 2020. High-resolution geological mapping towards an understanding of postglacial development and Holocene sedimentation processes in the eastern Gulf of Finland: an EMODnet Geology case study. *Geological Society, London, Special Publications*, **505**, https://doi.org/10.1144/SP505-2019-127

Sato, T., Furuyama, S., Komatsubara, J., Ozaki, M. and Yamaguchi, K. 2020. Bent incised valley formed in uplifting shelf facing subduction margin: case study off the eastern coast of the Boso Peninsula, central Japan. *Geological Society, London, Special Publications*, **505**, https://doi.org/10.1144/SP505-2019-117

Terrinha, P., Medialdea, T. *et al.* 2020. Integrated thematic geological mapping of the Atlantic Margin of Iberia. *Geological Society, London, Special Publications*, **505**, https://doi.org/10.1144/SP505-2019-90

Vallius, H.T.V., Kotilainen, A.T., Asch, K.C., Fiorentino, A., Judge, M., Stewart, H.A. and Pjetursson, B. 2020. Discovering Europe's seabed geology: the EMODnet concept of uniform collection and harmonization of marine data. *Geological Society, London, Special Publications*, **505**, https://doi.org/10.1144/SP505-2019-208

United Nations 2021*a* https://www.oceandecade.org

United Nations 2021*b* https://en.unesco.org/ocean-decade

Discovering Europe's seabed geology: the EMODnet concept of uniform collection and harmonization of marine data

Henry T. V. Vallius[1]*, Aarno T. Kotilainen[1], Kristine C. Asch[2], Andrea Fiorentino[3], Maria Judge[4], Heather A. Stewart[5] and Bjarni Pjetursson[6]

[1]Geological Survey of Finland (GTK)*, Vuorimiehentie 5, PO Box 96, FI-02151 Espoo, Finland

[2]Bundesanstalt für Geowissenschaften und Rohstoffe (BGR), Geozentrum Hannover, Stilleweg 2, D-30655 Hannover, Germany

[3]Istituto Superiore per la Protezione e la Ricerca Ambientale (ISPRA), via Vitaliano Brancati, 48 - 00144 Rome, Italy

[4]Geological Survey of Ireland (GSI), Beggars Bush, Haddington Road, Dublin D04K7X4, Ireland

[5]British Geological Survey (BGS), Lyell Centre, Research Avenue South, Edinburgh EH14 4AP, UK

[6]Geological Survey of Denmark and Greenland (GEUS), Øster Voldgade 10, DK-1350, Copenhagen, Denmark

KCA, 0000-0002-7423-7305
*Correspondence: henry.vallius@gtk.fi

Abstract: Maritime spatial planning, management of marine resources, environmental assessments and forecasting all require good seabed maps. Similarly there is a need to support the objectives to achieve Good Environmental Status in Europe's seas by 2020, set up by the European Commission's Marine Strategy Framework Directive. Hence the European Commission established the European Marine Observation and Data Network (EMODnet) programme in 2009, which is now in its fourth phase (2019–21). The programme is designed to assemble existing, but fragmented and partly inaccessible, marine data and to create contiguous and publicly available information layers which are interoperable and free of restrictions on use, and which encompass whole marine basins.

The EMODnet Geology project is delivering integrated geological map products that include seabed substrates, sedimentation rates, seafloor geology, Quaternary geology, geomorphology, coastal behaviour, geological events such as submarine landslides and earthquakes, and marine mineral occurrences. Additionally, as a new product during the ongoing and preceding phase of the project, map products on submerged landscapes of the European continental shelf have been compiled at various time frames. All new map products have a resolution of 1:100 000, although finer resolution is presented where the underlying data permit. A multi-scale approach is adopted whenever possible.

Numerous national seabed mapping programmes worldwide have demonstrated the necessity for proper knowledge of the seafloor. Acting on this, the European Commission established the European Marine Observation and Data Network (EMODnet) programme in 2009. The national geological survey organizations of Europe have a strong network of marine geological teams through the Marine Geology Expert Group of the association of European geological surveys (Eurogeosurveys). This network was the foundation of the EMODnet Geology consortium which today consists of the national geological surveys of Finland, the UK, Sweden, Norway, Denmark, Estonia, Latvia, Lithuania, Poland, The Netherlands, Belgium, France, Ireland, Spain, Italy, Slovenia, Croatia, Albania, Greece, Cyprus, Malta, Russia, Germany, Montenegro and Iceland, as well as marine teams of research organizations in Portugal (IPMA), Bulgaria (IO-BAS), Romania (GeoEcoMar), the UK (CEFAS), Greece (HCMR) and Ukraine (PSRGE, replaced in the fourth phase by Institute of Geological Sciences, NAS of Ukraine). The consortium is further strengthened with experts from six universities: Edge Hill University (UK), Sapienza University of Rome (Italy), University of Tartu (Estonia), University of Crete through FORTH-ICS, Institute of Marine Science and Technology of Dokuz Eylul University (Turkey), and EMCOL Research Centre of Istanbul Technical University – altogether, 30 partners and nine subcontractors. The EMODnet Geology programme is now in its fourth phase, which started in September 2019. In addition to geological information, the wider EMODnet programme aims to also bring together information from

From: Asch, K., Kitazato, H. and Vallius, H. (eds) 2022. *From Continental Shelf to Slope: Mapping the Oceanic Realm.* Geological Society, London, Special Publications, **505**, 7–18.
First published online July 23, 2020, https://doi.org/10.1144/SP505-2019-208

European seas on seabed habitats, physical properties, chemistry, biology, human activities and hydrography. This paper describes the EMODnet Geology project and the different end products which were delivered in the end of the third phase and will be further developed during the recent fourth phase of the project.

Project objectives

EMODnet Geology compiles marine geological information held primarily by the project partners with some additional datasets that are publicly available. The project outputs are delivered through the European Geological Data Infrastructure (EGDI) portal (http://www.emodnet-geology.eu/). The consortium delivers 'Operation, development and maintenance of a European Marine Observation and Data Network', collecting and harmonizing datasets on seabed substrates, sedimentation rates, seafloor (bedrock) lithology, seafloor (bedrock) stratigraphy, coastal behaviour (migration direction, rate and volume, resilience), mineral occurrences (e.g. oil and gas, aggregates, metallic minerals), geological events and probabilities (e.g. earthquakes, submarine landslides, volcanic centres). As a new output, included for the third phase of the project, information on the submerged landscapes of the European continental shelves have been compiled at various time frames (e.g. Last Glacial Maximum (LGM) and younger low sea-level stages). The submerged landscapes product includes: (1) shorelines and coastal environments and deposits (such as lagoons, dunes, estuaries and beachrocks); (2) valleys and riverbeds, terraces and associated deposits; (3) river deltas and delta clinoforms; (4) submerged water points (such as submerged springs and freshwater lakes); (5) thickness of Holocene deposits; and (6) flora and fauna on the submerged landscapes. A common classification process will be adopted for all EMODnet Geology data compilations, with the resolution of the map products at a scale of 1:100 000 but finer-resolution layers where the underlying data permit. The regional seas covered by the project are the Baltic Sea, the Barents Sea, the Bay of Biscay, the Celtic Sea, the Greater North Sea, the Iberian Coast, the Norwegian Sea, the White Sea, the North Atlantic Ocean (continental shelves around Iceland, the Faroe Islands and Macaronesia), the Mediterranean Sea (within waters of EU countries), the Black Sea and, from the beginning of the fourth phase, the Caspian Sea (Fig. 1)

Geological data and metadata

During the first two phases of EMODnet Geology (2009–12 and 2013–16), the project identified relevant data held by the project partners and other national organizations, specifically interpreted geological information. This involved a comprehensive audit and evaluation of national geological spatial datasets that could be compiled at the specified scales in all partner countries. During the first phase, all data were harmonized and delivered at 1:1 million scale, this was improved to 1:250 000 scale during the second phase. Subsequently, during the early stages of the third phase, all data that were available at a scale finer than 1:250 000 were compiled for addition to the harmonized EMODnet Geology datasets. The third phase (2017–19) delivered products at a scale of 1:100 000 or finer where the underlying data permitted, using the standards developed during the previous phases of the project.

The geographical scope of the project area has evolved from the first phase, which only included the northern European seas, with the inclusion of the other European seas from the start of the second phase. The Caspian Sea was included into EMODnet Geology from the beginning of the fourth phase.

The specification of the third phase also included the supply of metadata on multibeam echo-sounder and seismic surveys, and seabed cores to be made available. The EMODnet Geology project has collected such available metadata and delivered this through the European Geological Data Infrastructure (EGDI) portal as indexes with links to the owner of the original data. It is planned for this portal to continue to be maintained after the duration of the project.

Emodnet Geology deliverables and products

The deliverables for the EMODnet Geology project mainly comprise spatial data: that is, maps and data points open and freely accessible via a map viewer on the EMODnet Geology portal. These products are downloadable (e.g. as shapefiles), and are described here in detail.

Seabed substrate information and harmonization of European-wide data

The first map layer to be compiled comprised information on the seabed substrate. The information was harmonized by evaluating the different classification schemes used in each country, translating the submissions to a uniform sediment scheme (Kaskela *et al.* 2019), and generalizing those interpretations to 1:100 000 scale before combining them into a single seabed substrate dataset. The derived dataset was

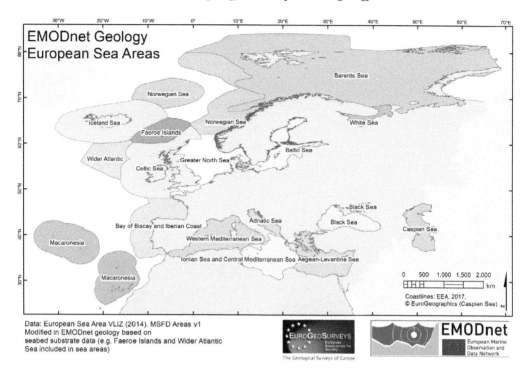

Fig. 1. The regional seas and sub-seas of Europe according to the Marine Strategy Framework Directive. Marine areas around Iceland, the Faroe Islands and the wider Atlantic Sea are included on the basis of the exclusive economic zone (EEZ) areas and data delivery in the EMODnet Geology phase III. The Caspian Sea is also included in the areas of EMODnet Geology during phase IV of the project.

delivered as geographic information system (GIS) shapefiles showing seabed substrate polygon features, with all maps provided in the WGS84 geographical coordinate system.

Owing to the vast heterogeneity in the different national classification schemes, a simplified reclassification scheme was adopted (Kaskela *et al.* 2019) that provides an estimate of the substrate representing sediments within 30 cm of the seafloor, whilst remaining ecologically relevant units.

It was agreed to follow the Folk sediment classification (Folk 1954) to include all 15 substrate classes, and also data on rock and boulders where possible. A nested hierarchy of Folk classifications was created with 16, seven and five classes (Kaskela *et al.* 2019). Thus, all classes could be translated into the simpler five class scheme to be used on the final map. The cutoff between 'Mud to muddy sand' and 'Sand' was changed from 4:1 to 9:1 during the first phase of the EMODnet Geology project.

Any submissions not originally provided using the Folk classification system were reclassified, with harmonization undertaken by the respective contributing partner to ensure the integrity of the interpretations were retained. The reclassification process required analysis of the surface material based on vast archives of seafloor samples, and an expert-based prediction. In each case, an attribute table that contains information related to the reclassification was created, with the reclassification validated by ground-truthing where possible. The final step was to combine the validated reclassified substrate maps, undertake quality control checks and to publish, for the first time, a Europe-wide coverage seabed substrate map (Fig. 2).

Sedimentation rates

The EMODnet Geology project has, since its first phase, collected all available data on sedimentation rates within European seas. Much of the data were compiled during the first phase of EMODnet Geology, although the dataset has been updated during the later phases, partly through further data mining and partly through input of new measurements, as well as through the inclusion of the European sea areas which were absent in the first phase of EMODnet Geology. The dataset now contains sedimentation rate (cm a^{-1}) data from over 1350 sites, with a clear difference in geographical coverage. The Baltic

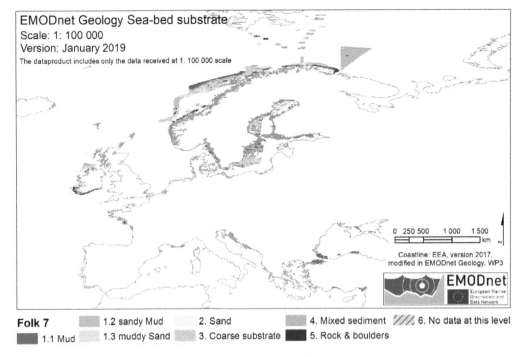

Fig. 2. EMODnet Geology seabed substrate data at a scale of 1:100 000 for the European seas hierarchy of seven Folk classes. The map shows one data layer (seabed substrate) as visualized in the EMODnet Geology map viewer on 1 April 2019. This open and freely-accessible product was made available by the EMODnet Geology project (http://www.emodnet-geology.eu/), implemented by EMODnet Geology phase III partners, and funded by the European Commission Directorate General for Maritime Affairs and Fisheries. These data were compiled by Anu Kaskela and Susanna Kihlman (GTK) from the EMODnet-3 Geology partners.

Sea, the Mediterranean Sea and, partly, also the Black Sea are well covered by sedimentation rate data, while there is a paucity of data from the nations bordering the North Atlantic Ocean, with the exception of areas offshore Norway and Iceland, as well as the Kattegat and Skagerrak straits. This is likely to be a result of local geological conditions as the coastal areas of the Atlantic Ocean are generally so highly dynamic that the accumulation of soft fine-grained sediments, which are best for accumulation measurements, do not occur in a comparable way to the enclosed seas such as the Baltic, the Mediterranean and the Black seas.

Pre-Quaternary geology and harmonization

During the first phase of EMODnet Geology, the offshore component of the harmonized 1:5 million-scale *International Geological Map of Europe and Adjacent Areas* (Asch 2005) was used as base to compile a pre-Quaternary geological map for the northern European sea areas. It was implemented as a Web Map Service (WMS) delivering information on both the lithology and chronostratigraphy

(labelled 'age' in the figure captions) of the seafloor. These datasets have since been updated during the subsequent phases of EMODnet Geology to cover all European sea areas, with improvements in both the detail of the rock descriptions and in the product resolution, so that today the best resolution available is 1:50 000 (Fig. 3). Currently, these data encompass not only information on the age and lithology of the seafloor in more detail but also on the genesis (event environment, event process) and structures (fault types) based on the INSPIRE Geology data specifications (https://inspire.ec.europa.eu/Themes/128/2892).

Quaternary geology

The Quaternary geology layer was added during the third phase of the EMODnet Geology project (Fig. 4). The aim was to compile all available Quaternary geology maps from each participating country, and resolve any major boundary issues to deliver information on the lithology, chronostratigraphy (labelled 'age' in the figure captions) and genesis at the best available resolution. Similarly to the

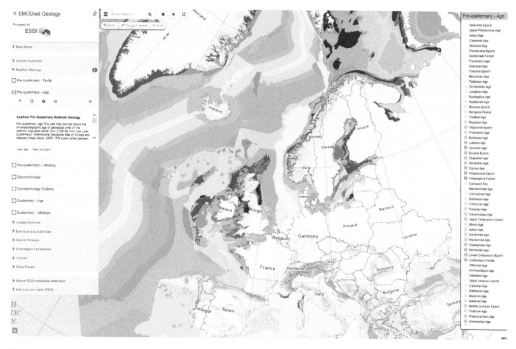

Fig. 3. Offshore pre-Quaternary geology as displayed on the EMODnet Geology portal. The map shows one data layer (seafloor age of the pre-Quaternary), as visualized in the EMODnet Geology map viewer on 1 April 2019. This open and freely-accessible product was made available by the EMODnet Geology project (http://www. emodnet-geology.eu/), implemented by EMODnet Geology phase III partners, and funded by the European Commission Directorate General for Maritime Affairs and Fisheries. These data were compiled by Kristine Asch (BGR) from the EMODnet-3 Geology partners.

pre-Quaternary dataset, the descriptions are based on the INSPIRE Geology specifications which may be further developed in phase IV as necessary.

Geomorphology

The geomorphology of the European seas was a new theme included in EMODnet Geology during phase III. A harmonized synthesis of geomorphological structures of the European seas did not exist prior to this project. Marine geomorphology in the framework of EMODnet Geology comprises delineated seafloor features, describing the submarine 'landscape', including information on their genesis.

Similarly to other themes within phase III of the project, this theme aimed to deliver information on the seafloor geomorphology at the best available resolution. This product utilized vocabulary and definitions developed in conjunction with the Commission for the Geological Map of the World/International Union for Quaternary Science CGMW/INQUA project of the *International Quaternary Map of Europe and Adjacent Areas*, produced at a scale of 1:2500 000 (IQUAME2500) (Asch 2019); work

which took place concurrent with phases II and III of the EMODnet Geology project.

The published data layer includes the major geological and geomorphological boundaries displayed at scales as detailed as possible as supplied by partner organizations (Fig. 5). Where information was not available at the stipulated 1:100 000 resolution, these data gaps were infilled using coarser-resolution products compiled during EMODnet Geology phases I and II.

Coastal behaviour

During the first phase of EMODnet Geology a GIS layer based on the EUROSION database on coastal erosion and sedimentation was supplemented and updated by the project partners. However, the dataset did not offer complete coverage of the European coasts and, moreover, the data were compiled from observations from different time periods, reducing the usefulness of the data. During the third phase of the project, a completely new approach was developed and applied. The new approach is based on remote sensing and comparison of satellite photographs over a time span of 10 years (2007–17)

Fig. 4. Offshore Quaternary geology as displayed on the EMODnet Geology portal. The map shows one data layer (chronostratigraphical 'age' of the Quaternary), as visualized in the EMODnet Geology map viewer on 10 February 2020. This open and freely-accessible product was made available by the EMODnet Geology project (http://www. emodnet-geology.eu/), implemented by EMODnet Geology phase III partners, and funded by the European Commission Directorate General for Maritime Affairs and Fisheries. These data were compiled by Kristine Asch (BGR) from the EMODnet-3 Geology partners.

with a spacing of 500 m between the observations. Importantly, for the first time, these observations were made with full European coverage. This approach gives very good resolution on both local and regional scales. The obtained results have been validated with field tests at a number of locations by project partners. The final outcome is a fully populated GIS layer and WMS of coastal behaviour information (Fig. 6). This will be followed up in the current phase by a GIS layer and WMS on coastal resilience, to be used as a pan-European tool for assessing the capacity to cope with the adverse effects of sea-level rise and coastal erosion. The resilience map then visualizes the socio-economic relevance of coastal–geological change as captured in the previous EMODnet Geology phases.

Geological events and probabilities

Data on geological events and probabilities have been collated from various sources but the major sources of information for this product were the national mapping programmes of the project partners. The web-page map-viewer provides fully populated GIS layers consisting of locations and, where available, additional attributes of features, such as

landslides, tectonics, fluid emissions, volcanoes and tsunamis, as polygons, lines and points (Fig. 7). Considering the diverse settings of European seas, it was necessary to compile an extensive and detailed list of attributes for the different features to represent the diverse characteristics of each occurrence (Battaglini *et al.* 2020). The EMODnet Geology portal links directly to the European–Mediterranean Seismological Centre (EMSC), which provides up-to-date information on earthquakes.

Minerals

As there has been a lack of compiled data on marine minerals within European waters, the aim of the minerals theme has been to identify areas of mineral occurrences (including aggregates, hydrocarbons and metalliferous minerals) both on and beneath the seabed. Each project partner collated and standardized known marine mineral occurrences from sources including publicly-available information (e.g. published scientific papers) for all EMODnet participating countries encompassing European marine regions and subregions.

During the third phase of the project the aim of delivering the first catalogue of marine minerals,

Fig. 5. Seafloor geomorphology of the European seas as displayed on the EMODnet Geology portal. The map shows one data layer (basic geomorphology), as visualized in the EMODnet Geology map viewer on 10 February 2020. This open and freely-accessible product was made available by the EMODnet Geology project (http://www. emodnet-geology.eu/), implemented by EMODnet Geology phase III partners, and funded by the European Commission Directorate General for Maritime Affairs and Fisheries. These data were compiled by Kristine Asch (BGR) from the EMODnet-3 Geology partners.

raw materials, hydrocarbons and metalliferous minerals offshore Europe was achieved. Now, standardized information on the spatial distribution of 12 mineral types across all European seas is available to download. The dataset and maps present information on: aggregates; cobalt-rich ferromanganese crusts; evaporites; gas hydrates; hydrocarbons; marine placers; metal-rich sediments; outcrops of rock, pegmatite and vein-hosted mineralization; polymetallic nodules; polymetallic sulfides; phosphorites; and sapropel (Fig. 8). It is important to understand that the marine mineral accumulations mapped are not just indicators of potential economic deposits but also indicators of palaeoenvironments like palaeobeaches. Take the Southern North Sea, for instance, where aggregate deposits indicate a glacial palaeoenvironment in the Quaternary, and evidence of evaporites in this area indicate salt deposits laid down during evaporation of a marine incursion during the Permian era. Extensive evaporites also have implications in terms of hydrocarbon trapping. While the profusion of cobalt-rich ferromanganese crust in the Canary Island Seamount Province indicates an environment dominated by longstanding volcanic precipitation of metals dissolved in seawater (Marino *et al.* 2017). Evaporites

are useful for carbon/methane sequestration, while sulfides, hydrates and nodules are endemic habitat indicators. Aggregates are valuable for offshore wind farm development and beach nourishment projects. Marine mineral information, together with other geological products, is a valuable resource for marine spatial planning.

Submerged landscapes

One of the new challenges for EMODnet Geology during the third phase was the request for data on reconstructions of the submerged landscapes of the European continental shelves at various time frames (e.g. LGM and younger low sea-level stages). Sea level is known to have fluctuated by more than 100 m over repeated glacial cycles, resulting in recurring exposure, inundation and migration of coastlines not only across Europe but worldwide. Landscape response to these changes in sea level, and the preservation of these features on continental shelves around Europe, are an invaluable resource for improving our understanding of human history and environmental change over geological time, while also providing data for potential use in examining future sea-level rise scenarios.

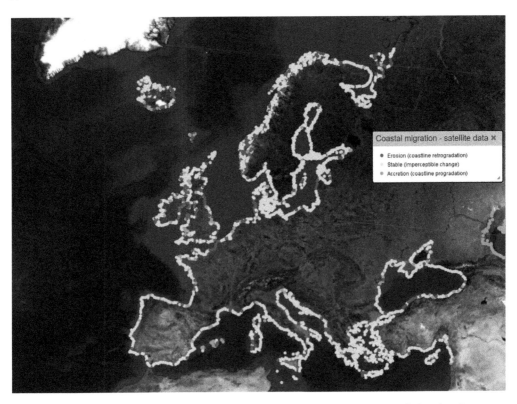

Fig. 6. Pan-European coastal behaviour based on satellite imagery from the years 2007–17 as displayed on the EMODnet Geology portal. The different colours mark erosion, stable coast or accretion. This open and freely-accessible product was made available by the EMODnet Geology project (http://www.emodnet-geology.eu/), implemented by EMODnet Geology phase III partners, and funded by the European Commission Directorate General for Maritime Affairs and Fisheries. These data were compiled by Deltares and Sytze Van Heteren (TNO). Base map ESRI World imagery.

The work package aimed to compile and harmonize available information on submerged landscape features by integrating existing records of palaeoenvironmental indicators with interpretations of geomorphology, stratigraphy and type of sediment (Fig. 9). The fully attributed GIS layer will be used to underpin palaeogeographical reconstructions across various time frames during the fourth phase of the project.

More than 10 000 features representing 26 classes of submerged landscape and palaeoenvironmental indicators ranging from mapped and modelled palaeocoastlines, evidence for submerged forests and peats, thickness of post-LGM sediments and submerged freshwater springs across all European seas have been collated and delivered on the EGDI portal for the first time. Building on the work of other projects, such as the COST Action SPLASHCOS project (http://www.splashcos.org/), the EU FP-7 project SASMAP (http://sasmap.eu/); and the MEDFLOOD project, this theme aimed at meeting the

recommendations of the European Marine Board SUBLAND group (http://www.marineboard.eu/continental-shelf-prehistoric-research-wg-subland). Recent advances in both data acquisition and availability over the last two decades has enabled researchers to more accurately reconstruct the extent and dynamics of fluctuating palaeocoastlines. High-resolution multibeam bathymetry and sub-bottom seismic data, in particular, have resulted in a step change in our understanding of palaeoshorelines and other traces of the original landscape topography and sediments. With preservation of these now submerged features under threat from commercial activities and natural erosion, bringing together existing knowledge through delivery of this work package is timely.

As this theme was new in the third phase of the project, data compilation and harmonization is ongoing with not all jurisdictional waters represented as yet. Furthermore, visualization of these data and interpretations at the European scale was challenging

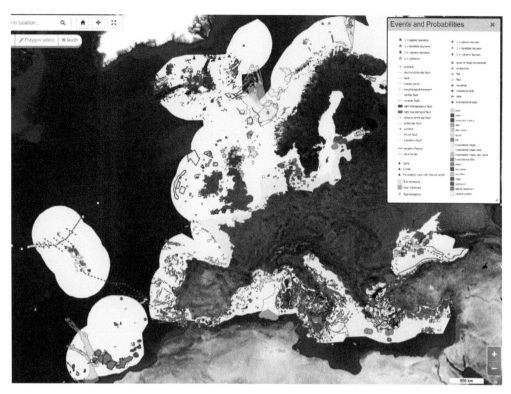

Fig. 7. Geological events and probabilities of the European seas as displayed on the EMODnet Geology portal, as visualized in the EMODnet Geology map viewer on 1 April 2019. This open and freely-accessible product was made available by the EMODnet Geology project (http://www.emodnet-geology.eu/), implemented by EMODnet Geology phase III partners, and funded by the European Commission Directorate General for Maritime Affairs and Fisheries. These data were compiled by ISPRA from submissions from the EMODnet-3 Geology partners.

as much of the obtained data are restricted in both geological time and spatial coverage, and therefore not easy to correlate across a whole continent. Future work will look to add and develop these data layers, incorporating feedback from the wider community and work on reconstructions at key snapshots in time.

Challenges of map production

There are a number of major challenges involved in the production of pan-European maps. A main issue is the scattering of data, which in fact was one of the main reasons for starting the EMODnet programme. It has been known for a long time that there are a lot of data collected from the European seas but by many different organizations, in many different countries and stored in many different ways, often inaccessible for the public. That being said, a strength of the EMODnet Geology consortium was access to key data repositories, as the project partners are the geological survey organizations of all European maritime nations, themselves the

major geological data owners in Europe. After the initial phase of collecting available data, the first drawbacks – the data gaps – were observed. It was known from the outset that a large number of data gaps existed, which was confirmed during the execution of the EMODnet Geology project. One of the main issues of importance for the European Union and member states should be the recognition of these knowledge gaps which should be prioritized so they can be strategically addressed in the future, based on environmental values and level of (conflicting) interests in the respective European sea areas.

After, collection of all available data it was clear that there were a number of different classification schemes describing the data made available from various partners. This highlighted the necessary steps in harmonizing the data, often a rather time-consuming process which can, unfortunately, also impact the quality of the result. For example, the compilation of seabed substrate data and their harmonization has been discussed here as an example of this complex process.

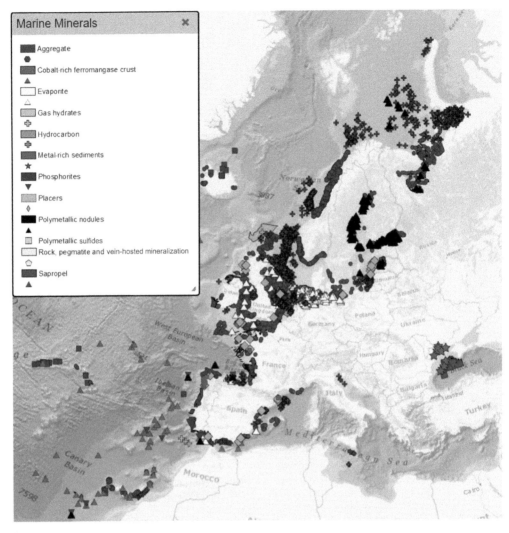

Fig. 8. Screenshot from the EMODnet Geology map viewer (on 1 February 2020) of the marine minerals map including 12 different mineral types, displayed here in point format. These points represent the single-point location for occurrences such as polymetallic sulfides. Where occurrences, such as aggregates, span a larger geographical footprint the central point of this geographical area is illustrated as the unique point representing that unique dataset. Polygons that illustrate the geographical distribution for each of these datasets are also available. This open and freely-accessible product was made available by the EMODnet Geology project (http://www.emodnet-geology.eu/), implemented by EMODnet Geology phase III partners, and funded by the European Commission Directorate General for Maritime Affairs and Fisheries. These data were compiled by Maria Judge (GSI).

Another issue affecting the value of the resulting maps is the quality of the original data. The quality does not concern only the measurements but also the data acquisition and, in the earlier days, accurate positional data. For instance, the low-resolution data on which many of the national geological interpretations are based have highlighted the data gaps and deficiencies. Sometimes, a cross-border connection of the maps has been difficult to reconcile due to such issues. The high-resolution datasets of today, using the latest acquisition and imaging techniques, are of a completely different standard to the archived datasets, which can make harmonization difficult.

Web portal and presentation of data products

The Geological Survey of Denmark and Greenland (GEUS) has, since the beginning of phase III of the project, been responsible for hosting and upgrading the portal, managing data products and allowing

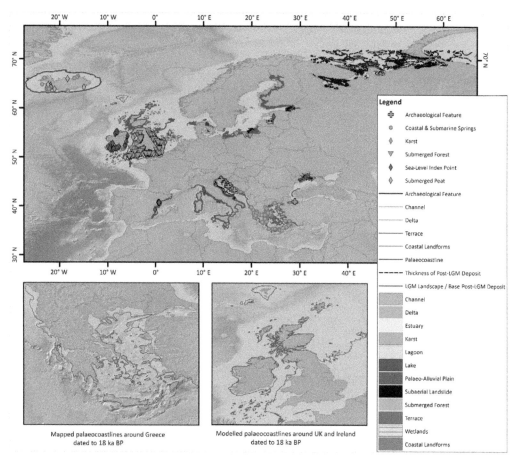

Fig. 9. Submerged landscapes of European seas as displayed on the EMODnet Geology portal (main panel). Mapped palaeocoastlines around Greece (lower left panel) and modelled palaeocoastlines around the UK and the Republic of Ireland (lower right) dated to 18 ka BP are displayed. These data were visualized in the EMODnet Geology map viewer on 1 April 2019. This open and freely-accessible product was made available by the EMODnet Geology project (http://www.emodnet-geology.eu/), implemented by EMODnet Geology phase III partners, and funded by the European Commission Directorate General for Maritime Affairs and Fisheries. These data were compiled by Heather Stewart (BGS) from the EMODnet-3 Geology partners.

users easy access to data products. Access includes downloading, machine-to-machine services and interactive maps. Synergies were found with other EU projects by adopting the technical infrastructure of Eurogeosurveys, called EGDI (see http://www.europe-geology.eu). Using this approach, a suite of functionality came 'out-of-the-box' and reduced initial costs. The service includes access to all EMODnet Geology products and, in case of technical queries regarding data access or contributions, a help desk is available during office hours.

Acknowledgements A large number of marine scientists, engineers and data managers have contributed to the information that has been compiled during the EMODnet Geology project. Staff of the EMODnet Geology partner organizations and subcontractors who have been directly involved in the project and who have made a major contribution are: Teresa Medialdea Cela, Pedro Terrinha, Slobodan Miko, Bogomir Celarc, Irene Zananiri, Dimitris Sakellariou, Lyubomir Dimitrov, Jørgen O. Leth, David Tappin, Heather Stewart, Rhys Cooper, Sophie Green, Joana Gafeira, Dayton Dove, Albert Caruana, Vera van Lancker, Kristine Asch, Laure Simplet, Axel Ehrhold, Wojciech Jeglinski, Dorota Kaulbartz, Slobodan Radusinović, Daiga Pipira, Zomenia Zomeni, Cherith Moses, Sten Suuroja, Anu Veski, Andres Kask, Lis Mortensen, Fabien Paquet, Agnes Tellez-Arenas, Isabelle Thinon, Jolanta Čyžienė, Arni Hjartarson, Reidulv Boe, Szymon Uscinowicz, Koen Verbruggen, Charise McKeon, Sokol Marku, Marenglen Gjoka, Joni Topulli, Gabriel Ion, Daria Ryabchuk, Lutz Reinhardt, Lovisa Zillen-Snowball,

Terje Thorsnes, Mikael Pedersen, Xavier Monteys, Markus Diesing, Anu Kaskela, Susanna Kihlman, Lars Kint, Ivana Raznatovic, Elena Borisova, Sytze van Heteren, Tamara van de Ven, Charles Galea, Gülsen Ucarkus, Kadir Eris, Murat Alyaz, Rekay Sergin, Ola Hallberg, Duncan Hume, Simeon Archer, Peter Mitchell, Mateusz Damrat, Agnese Jansone, Ieva Bukovska, Lauris Laiko, Johan Nyberg, Rokitskyi Valerii, Sonja Breuer, Alexander Müller, Pawel Gdaniec, Odd Harald Selboskar, Spela Kumelj, Ricardo Leon, Nuno Lourenco, Luis Batista, Aspasia Zalachori, Paraskevi Drakopoulou, Diana Perşa, Vladimir Zhamoida, Olga Kovaleva, Luis Somoza, Janine Guinan, Namik Çagatay, Alar Rosentau, Francesco Chiocci, Daniele Casalbore, Cathal Jordan, Jenny Hettelaar, Loredana Battaglini, Silvana D'Angelo, Marco Pantaloni, Ulla Alanen, Ozren Hasan, Nikolina Ilijanić, Vivi Drakopoulou, Alexandra Zavitsanou, Vangelis Zimianitis, Bogdan Prodanov, Eglė Šinkūnė, Mustafa Ergun, Gunay Cifci, Zoran Janković, Božica Jovanović, Susana Reino, Tatjana Durn, Olag Kovaleva, Viktor Snezhko, Bartal Højgaard, Vytautas Minkevičius, Skuli Vikingason, Špela Kumelj, Stefan Marincea, Delia Dumitras, Nilhan Kizildag, Eleni Georgiou Morisseau, Victor Snezhko, Ögmundur Erlendsson, Konstantina Tsoumparaki-Kraounaki, Julia Gimenez Moreno, Javier González, Eoin Mac Craith, Lars Kint, Nena Galanidou, Henk Weerts, Susana Muinos and Seda Okay. Additional valuable help has been provided by non-partners Manfred Zeiler and Jennifer Valerius (both BSH), Gerben Hagenaars (Deltares), Ranko Crmaric, and Ivan Petričević (HHI).

The authors wish to express their gratitude to the two anonymous reviewers who made valuable suggestions and improved the manuscript.

Funding Phases III and IV of the EMODnet Geology project have been funded by the Executive Agency for Small- and Medium-sized Enterprises (EASME) through contracts EASME/EMFF/2016/1.3.1.2 - Lot 1/ SI2.750862–EMODnet Geology and EASME/EMFF/ 2018/1.3.1.8 - Lot 1/SI2.811048–EMODnet Geology.

Author contributions HTVV: project administration (lead); **ATK**: investigation (equal), methodology (equal), validation (equal), writing – original draft (equal), writing – review & editing (equal); **KCA**: investigation (equal), methodology (equal), writing – original draft (equal), writing – review & editing (equal); **AF**: investigation (equal), methodology (equal), validation (equal), writing – original draft (equal), writing – review & editing (equal); **MJ**: methodology (equal), writing – original draft (equal); **HAS**: methodology (equal), writing – original draft (equal), writing – review & editing (equal); **BP**: methodology (equal), software (equal), writing – original draft (supporting).

Data availability statement The datasets generated during and/or analysed during the current study are available in the EMODnet repository, https://www.emod net.eu/.

References

Asch, K. 2005. *The 1: 5 Million International Geological Map of Europe and Adjacent Areas*. Bundesanstalt für Geowissenschaften und Rohstoffe (BGR), Hannover, Germany.

Asch, K. 2019. Assembling a jigsaw puzzle: an introduction to the International Quaternary Map of Europe Project (IQUAME2500). Paper presented at the 20th Congress of the International Union for Quaternary Research (INQUA), 25–31 July 2019, Dublin, Ireland.

Battaglini, L., D'Angelo, S. and Fiorentino, A. 2020. Digital mapping of geological events in European seas. *Geological Field Trips and Maps*, **12**(1.1), https:// doi.org/10.3301/GFT.2020.01

Folk, R.L. 1954. The distinction between grain size and mineral composition in sedimentary rock nomenclature. *The Journal of Geology*, **62**, 344–359, https:// doi.org/10.1086/626171

Kaskela, A.M., Kotilainen, A.T. *et al.* 2019. Picking up the pieces – harmonising and collating seabed substrate data for European maritime areas. *Geosciences*, **9**, 84, https://doi.org/10.3390/geosciences9020084

Marino, E., González, F.J. *et al.* 2017. Strategic and rare elements in Cretaceous–Cenozoic cobalt-rich ferromanganese crusts from seamounts in the Canary Island Seamount Province (northeastern tropical Atlantic). *Ore Geology Reviews*, **87**, 41–61, https://doi.org/10. 1016/j.oregeorev.2016.10.005

High-resolution geological mapping towards an understanding of post-glacial development and Holocene sedimentation processes in the eastern Gulf of Finland: an EMODnet Geology case study

Daria Ryabchuk[1]*, Alexander Sergeev[1], Vladimir Zhamoida[1], Leonid Budanov[1], Alexander Krek[2], Igor Neevin[1], Ekaterina Bubnova[2], Aleksandr Danchenkov[2,3] and Olga Kovaleva[1]

[1]A.P. Karpinsky Russian Geological Research Institute 74, Sredny Prospect, Saint Petersburg 199106, Russia

[2]Shirshov Institute of Oceanology, Russian Academy of Sciences, 36 Nahimovskiy Prospekt, Moscow 117997, Russia

[3]Immanuel Kant Baltic Federal University, 14 Nevskogo Alexandre Street, Kaliningrad 236016, Russia

EB, 0000-0002-8168-2984; AD, 0000-0002-1710-3757; OK, 0000-0002-9258-815X
*Correspondence: Daria_Ryabchuk@mail.ru

Abstract: Analyses of high-resolution multibeam and sub-bottom profiling data, acquired during marine geological field cruises between 2017 and 2019 in the eastern Gulf of Finland (Baltic Sea), enabled the detailed mapping of Quaternary deposits, and revealed diverse submerged glacial and post-glacial landforms (e.g. streamlined moraine ridges, large retreat moraine ridges, De Geer moraines and kettle holes). The morphology of these glacial features provides evidence of the ice-sheet retreat direction and rate throughout the deglaciation of the region, which occurred between 13.8 and 13.3 ka BP (Pandivere–Neva Stage) and 12.25 ka BP (Salpausselkä I Stage). Analysis of sub-bottom profiling, supported by piston long-core sampling, indicates periods of bottom erosion/non-deposition during the Holocene caused by relative water-level regressions. Significant negative relief features are also observed in the area for the first time. These linear and curved V-shaped furrows are several kilometres long and 5 m deep, and are tentatively ascribed to bottom current and gas-seepage processes.

Multiscale geological and geomorphological seabed mapping has become an important tool for addressing a wide spectrum of scientific challenges, including: reconstruction of geological and palaeoenvironmental histories, biological evolution and biodiversity assessment; and climate change operating over different timescales – from millions of years to annual variations (Petrov and Miletenko 2014; Smerlor 2020). These analyses then play an important practical role, informing marine spatial planning, maritime activity safety, anthropogenic impact evaluation, pollution assessment and sustainable use of marine resources (Dorschel *et al.* 2010; Diesing *et al.* 2014; Harris and Baker 2019).

Broad-scale geological mapping produces a global and regional understanding of geological structures, the potential of mineral resources, and the probability of natural hazardous processes. Whilst detailed mapping, based on the new and continuously improving technologies of both proxy studies and GIS modelling, provides high-resolution data on seafloor structure, topography, surficial sediments and the anthropogenic impact.

During its first two phases, the EMODnet Geology project (https://www.emodnet-geology.eu) was mainly targeted on compilations of broad-scale seafloor maps of European seas, which were based on harmonized multilateral cross-border mapping approaches, legends and classifications. During its third phase, the project processed and presented existing data products at a higher resolution, using geological sampling and submarine video-survey data, seismic survey data, side-scan sonar, and multibeam survey data.

This paper presents the results of high-resolution geological studies in several key areas of the eastern Gulf of Finland (northeastern Baltic Sea) (Fig. 1e). Despite a long history of marine geological studies, there are still important unsolved problems, of which the main ones include:

(i) the location of end moraines and glaciofluvial deposits in the Gulf of Finland;

From: Asch, K., Kitazato, H. and Vallius, H. (eds) 2022. *From Continental Shelf to Slope: Mapping the Oceanic Realm.* Geological Society, London, Special Publications, **505**, 19–38.
First published online July 21, 2020, https://doi.org/10.1144/SP505-2019-127

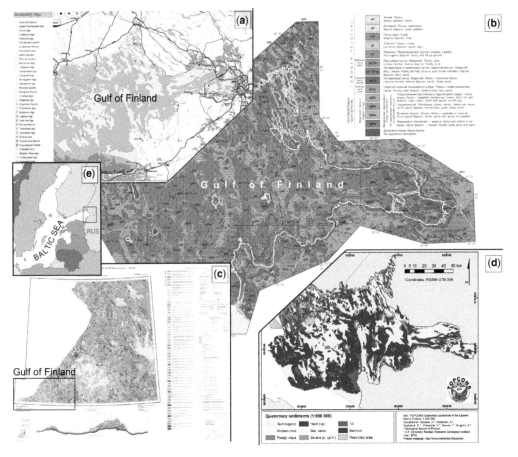

Fig. 1. Samples of broad-scale maps of Quaternary deposits of the eastern Gulf of Finland: (**a**) EMODnet Geology (2018) (https://www.emodnet-geology.eu/map-viewer/?bmagic=y&baslay=baseMapEEA,baseMapGEUS&optlay= &extent=-2179400,-295790,7283560,5318790&layers=emodnet_substrate_1m); (**b**) results of 1:200 000-scale mapping (Petrov 2010); (**c**) 1:1 000 000 scale State Geological Map (Legkhova *et al.* 2000); and (**d**) transboundary map compiled in the framework of the TOPCONS project (Ryabchuk *et al.* 2019). (**e**) Location of the study area on the map of the Baltic Sea region.

(ii) the age of deglaciation and rate of glacial front retreat;

(iii) the number of Holocene sea-level fluctuations and the amplitude of relative sea-level regressions (Virtasalo *et al.* 2014; Ryabchuk *et al.* 2016).

Due to the occurrence of well-preserved end-moraine complexes onshore (Pandivere–Neva and Palivere stages at the southern coast of the Gulf of Finland, and the Salpausselkä I and II stages at the northern coast of southeastern Finland), palaeore-constructions of deglaciation have been the focus of many studies (Zarina 1970; Krasnov *et al.* 1971; Kvasov 1975; Raukas and Hyvärinen 1991; Donner 1995; Saarnisto and Saarinen 2001; Hang 2003;

Kalm 2006; Subetto 2009; Vassiljev *et al.* 2011; Vassiljev and Saarse 2013). All of these reconstructions are based on terrestrial investigations alone, and so the location of ice-sheet margins within the Gulf of Finland are generally shown on palaeoreconstructions as dashed lines (inferred) or attributed as unknown (Hughes *et al.* 2016).

Another subject of active discussion is the problem of Holocene sea-level fluctuations within different parts of the Baltic Sea. Levels of relative regression are still under dispute. The eastern Gulf of Finland area is characterized by $+8 \text{ mm a}^{-1}$ rates of glacio-isostatic rebound during the Early Holocene (Ekman 1996; Poutanen and Steffen 2015), very low (from $+1$ to $+3 \text{ mm a}^{-1}$) (Harff *et al.* 2017) rates of recent uplift (at a near zero-rate of recent sea-level change) and Holocene transgressions recorded onshore as numerous distinct morphological traces (relict accretion formations, spits, coastal dunes and erosion escarpments). In addition, published relative sea-level change curves for the different onshore areas surrounding the Gulf of Finland are contradictory (Sandgren *et al.* 2004; Miettinen *et al.* 2007; Rosentau *et al.* 2013). The dating of morphological traces of Holocene transgressions are, in many cases, supported by archaeological evidence (Rosentau *et al.* 2013; Kriiska and Gerasimov 2014). Coastal lakes, which were bays during transgressive phases, also provide information about Holocene transgressions (Miettinen *et al.* 2007; Subetto 2009). Proxy evidence of maximum regression levels can be only determined by marine geological studies, so this question is still under dispute (Rosentau *et al.* 2013; Ryabchuk *et al.* 2016; Nirgi *et al.* 2019). Important remaining questions include: what are the rates of recent sedimentation and submarine erosion? What are the directions and intensities of near-bottom currents and sediment flows? Lastly, what are the sources and intensities of the mechanisms of gas seepage and submarine ground-water discharge?

The EMODnet Geology project has led to a reinterpretation of available broad-scale geological and geophysical data, and to the production of both a map of Quaternary deposits and a geomorphological map (WP3). As a result of this, several hypotheses of submerged glacial and possible post-glacial relief locations were suggested, which were then tested during the field cruises of 2017–19 (Table 1). The results of this research are presented in this paper.

It is worth noting that a proper understanding of the geological structures, morphology and modern sedimentation processes of the eastern Gulf of Finland is very important from practical point of view as it is one of the most intensely anthropogenically impacted marine areas of the Russian Federation (Vallius 2016; Raateoja and Setälä 2016; Kosyan 2017). St Petersburg, Russia's second largest city, occupies the coastal area where the Neva River debouches into the Neva Bay – the easternmost part of the Gulf (Fig. 1e). St Petersburg has a protracted history of industrial-, transportation- and urban-related activity that has affected the eastern Gulf of Finland (Ryabchuk *et al.* 2017). Since 2000, several large harbours have been constructed or renovated, intense dredging and dumping activity has been undertaken, and the Nord-Stream gas pipeline has been build.

Study area and previous research

State geological mapping (1:200 000 scale) of the eastern Gulf of Finland seabed was carried out by the Department of Marine and Environmental Geology of Russian Research Geological Institute (VSEGEI) in 1986–2000. The total length of acoustic sub-bottom profiling (SBP) completed at that time using sparker (500 Hz) and piezoceramic transmitters (7.5 kHz) was approximately 8000 km. The profiles were mainly orientated parallel to meridians with an approximate 2 km distance between the lines. For the interpretation of the geophysical data, more than 6000 sediment samples including gravity cores and vibro-cores were collected at that time. As a result, several sets of geological maps (1:200 000 (Fig. 1b) and 1:1 000 000 scale (Fig. 1c)) were published, including ones for the eastern Gulf of Finland and the *Atlas of Geological and Environmental Geological Maps of the Russian Area of the Baltic Sea* (1:700 000 scale) (Petrov 2010) (Fig. 1b).

In 2004–06, as a result of the Finnish–Russian project SAMAGOL (Sediment Geochemistry and Natural and Anthropogenic Hazards in the Marine Environments of the Gulf of Finland'), a joint effort to study and map Quaternary deposits using concerted methods was carried out (Spiridonov *et al.* 2007). In 2012–14, the joint Finnish–Russian project TOPCONS ('Transboundary Tool for Spatial Planning and Conservation of the Gulf of Finland') of the South-East Finland–Russian European Neighbourhood Partnership Instrument Cross-Border Cooperation (ENPI CBC) Programme was undertaken. As part of this project, broad-scale transboundary geological (maps of Quaternary deposits and of seabed sediments) (Fig. 1d) and habitat maps were compiled (Kaskela *et al.* 2017). The geological map compilation was preceded by a harmonization of sub-bottom profiling interpretation methods, grain-size classifications and GIS mapping approaches used by Finnish and Russian specialists. The first multibeam survey for key areas of the Russian part of the Gulf of Finland was also carried out during the TOPCONS project (Fig. 2, areas c, d, e, f and g).

Within the EMODnet Geology project, the Quaternary geological map and the map of bottom sediments of the eastern Gulf of Finland compiled by the VSEGEI were modified according to the unified rules and guidelines of the project using INSPIRE dictionaries (Asch and Müller 2017; Heteren *et al.* 2017). As a result of the harmonization and generalization of the geological data according to the EMODnet Geology legends (semantic vocabulary, guidelines and harmonized classifications, including the Folk grain-size classification), the initial Quaternary geological map (Fig. 1a) and the map of bottom substrate (bottom sediments) of the eastern Gulf of Finland became essentially simplified.

In general, the Quaternary geological structure and bottom morphology of the Gulf of Finland has been well known since the 1980s (Spiridonov *et al.* 1988, 2007; Winterhalter 1992). Acoustic–seismic units, identified in sub-bottom profiling (SBP),

Table 1. *Volume of multibeam echo sounder (MBES) and sub-bottom profiles (SBP) within the key studied areas*

Key area	Research vessel, year, cruise number	No. of MBES and SBP profiles	Length of the MBES and SBP profiles (km)	Square of multibeam coverage (km^2)	Sediment sampling sites
Moschny Island (area a, Fig. 2)	R/V *Academic Nikolaj Strakhov*, cruise 35, 21–24 July 2017 R/V *SN 1303*, 9–11 September 2017 R/V *SN 1303*, 23–28 June 2018	33	119	6.8	29
Vyborg Bay (area b, Fig. 2)	R/V *Academic Nikolaj Strakhov*, cruise 35, 21–24 July 2017 R/V *SN 1303*, 23–28 June 2018	39	164	9	15
Gogland Island (area h, Fig. 2)	R/V *Academic Nikolaj Strakhov*, cruise 35, 2017 R/V *SN 1303*, 9–11 September 2017 R/V *SN 1303*, 23–28 June 2018	34	697	58.5	66
Virgin Islands (area i, Fig. 2)	R/V *Academic Nikolaj Strakhov*, cruise 39, 2019	31	164	15.4	9
Sommers Island (area j, Fig. 2)	R/V *Academic Nikolaj Strakhov*, cruise 39, 2019	21	134	18.5	11

correspond to the concept of deglaciation and postglacial development of the Baltic Sea. The lowest unit is represented by the acoustic transparent zone that lacks reflections; it is interpreted as dense boulder sandy or clayey loam deposits of Late Pleistocene glacial till which usually (with exception of palaeovalleys) form the lowermost part of the Quaternary sequence (Fig. 3). The upper till boundary is characterized by discontinuous high-amplitude reflections lacking an explicit in-phase correlation. The next unit exhibits parallel rhythmic reflections with in-phase axes of varying amplitudes in the seismic-reflection profiles that conform with the surface of the underlying acoustic unit. It represents varved clays and homogenous clays of the Baltic Ice Lake. The upper acoustic unit is characterized by subhorizontal in-phase axes of variable amplitude reflections, and it represents the Holocene Ancylus Lake and the marine Littorina and Post-Littorina silty-clayey mud (Fig. 3). The boundary between the Ancylus and Littorina deposits is not easily traced on acoustic profiles; for the subdivision of the Holocene sequence, the long core interpretation is used (Spiridonov *et al.* 1988, 2007; Winterhalter 1992).

The primary geophysical (paper seismic-reflection profiles) and geological (core descriptions)

data collected in the eastern Gulf between 1986 and 2000 by the VSEGEI Department of Marine and Environmental Geology were used, after digitizing and reinterpretation, for a compilation of a broadscale maps of Quaternary deposits, and 3D models of pre-Quaternary relief, moraine and Late Pleistocene; and for calculating the thickness of the till, glaciolacustrine and Holocene sediments (Ryabchuk *et al.* 2018).

The upper boundaries of the till and glaciolacustrine clays were easily traced. The 3D models of pre-Quaternary relief, moraine and Late Pleistocene surfaces were compiled; the thickness of the till, glaciolacustrine and Holocene sediments were calculated based on interpretations of the seismic-reflection profiles collected between 1986 and 2000. A broad-scale bathymetric model of the eastern Gulf of Finland bottom was analysed using traditional GIS instruments (aspect, slope, terrain ruggedness and bathymetric position index (BPI) using the ArcGIS Spatial Analyst and the Benthic Terrain Modeler toolbox). This allowed us to localize the submarine landforms, which were further interpreted using all available geological and geophysical information (Ryabchuk *et al.* 2018).

Materials and methods

GIS analyses of broad-scale maps and a reinterpretation of primary proxy data were used to choose key areas for detailed studies and mapping with the aim of collecting new data on the eastern Gulf Finland deglaciation and post-glacial sedimentation processes. In 2016–19, during VSEGEI cruises, side-scan sonar profiling was conducted using a Klein 3000 (100 and 500 kHz) side-scan sonar on the Finnish side, and a CM2 (C-MAX Ltd, UK0 side-scan sonar with a working acoustic frequency of 325 kHz on the Russian side. Positioning was carried out using the differential global positioning system (DGPS) Trimble AgGPS132. The SBP survey was performed using a GEONT-HRP 'Spektr-Geophysika' Ltd (Russian) sparker operated at a working frequency range of 0.03–2 kHz. The vessel continuously recorded positions using a DGPS Furuno

GP7000F system in combination with a Vector VS330 Hemishere GNSS (USA).

In 2017 and 2019, during the cruises of the R/V *Academic Nikolaj Strakhov*, five key areas in the eastern Gulf of Finland were studied using a multibeam echo sounder (MBES) with a Teledyne RESON Seabat 8111-H, E208-3F66 Dry MBES system and an acoustic–seismic survey with Edge-Tech 3300-HM sub-bottom profiler (with Discover Sub-Bottom v3.36) (Table 1; Fig. 2, areas a, b, h, i and j). Three additional seismic-reflection profiles were collected in the Gogland Island key area onboard R/V *SN 1303* in September 2017 using a GEONT-HRP 'Spektr-Geophysika' Ltd (Russian) sparker operated at a working frequency range of 0.03–2 kHz.

A box-corer sampler was used for most of the sampling stations. Long cores were taken using a 3 m gravity corer (VSEGEI cruises of 2016–19)

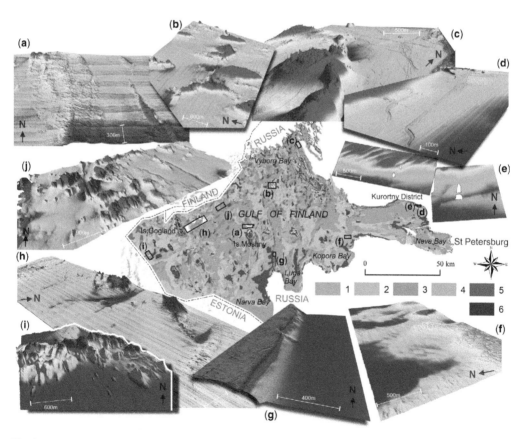

Fig. 2. Map of Quaternary deposits of the eastern Gulf of Finland (Petrov 2010): Holocene: 1, Ancylus, Littorina and Post-Littorina clays; Pleistocene: 2, Baltic Ice Lake clays; 3, varved clays; 4, glacio-fluvial deposits; 5, till; 6, bedrocks. Key areas of detailed mapping (scale of height 10 times larger than horizontal): (**a**) Moschny Island (2017); (**b**) Vyborg Bay (2017); (**c**) Kipperort Peninsula (2013); (**d**) Repino 1 (2012); (**e**) Repino 2 (2015); (**f**) Kopora Bay (2012); (**g**) Kurgalsky Reef; (**h**) Gogland Island (2017); (**i**) Virgin Islands (2019); and (**j**) Sommers Island (2019).

and a 6 m gravity corer (R/V *Academic Nikolaj Strakhov* cruises). In addition submarine video-surveys within key areas were carried out using an underwater remotely operated vehicle (microROV *Super-GNOM*, IO RAS, Russia) modernized by a GoPro HD HERO2 videocamera.

The MBES data were collected using an operating frequency of 100 kHz. The bottom surface was covered by 101 beams at 1.5° × 1.5°, 150° perpendicular to the direction of travel and 1.5° in the direction of travel. The resolution of each beam, regardless of the distance from the nadir, was 3.7 cm. The Edge-Tech 3300-HM fundamental frequency range was 2–10 kHz with a pulse width of 5–100 ms at a sampling rate of 20, 25, 40 and 50 kHz, depending on the pulse of the higher frequency. To calculate depths and positions, data from external GPS and motion sensors fixed to the vessel were used. The position of these sensors and the GPS antennae were located at the same reference point corresponding to the point of the OCTANS sensor arrangement, which was installed at the waterline as close as possible to the *x*-axis of the vessel. Georeferencing of data

received from the sensors was carried using the WGS84 coordinate system.

Three main stages of data processing were designed using the PDS2000 and ArcGIS 10.2 software packages. Subsequent processing and calculation of seafloor attributes led to the construction of geomorphological and substrate maps. The compilation of the Quaternary deposits distribution map for each key area was performed using ArcGIS. The seismic stratigraphic unit exposures at the seabed were traced and presented in GIS as a continuous line formed by set of points corresponding to each seismic trace. These exposures were interpolated into the polygon objects taking into account the bottom relief, which is diverse for different types of deposits.

The results of the GIS analyses of geophysical data collected in 2017 for key areas A (Moschny Island) and B (Vyborg Bay) have already been published (Ryabchuk *et al.* 2018). The data revealed well-defined elongated SSE–NNW (160°–170°) linear ridges up to 1000 m long, 100–170 m wide and 15–20 m high observed on the till surface. Within

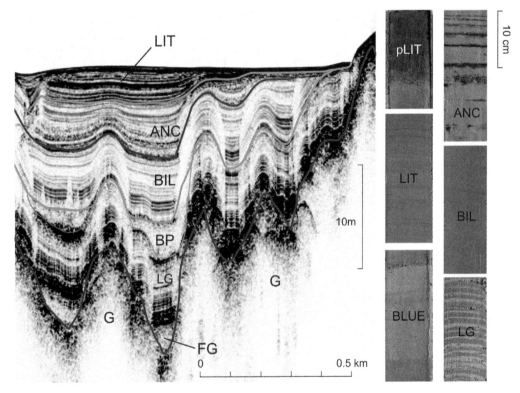

Fig. 3. Typical acoustic–seismic profile of post-glacial deposits of the eastern Gulf of Finland and photographs of cores of selected units. Glacial deposits (moraine) (unit G); fluvio-glacial deposits (unit FG); varved clays (unit LG); deposits of Baltic Ice Lake (units BP and BIL); Ancylus lacustrine clays (unit ANC); blue clays (unit BLUE); Littorina (unit LIT); and Post-Littorina (unit pLIT) muds.

the Moschny Island area, the following were observed: (i) a NNE–SSW (10°) trend for a less well-defined (than that observed for the primary orientation) bottom relief, representing linear elongations up to 1200 m long, 20–60 m wide and 0.5–1 m high (only observed within the Moschny Island area); and (ii) a SE–NW (120°) trend for small (0.5–1.5 m, with some up to 2 m, high), 8–10 m wide and up to 1300 m long rhythmic parallel ridges spaced 50–150 m apart, which occurred within relatively high bathymetric areas. In the Vyborg Bay area, the main relief feature was the >4300 m-long curved bedform, which is orientated roughly NE–SW to SE–NW (65°–100°) and can be traced on the till surface. Morphologically, this orientation corresponds to relatively high (10–20 m) and wide (70–200 to 300–1000 m) ridges. Small rhythmically curved-lined bedforms of similar orientation are observed on the ridges' surfaces. These bedforms are 0.5–1.5 m (up to 2 m) high, 8–10 m wide and up to 300 m long. These features, shown on the resultant thematic maps, were interpreted as streamlined till ridges, end-moraine ridges and De Geer moraines, and are used for reconstruction of the deglaciation in the eastern Gulf of Finland. This deglaciation occurred between 13.8 and 13.3 ka BP (Pandivere–Neva Stage) and 12.25 ka BP (Salpausselkä I Stage) (Ryabchuk et al. 2018).

The results of research and the mapping of three other areas (from west to east) – i (Virgin Islands), h (Gogland Island) and j (Sommers Island) (Fig. 2) – are discussed further in the following section.

Results

Acoustic–seismic stratigraphy

All of the described key areas are located within the Baltic Crystalline Shield where Vendian and Cambrian sedimentary rocks were eroded, so that Proterozoic igneous and supracrustal rocks outcrop on the pre-Quaternary surface (Petrov 2010). The thickness of the Quaternary deposits varies from 0–2 to 45 m near the Virgin Islands, from 2 to 45 m to the east of Gogland Island and from 5 to 35 m near Sommers Island.

According to the results of the geological survey, within the Virgin Islands area the surface of the Proterozoic rocks is raised and so crops out onshore in the islands and locally on the seafloor (including a part of the key study area). Within the other two study areas, the crystalline rocks are completely covered by Quaternary deposits.

Within all studied key areas, several of the characteristics of the exposed Quaternary deposits acoustic units are very similar (Figs 4d, 5d & 6d). An acoustic transparent zone lacking reflections lies below the basal acoustic surface of unit G; it is interpreted as

the upper surface of till (Figs 4d, 5d & 6d). This surface is characterized by a knob-and-kettle topography with numerous ridges and depressions. The maximum relative elevation changes of the unit G surface achieves 30 m. Rarely, for example in the Virgin Islands key area, is it possible to find deposits forming acoustic units LG_1 and BP_1 infilling local depressions between ridges; although the surface of acoustic unit G is usually covered by deposits of unit LG. Unit LG is characterized by a series of parallel reflections of varying amplitudes, conformable with the reflections of this surface. Varved clays are well identified in the seismic profiles due to their parallel rhythmic reflections and 'clothing' bedding. The overlying acoustic unit BIL, which is also characterized by 'clothing' bedding, is represented according to sediment core data, by a thin laminated clay in the lower part of the unit and an homogeneous glaciolacustrine deposit in its upper part. Sometimes it is impossible to differentiate between these two acoustic units. However, near the high morainic ridges in the Sommers and Virgin Islands key areas, an additional acoustic BP unit occurs between the LG and BIL units. This BP unit differs from other units by its acoustical transparency and very limited extension, usually no more than 500–900 m. The deposits forming unit BP reach a thickness of 6 m. The acoustic unit BIL overlying units BP and LG can be traced all over the eastern Gulf of Finland, and at the bottom surface its deposits are exposed in areas surrounding glacial till ridges and shallows characterized by submarine erosion or non-sedimentation. The thickness of the deposits forming the acoustic unit BIL rarely exceeds 4–5.5 m.

The acoustic unit ANC overlying unit BIL is traced mainly as the base layer of the Holocene deposits. The surface between units BIL and ANC is usually sharp due to the partly erosive nature of this contact. The areas of deposits that crop out and form the ANC acoustic unit are usually situated on the margins of local basins of silty–clay mud sedimentation.

The LIT acoustic unit forms the upper part of the geological section and is found within the in the deepest part of local sedimentary basins, where its thickness achieves 8–10 m. Often within the section of the LIT unit, several sharp steady sub-parallel contacts can be traced. On the slopes, the thickness of the LIT unit can be reduced by up to tens of centimetres. In the Sommers key area it is possible to distinguish the additional acoustic unit pLIT, with deposits that fill very locally as some elongated narrow runnels dissecting several acoustic units.

The interpretation of acoustic units permits us to construct schemes of acoustic unit outcrops (Figs 3b, 4b & 5b), 3D schemes of the till surface and Late Pleistocene clays surface (Figs 4c, 5c & 6c), on the basis of which maps of Quaternary deposits (Figs 4a, 5a & 6a) were compiled.

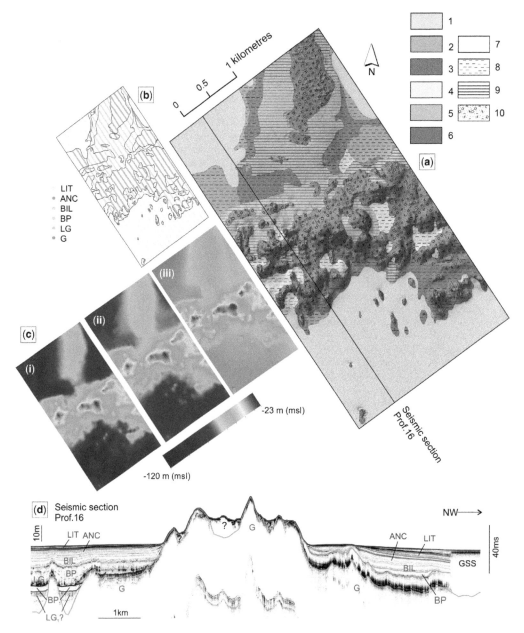

Fig. 4. (**a**) Map of Quaternary deposits of the Virgin Islands key area: 1, Littorina (unit LIT) and Post-Littorina (unit pLIT) muds; 2, Ancylus lacustrine clays (unit ANC); 3, glacio-lacustrine deposits of the Baltic Ice Lake (unit BIL); 4, homogenous sediment (unit BP); 5, varved glacio-lacustrine clays (unit LG); 6, glacial deposits (moraine) (unit G); 7, mud; 8, clay; 9, varved clays; 10, till. (**b**) Geophysical profiling and unit outcrop locations. (**c**) The surface of: (i) glacial deposits: (ii) glacio-lacustrine deposits; and (iii) the sea bottom. (**d**) Interpretation of the acoustic seismic profile. GSS, gas-saturated sediments. msl, mean sea level; Prof. 16, profile 16.

Seabed morphology

The best-defined seabed feature, observed in the central part of the Virgin Islands key area, is a ridge of sublatitudinal direction (east–west) formed by the elevation of the acoustic fundament surface from absolute depth −90 m to −30 m. The total width of the ridge on the seafloor surface varies from 500

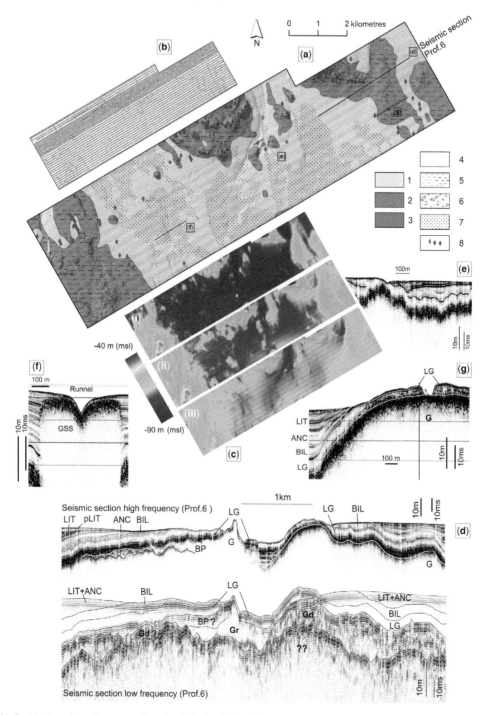

Fig. 5. (**a**) Map of the Quaternary deposits of Gogland Island key area: 1, Littorina (unit LIT) and Post-Littorina (unit pLIT) muds; 2, glacio-lacustrine deposits of Baltic Ice Lake (unit BIL); 3, glacial deposits (moraine); 4, mud; 5, clay; 6, till; 7, gas-saturated sediments (GSS); 8, buried furrow. (**b**) Geophysical profiling locations. (**c**) The surface of: (i) glacial deposits; (ii) glacio-lacustrine deposits; and (iii) the sea bottom. (**d**) Interpretation of the seismic profile (Gr, end moraine; Gd, drumlin). (**e**) Seismic section of buried furrow. (**f**) Seismic section of furrow. (**g**) Seismic section of 'crater'. msl, mean sea level; Prof. 6, profile 6.

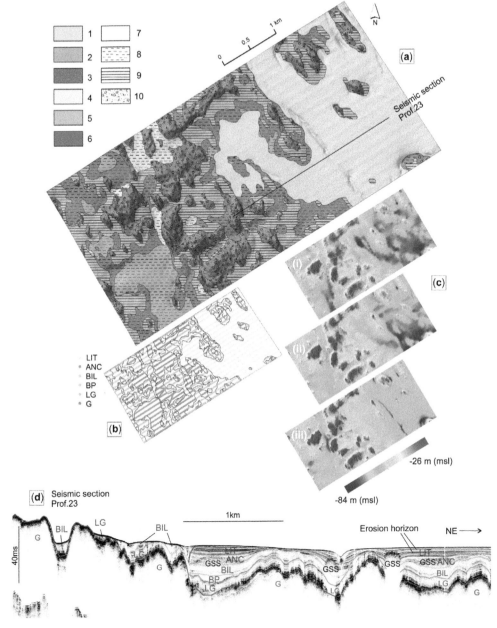

Fig. 6. (a) Map of Quaternary deposits of the Sommers Island key area: 1, Littorina (unit LIT) and Post-Littorina (unit pLIT) muds; 2, Ancylus lacustrine clays (unit ANC); 3, glacio-lacustrine deposits of Baltic Ice Lake (unit BIL); 4, homogenous sediment (unit BP); 5, varved glacio-lacustrine clays (unit LG); 6, glacial deposits (moraine); 7, mud; 8, clay; 9, varved clays; 10, till. (b) Geophysical profiling and unit outcrop locations. (c) The surface of: (i) glacial deposits; (ii) glacio-lacustrine deposits; and (iii) the sea bottom. (d) Interpretation of the seismic profile. GSS, gas-saturated sediments. msl, mean sea level; Prof. 23, profile 23.

to 1000 m, it crosses the entire study area from east to west, and, according to broad-scale maps, continues 2 km ESE and up to 10 km WNW (Table 2; Fig. 4).

According to results of a previous geological survey (Raukas and Hyvärinen 1991), the crystalline rocks' surface rose within the same area from −110

Table 2. *Results of morphometric analyses of MBES and SBP data (positive relief forms)*

Location	Shape	Direction (azimuth)	Height above the seafloor (m)	Height of ridge above its base (m)	Width of the ridge base (m)	Length of the ridge (m)	Slope (°)	Crests interval (m)	Geological interpretation
Virgin Islands	Linear (oval)	North-south (175°)	3–15 or 3–5	10–15 or 25	60–140 or 300–500	100–300 or 2000	10–25 (north, western and eastern), 5–10 (south)	–	Streamlined moraine ridges
	Linear	East-west (71°)	20–30	40	500–1000	more than 3500	10–25 (north, western and eastern) 5–10 (south)	–	End-moraine ridges with glaciofluvial complex
	Linear or crescent	East-west (76°)	1–2	1–3	10–60	100–600	5–10	50–150	De Geer moraine
Gogland Island	Linear (oval)	North-south (172°)	2–5 or 10–15	20–25	40–140 or 200–350	200–300 or 1500–2000	2–5	–	Streamlined moraine ridges
	Linear	East-west (97°)	5–10	20–30	100–300	more than 2000	10–25	–	Retreated moraine
Sommers Island	Crescent	East-west (86°)	1–2	1–3	5–10	130–400	10–25	50–150	De Geer moraine
	Linear (oval)	SSE–NNW (144°)	2–20	20–25	50–150	130–500	10–25 (NW), 5–10 (SE)	–	Streamlined moraine ridges
	Linear	WSW–ENE (56°)	8–20	25–40	300–500	800–2000	10–25 (NW), 5–10 (SE)	–	Retreated moraine
	Linear or crescent	WSW–ENE (53°)	1–2	1–2	40–60	500–600	10–25	50–150	De Geer moraine
Moschny Island (Ryabchuk et al. 2018)	Linear	NNE–SSW (10°)	0.5–1	–	20–60	1200	1–3	–	Glacial erosion ridges
	Linear (oval)	SSE–NNW (160°)	5–8	15–20	100	1000	5–20	–	Streamlined moraine ridges
	Linear	SE–NW (120°)	0.5–1.5	1–2	8–10	1300	5–15	50–150 (average 85)	De Geer moraine
Vyborg Bay (Ryabchuk et al. 2018)	Linear (oval)	SSE–NNW (170°)	10–15	15–20	130–170	1000	5–20	–	Streamlined moraine ridges
	Crescent	NE–SW and SE–NW (65° and 100°)	10–20	10–25	from 70 to 200 to 300–1000	more than 4300	3–4 (north), 10 (south)	–	End-moraine ridges
	Crescent	NE–SW and SE–NW (65° and 100°)	0.5–1.5	1–2	8–10	300	5–20	50	De Geer moraine

to −50 m. The height of the ridge above the surrounding seafloor is 20–30 m. The inner structure of the ridge is very complicated, it consists of smaller ridges orientated in an east–west direction (500–700 m wide and 1000–1200 m long) and a north–south (175°) direction, (with steeper (10°–25°) northern, eastern and western slopes, and smoother (5–10°) southern slopes). These features .are 60–140 m wide and 100–300 m long. The slopes of ridge are covered by post-glacial deposits. Within local areas on the ridge surface, tiny (0.5–1 m high, 8–10 m wide and 200–700 m long) linear or crescent sub-parallel ridges occur. In the northern part of study area, the larger ridge with a flat top (300–500 m wide and 2000 m long) running in a north–south direction occurs. It is poorly defined in recent bottom relief as its base is located at the absolute depth of 80 m and it is covered by glacio-lacustrine clays. On its surface, small sub-parallel ridges are observed, along with several round holes about 100 m in diameter. Ridges located within areas of Holocene mud accumulation are surrounded by V-shape furrows, which are mainly directed submeridionally (Table 3; Fig. 7).

The majority of the Gogland Island key area is located within a sedimentary basin filled with Holocene marine clayey mud with a very plain bottom relief (Fig. 8). On the mud surface, V-shaped furrows of two types (crescent, located around submarine highs; and linear, mostly submeridional) can be observed (Fig. 5f). The relative depths of furrows reach 3 and 5 m, with lengths of up to 4000 and 1500 m, respectively. Positive seabed features are represented by ridges of acoustic fundament of two perpendicular directions: north–south (172°) and east–west (97°). The ridges of the first direction are about 20–25 m in true height, and range from 40 to 350 m in width and from 200 to 2000 m in length. The slopes of the ridges are gentle (2°–5°). On the ridge surface, round holes 100 m in diameter and up to 5 m deep with steep (10°–25°) slopes and small (1–3 m high) sub-parallel ridges elongated east–west (86°) can be observed. The elongated elevations of the second direction are represented by larger sublatitudinal ridges (20–30 m high, 100–300 m wide and up to 2000 m long) and can be observed in the northern part of study area (Table 2; Fig. 8).

Table 3. *Results of morphometric analyses of MBES and SBP data (negative relief forms)*

Location	Shape (plane)	Direction (azimuth)	Shape (vertical)	Depth (m)	Width (m)	Length (m)	Sediment substrate type	Interpretation
Virgin Islands	Crescent (around submarine highs)	Approximately north–south	V-shape	3	20	100–200	Holocene mud	Current runnels
	Round	–	U-shape	3	100	100	Late Pleistocene clays	Kettle-holes of ice or water impact during deglaciation or meteorite impact
Gogland Island	Crescent (around submarine highs)	Approximately north–south	V-shape	3	20	Up to 4000	Holocene mud	Current runnels
	Linear	Approximately north–south	V-shape	5	20	Up to 1500	Gas saturated Holocene mud	Gas seepage or current runnels
	Round	–	U-shape	3	100	100	Late Pleistocene clays	Kettle-holes or water impact during deglaciation or (?) meteorite impact
Sommers Island	Crescent (around submarine highs)	All directions	V-shape	3–10	20	Up to 1000	Holocene mud	Current runnels
	Linear	Approximately SE–NW	V-shape	5–9	20	Up to 2000	Gas saturated Holocene mud	Gas seepage or (?) current runnels

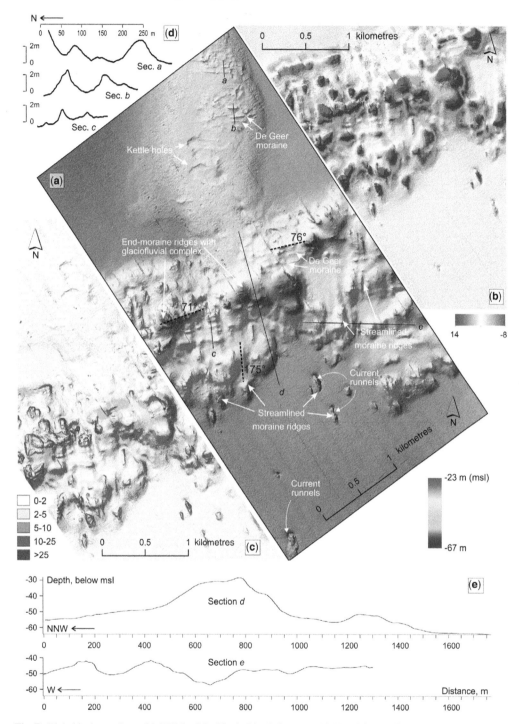

Fig. 7. Digital bathymetric model (DBM) of the Virgin Islands key area. (**a**) MBES image. (**b**) BPI map. (**c**) The slope in degrees. (**d**) and (**e**) Bathymetric profiles constructed on the DBM. msl, mean sea level.

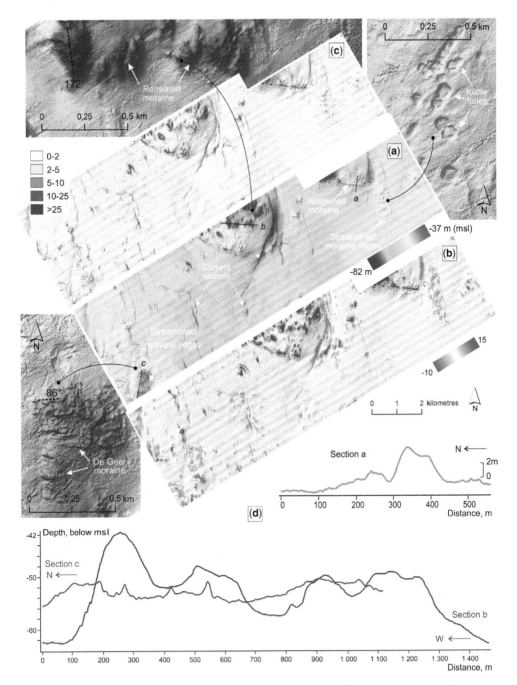

Fig. 8. Digital bathymetric model (DBM) of the Gogland Island key area. (**a**) MBES image with more detailed fragments. (**b**) BPI map. (**c**) The slope in degrees. (**d**) Bathymetric profiles constructed on the DBM (**d**).

The surface of the acoustic fundament within the Sommers Island key area rises from an absolute depth of −80 to −70 m in the eastern part to up to 20 m in the western part (Fig. 9). In the western part of the area, the ridges (25–40 m in true height, 8–20 m in height above the bottom surface, 300–500 m in width and 800–2000 m long) are elongated from SSE to NNW (144°) and crop out on the seafloor

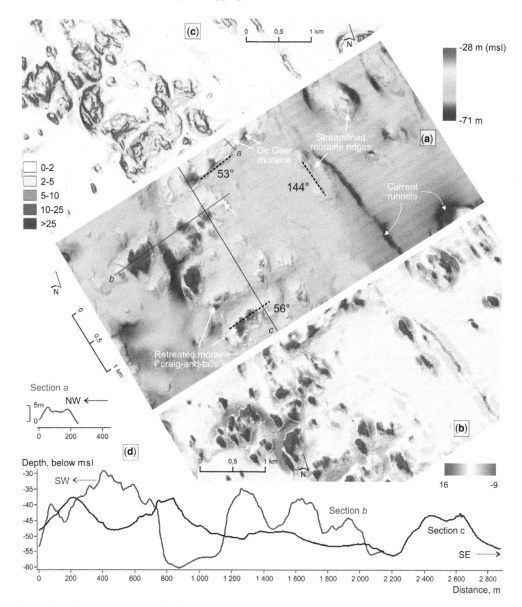

Fig. 9. Digital bathymetric model (DBM) of the Sommers Island key area. (**a**) MBES image. (**b**) BPI map. (**c**) The slope in degrees. (**d**) Bathymetric profiles constructed on the DBM.

surface. The ridges of the second type are directed WSW–ENE (56°), have a true height of 20–25 m, a width of 50–150 m and a length of 130–500 m. The ridges show steep northwestern (10°–25° and steeper) and more gentle (2°–10°) southeastern slopes. Within the bathymetrically high areas, small regular sub-parallel ridges (1–2 m high, 40–60 m wide and 500–600 m long, running in a WSW–ENE direction (53°)) with steep slopes are locally observed on the bottom surface. In the eastern part of the described

key area, a 2000 m-long linear V-shape furrow and several 1000 m runnels surrounding the till ridges can also be observed (Table 3; Fig. 9).

Discussion

Analysis of the shape and structure of the till surface (Figs 4d, 5d & 6d), as well as some previous seismic acoustic data collected during the geological survey of these and adjacent areas, indicate that some sharp hill crests covered by sediments are formed by pre-Quaternary rocks. Till, identified within all investigated key areas, is partly exposed at the sea bottom where it is eroded. However, the till is largely overlain by deposits of acoustic unit LG, which was identified as a layer of typical Late Pleistocene varved glacio-lacustrine clay accumulated in the ice marginal lakes close to the edge of the ice sheet at the beginning of deglaciation (Spiridonov *et al.* 2007).

Varved clays and the overlying unit BIL sediments were accumulated during a stage of the Baltic Ice Lake development. Generally, the boundary between typical varved clays and the overlying laminated and homogeneous clay is a facies transition. However, locally, in the troughs of the moraine surface between the layer of varved clays (unit LG) and the overlying glacio-lacustrine deposits (unit BIL), there is a layer of homogenous sediment (unit BP), which is characterized by acoustical transparency and limited extension. Such acoustic properties of unit BP and its position within the glacio-lacustrine deposits indicate that the accumulation of these homogenous sediments took place as a result of rapid single-step processes, most probably glacio-fluvial. The composition of unit BP sediments has been an ongoing question. One initial suggestion was that the BP sediments are composed of sand; however, a core sample collected in the Sommers key area has shown that homogenous sediments are represented by clay.

The upper boundary of the BIL unit is sometimes easily traced on the acoustic profiles due to a rather sharp erosive contact between Late Pleistocene glacio-lacustrine clays and the overlying Early Holocene sediments accumulated in the Ancylus Lake stage. Sometimes, within the relatively deep sedimentary basins, the contact is transitional. The situation is similar at the border between deposits of the ANC unit (Ancylus lacustrine clays) and the LIT unit, which is formed by marine Littorina and Post-Littorina muds. Sometimes, this border is easily traced due to an erosive contact and sometimes this contact is transitional, as evidenced by core sampling. Distinct erosion contacts mark periods when the water level has dropped during relative regressions.

Analysis of seabed relief features and the geological sequence of the study areas, and comparison of these data with results of the previous research (Ryabchuk *et al.* 2018), have shown that, within each polygon, the long axes of moraine ridges are elongated in two perpendicular directions. These can be interpreted as ice-sheet streamlined and glacial edge parallel, respectively. These directions are slightly different from one study area to another.

Streamlined features have north–south (Virgin Islands, Vyborg Bay and Gogland Island: azimuth 170°–175°) or NNE–SSW (Moschny Island and Somers Island: azimuth 144°–160°) directions. Ridges are characterized by steep northern slopes (10°–25° or steeper), while usually the southern slope is smoother, which is a feature typical of drumlin-like subglacial relief forms (Figs 7–9). Within all study areas, these bedforms vary from 50 to 150 m in width, from 100 to 300–500 m in length and from 20 to 25 m in true height. In the Virgin Islands and the Gogland Island key areas, the ridges extend in the same direction but are larger in size (350–500 m wide and 1500–2000 m long) (Figs 7 & 8). Within the Virgin Islands key area, the ridge slopes, according to SBP, are covered by glacial–lacustrine deposits (up to 10 m thick). Streamlined ridges are interpreted as drumlins, or drumlinoides, formed beneath an ice sheet, which are typical for areas of Scandinavian Ice Sheet deglaciation (Breilin *et al.* 2004, 2005). Conversely, some streamlined bedforms within the Virgin Islands key area look very similar to megascale glacial lineations, caused by glacial erosion.

On the top of the southwestern part of the ridge, glacial–lacustrine clays are partly eroded (with an average thickness of around 2–3 m); in addition, 'holes' or 'craters' about 100 m in diameter and about 4 m deep can also be observed here (Table 2; Figs 5 & 8a). The same round-shaped bedforms were revealed on the tops of large streamlined glacial ridges in the Gogland Island key area. Some of these are partly buried under recent marine mud but most are practically cropped out at the sea bottom. Acoustic profiles that cross the craters show that they are cut out in lower part of laminated sediment section (glacial–lacustrine or glacial–fluvial deposits are eroded) and that their bottom is formed by till. Accordingly, the craters cannot be older than 13.8–13.3 ka BP. These features differ from typical eastern Gulf of Finland pockmark features, which are much smaller (about 15–20 m in diameter), and their occurrence on the top of till ridges excludes any link for their genesis with gas seepages or submarine groundwater discharge. However, technogenic processes (such as recent or past naval activity) cannot be considered as a possible cause for the development of the described bedforms, as on the slopes of the 'holes' large Fe–Mn concretions and crusts can be observed. The surface sediment collected from the crater's bottom and slopes consist of silty–clayey mud (up to 15–20 cm thick), including sandy particles and coarse-grained granitic debris in the lower part of the layer. The sediment samples also contain irregular ferromanganese cores and large (up to 12–15 cm in diameter) flat concretions. A previous investigation of ferromanganese concretions collected in the eastern Gulf of Finland concluded that the age of such large flat concretion is over 1–1.5 kyr (Grigoriev *et al.* 2013), accordingly the craters cannot be younger.

There are several possible mechanisms for the formation of the round-shaped depressions caused by ice impact.

There was no submarine permafrost during deglaciation within the study area. After glaciolacustrine sedimentation had started and the recent seafloor at a depth of 50–60 m water depth was never dried, the kettle holes were most likely to have formed as a result of melting ground ice buried in glacial lacustrine clays during an early stage of the Baltic Ice Lake development, when the ice-sheet margin was not far from study bottom area.

Landforms that are parallel to the ice-sheet edge provide the most important evidence in respect to glacier recession. Within the key study areas – Virgin Islands, Gogland Island and Sommers Island, as well as within the previously investigated Viborg Bay and Moschny Island – these features are represented by large retreat moraines and smaller transversal ridges (De Geer moraine) (Table 2). Both landform types are directed east–west (SSE–NNW) (for retreat moraines and transverse ridges, azimuths vary between 56° and 53° in the Sommers Island area, between 71° and 76° near the Virgin Islands, between 75° and 80° in the Vyborg Bay, and between 86° and 97° near Gogland Island). A slight variation in the ridge direction indicates differences in the ice-sheet edge position. Deposits of transversal formations cover the streamlined ridges. Morphological parameters of small ridges can be interpreted as De Geer moraines and correspond very well to the results reported for other areas (Finlayson et al. 2007). It is worthy of note that the De Geer moraine, found for the first time within the study area in 2017 (Ryabchuk et al. 2018), seems to be a widespread relief feature for submarine glacial heights of the eastern Gulf of Finland. This indicates areas of ice-sheet contact with the surface in the ice-dammed lake. Within the Virgin Islands and Gogland Island key areas, De Geer moraine ridges are located close to the 'kettle holes'.

Retreat moraines are 25–40 m in true height (the height of the ridge above its base), 100–500 m wide and can be traced through all the key study areas; the revealed fragments reached a length of 2000–4300 m. Within the Virgin Islands key area, a zone of several sub-parallel ridges occurs and can most probably be interpreted as a terminal moraine. According to broad-scale map analyses in the context of this research, we have studied just a small western fragment of this formation. It is important to mention that the large ridge of latitudinal direction, found earlier in Viborg Bay (Ryabchuk et al. 2018), shows the features typical of an end-moraine formation: its asymmetrical shape, with a steeper (10° dipping) southern (lee-side) slope and more gentle (3–4°) northern (stoss-side) slope is characteristic of a typical end-moraine morphology (Breilin

et al. 2004, 2005; Kotilainen and Kaskela 2017). Analyses of SBP and the till-surface model of the Virgin Islands and Sommers Island key areas have shown that the same specific features of the transversal formations are not shown here. In the Virgin Islands key area, the till ridges show steeper northern (stoss-side) slopes and more gentle southern (lee-side) slopes. In the Sommers Island key area, the retreat moraines formation consists of a series of ridges; some profiles perpendicular to the ridges crests demonstrate that three ridges from five have steeper NW slopes, while two ridges are characterized by opposite morphological features with steeper SE slopes. This can be explained by pre-Quaternary surface morphology because, according to previous investigations (Raukas and Hyvärinen 1991), the bedrock's surface is represented by a series of cuestas with very steep northern slopes. The till ridge overlies the bedrock high of the same direction (east–west), locally cropping out onshore of the Virgins Islands. Till-ridge thickness and morphology allow us to interpret it as a part of a terminal moraine, as previously suggested (Raukas and Hyvärinen 1991); however, future investigations with low-frequency SBP and drilling are needed. Within the Gogland area, low-frequency acoustic profiling very clearly shows the overlaying of basal till and drumlins by the retreat moraine ridge (Fig. 8d). An interpretation of the layers underneath is very difficult without drilling boreholes.

The genesis of linear and crescent V-shape furrows is not yet quite clear. Furrows surround submarine highs, which are most probably caused by near bottom currents. Conversely, some of furrows are spatially linked with gas-saturated sediments, and so are at least partly connected with gas-seepage processes (Fig. 8e, f).

The distribution and thickness of the Holocene silty–clayey mud within the key study areas indicate a drastic change in the sedimentation processes from the beginning of the Holocene, as with more eastern parts of the Gulf (Ryabchuk et al. 2018). Accumulation occurred within local sedimentary basins, while large areas of the Gulf of Finland were sediment-starved or experiencing erosion. In addition, SBP analyses revealed traces of several deep regressions, resulting in the formation of erosion layers that experienced no deposition of sandy and silty material in the pre-Ancylus Stage (11.7–10.7 ka BP) (Andren et al. 2011) and pre-Littorina time (c. 8.5 ka PB). The conclusion of a lower water level in pre-Littorinian time is supported by high-resolution sediment core investigations (Virtasalo et al. 2014), modelling of the submarine terrace formation (Amantov et al. 2012), onshore geoarchaeological research (Sergeev et al. 2014) and the study of a submerged river valley in Parnu Bay (Nirgi et al. 2019). Interesting evidence of regression was found via analyses of furrow distribution in the Gogland study area. This study has identified around 40 buried hollows in the Holocene mud sediment sequence in a sediment thickness of 5–10 m, this marked the depth of relict furrows formed during lower water-level stand. The dating of these buried relief forms is an important task of future investigations.

Conclusions

Based on high-resolution multibeam and acoustic profiling surveys, detailed maps of Quaternary deposits were compiled and geomorphological analyses were undertaken.

Geophysical research was conducted in key areas of the Virgins Islands, Gogland Island and Sommers Island at well-defined streamlined moraine ridges, large retreat moraine ridges and De Geer moraines. The morphology of these glacial features indicates the ice-sheet margins during different stages of deglaciation in the eastern (Russian) Gulf of Finland.

For the first time, negative bottom-relief features – linear and curved V-shape furrows (both on recent seafloor and that buried by Holocene mud) and huge (about 100 m in diameter) round craters in the Late Pleistocene clays – were revealed. The genesis of the linear and crescent V-shape furrows is not yet quite clear, however. Furrows surrounding submarine highs were most probably caused by near bottom currents. However, some of the furrows are spatially linked with gas-saturated sediments, so they are at least partly connected with gas-seepage processes. Round craters in Late Pleistocene deposits are interpreted as kettle holes, which formed during deglaciation.

Acknowledgements Authors thank the captain and crew of R/V *Academic Nikolaj Strakhov*. The authors thank the Reviewers – Dr. Kristine Asch and Dr. Dayton Dove for a careful revision of our manuscript, and very helpful and important comments.

Funding Multibeam data processing was supported by the state assignment Theme No. 0149-2019-0013. Acoustic data analyses and interpretation were carried out under project No. 17-7720041 of the Russian Science Foundation. Multibeam data analyses and interpretation were undertaken with support of the Russian Foundation for Basic Research No. 19-05-00768.

Author contributions DR: methodology (lead), supervision (lead), writing – original draft (lead); **AS**: investigation (lead), software (lead), visualization (lead), writing – original draft (lead); **VZ**: investigation (lead), supervision (lead), writing – original draft (lead); **LB**: investigation (equal), software (equal), visualization (equal); **AK**: investigation (lead), software (lead); **IN**: investigation (equal), software (equal); **EB**: investigation (equal), software (equal); **AD**: investigation (equal), software (equal), visualization (equal); **OK**: software (equal).

Data availability statement The multibeam datasets generated during and/or analysed during the current study are not available due to national restrictions. Subbotom profiling and sediment sampling metadata are available at https://www.emodnet-geology.eu/map-viewer (indexes).

References

Amantov, A.V., Zhamoida, V.A., Ryabchuk, D.V., Spiridonov, M.A. and Sapelko, T.V. 2012. Geological structure of submarine terraces of the eastern Gulf of Finland and modeling of their development during postglacial time. *Regional Geology and Metallogeny*, **50**, 15–27 (in Russian).

Andren, T., Björck, S., Andren, E., Conley, L.Z. and Anjar, J. 2011. The development of the Baltic Sea Basin during the last 130 ka. *In*: Harff, J., Björc, K.S. and Hoth, P. (eds) *The Baltic Sea Basin*. Springer, Berlin, 75–97.

Asch, K. and Müller, A. 2017. *Guidelines and Technical Guidance. Work Package 4: Sea-Floor Geology/Geomorphology. EMODnet-3*. Bundesanstalt für Geowissenschaften und Rohstoffe (BGR), Hannover, Germany.

Breilin, O., Kotilainen, A., Nenonen, K., Virransalo, P., Ojalainen, J. and Stén, C.-G. 2004. *Geology of the Kvarken Archipelago*. Geological Survey of Finland, Espoo, Finland.

Breilin, O., Kotilainen, A., Nenonen, K. and Räsänen, M. 2005. The unique moraine morphology, stratotypes and ongoing geological processes at the Kvarken Archipelago on the land uplift area in the Western coast of Finland. *Geological Survey of Finland Special Papers*, **40**, 97–111.

Diesing, M., Green, S.L., Stephens, D., Lark, R.M., Stewart, H.A. and Dove, D. 2014. Mapping seabed sediments: Comparison of manual, geostatistical, object-based image analysis and machine learning approaches. *Continental Shelf Research*, **84**, 107–119, https://doi.org/10.1016/j.csr.2014.05.004

Donner, J. 1995. *The Quaternary History of Scandinavia*. Cambridge University Press, London.

Dorschel, B., Wheeler, A.J., Monteys, X. and Verbruggen, K. 2010. On the Irish seabed. *In*: *Atlas of the Deep-Water Seabed: Ireland*. Springer, Dordrecht, The Netherlands, 21–24, https://doi.org/10.1007/978-90-481-9376-1_5

Ekman, M. 1996. A consistent map of the postglacial uplift of Fennoscandia. *Terra Nova*, **8**, 158–165, https://doi.org/10.1111/j.1365-3121.1996.tb00739.x

Finlayson, A., Bradwell, T., Golledge, N. and Merritt, J. 2007. Morphology and significance of transverse ridges (De Geer Moraines) adjacent to the Moray Firth, NE Scotland. *Scottish Geographical Journal*, **123**, 257–270, https://doi.org/10.1080/1470254080 1968477

Grigoriev, A.G., Zhamoida, V.A., Gruzdov, K.A. and Krymsky, R.S. 2013. Age and growth rates of ferromanganese concretions from the Gulf of Finland derived from ^{210}Pb measurements. *Oceanology*, **53**, 345–351, https://doi.org/10.1134/s00014370130 30041

Hang, T. 2003. A local clay-varve chronology and proglacial sedimentary environment in glacial Lake Peipsi, eastern Estonia. *Boreas*, **32**, 416–426, https://doi.org/10.1111/j.1502-3885.2003.tb01094.x

Harff, J., Deng, J. *et al.* 2017. What determines the change of coastlines in the Baltic Sea? *In*: Harff, J., Furmanczyk, K. and Von Storch, H. (eds) *Coastline Changes of the Baltic Sea from South to East: Past and Future Projection*. Springer, Berlin, 15–35.

Harris, P. and Baker, E. (eds). 2019. *Seafloor Geomorphology as Benthic Habitat: GeoHab Atlas of Seafloor Geomorphic Features and Benthic Habitats*. 2nd edn. Elsevier Science, Amsterdam.

Heteren, S., van de Ven, T. and van Moses, C.A. 2017. *EMODnet 3 Geology. WP5: Coastal Behavior*. Task Guide for Internal Use.

Hughes, A.L.C., Gyllencreutz, R., Lohne, Ø.S., Mangerud, J. and Svendsen, J.I. 2016. The last Eurasian ice sheets – a chronological database and time-slice reconstruction, DATED-1. *Boreas*, **45**, 1–45, https://doi.org/10.1111/bor.12142

Kalm, V. 2006. Pleistocene chronostratigraphy in Estonia, southeastern sector of the Scandinavian glaciation. *Quaternary Science Reviews*, **25**, 960–975, https://doi.org/10.1016/j.quascirev.2005.08.005

Kaskela, A.M., Rousi, H. *et al.* 2017. Linkages between benthic assemblages and physical environmental factors: The role of geodiversity in Eastern Gulf of Finland ecosystems. *Continental Shelf Research*, **142**, 1–13, https://doi.org/10.1016/j.csr.2017.05.013

Kosyan, R. (ed.). 2017. *The Diversity of Russian Estuaries and Lagoons Exposed to Human Influence. Estuaries of the World*. Springer International, Cham, Switzerland.

Kotilainen, A.T. and Kaskela, A.M. 2017. Comparison of airborne LiDAR and shipboard acoustic data in complex shallow water environments: Filling in the white ribbon zone. *Marine Geology*, **385**, 250–259, https://doi.org/10.1016/j.margeo.2017.02.005

Krasnov, I.I., Duphorn, K. and Voges, A. 1971. *International Quaternary Map of Europe 1:2 500 000. Sheet 3, Nordkapp*. Bundesanstalt für Geowissenschaften und Rohstoffe (BGR), Hannover, Germany.

Kriiska, A. and Gerasimov, D.V. 2014. Late Mesolithic period in the Eastern Baltic: formation of the coastal settlement system from the Riga Bay till the Vyborg Bay. *In: From the Baltic to Urals: Essays on the Stone Age Archaeology*. Syktyvkar, Russia, Komi Scientific Centre of Russian Academy of Science, 5–36 (in Russian with English abstract).

Kvasov, D.D. 1975. *Late Quaternary History of the Large Lakes and Inner Seas of the Eastern Europe*. Nauka, Leningrad, USSR.

Legkhova, V.G., Ydachona, O.N., Zatulskaya, T.Y., Rybalko, A.E., Moskalenko, P.E. and Spiridonov, M.A. 2000. *Map of Quaternary Deposits / State Geological Map of Russian Federation P (35)37, Scale 1:1 000 000*. VSEGEI, St Petersburg, Russia.

Miettinen, A., Savelieva, L., Subetto, D., Dzhinoridze, R., Arslanov, K. and Hyvärinen, H. 2007. Palaeoenvironment

of the Karelian Isthmus, the easternmost part of the Gulf of Finland, during the Litorina Sea stage of the Baltic Sea history. *Boreas*, **36**, 441–458, https://doi.org/10.1080/03009480701259284

Nirgi, T., Rosentau, A. *et al.* 2019. Holocene relative shore-level changes and Stone Age palaeogeography of the Pärnu Bay area, easten Baltic Sea. *The Holocene*, **30**, 37–52, https://doi.org/10.1177%2F095968361986 5603

Petrov, O.V. (ed.). 2010. *Atlas of Geological and Environmental Geological Maps of the Russian Area of the Baltic Sea*. VSEGEI, St Petersburg, Russia.

Petrov, O.V. and Miletenko, N.V. (eds). 2014. *Priority Areas of Geological Exploration. On Materials of the 34th Session of the International Geological Congress*. A.P. Karpinsky Russian Geological Research Institute, St Petersburg, Russia.

Poutanen, M. and Steffen, H. 2015. Land uplift at Kvarken Archipelago/high coast UNESCO World Heritage area. *Geophysica*, **50**, 49–64.

Raateoja, M. and Setälä, O. (eds). 2016. *The Gulf of Finland Assessment*. Reports of the Finnish Environment Institute, **27/2016**.

Raukas, A. and Hyvärinen, H. (eds). 1991. *Geology of the Gulf of Finland*. Valgus, Tallinn.

Rosentau, A., Muru, M. *et al.* 2013. Stone Age settlement and Holocene shore displacement in the Narva-Luga Klint Bay area, eastern Gulf of Finland. *Boreas*, **42**, 912–931, https://doi.org/10.1111/bor.12004

Ryabchuk, D., Zhamoida, V. *et al.* 2016. Development of the coastal systems of the easternmost Gulf of Finland, and their links with Neolithic–Bronze and Iron Age settlements. *Geological Society, London, Special Publications*, **411**, 51–76, https://doi.org/10.1144/SP411.5

Ryabchuk, D., Vallius, H. *et al.* 2017. Pollution history of Neva Bay bottom sediments (eastern Gulf of Finland, Baltic Sea). *Baltica*, **30**, 31–46, https://doi.org/10.5200/baltica.2017.30.04

Ryabchuk, D.V., Yu Sergeev, A. *et al.* 2018. New data on the postglacial development of Narva-Luga Klint Bay (eastern Gulf of Finland): results of geoarchaeological research. *Journal of Coastal Conservation*, **23**, 727–746, https://doi.org/10.1007/s11852-018-0670-5

Ryabchuk, D., Orlova, M. *et al.* 2019. The eastern Gulf of Finland (Baltic Sea) landscapes – brackish water estuary under natural conditions and anthropogenic stress. *In*: Harris, P. and Baker, E. (eds) *Seafloor Geomorphology as Benthic Habitat: GeoHab Atlas of Seafloor Geomorphic Features and Benthic Habitats*. 2nd edn. Elsevier Science, Amsterdam, 281–301.

Saarnisto, M. and Saarinen, T. 2001. Deglaciation chronology of the Scandinavian ice sheet from the Lake Onega basin to the Salpausselkä end moraines. *Global and Planetary Change*, **31**, 387–405, https://doi.org/10.1016/S0921-8181(01)00131-X

Sandgren, P., Subetto, D.A., Berglund, B.E., Davydova, N.N. and Savelieva, L.A. 2004. Mid-Holocene Littorina Sea transgressions based on stratigraphic studies in coastal lakes of NW Russia. *GFF*, **126**, 363–380, https://doi.org/10.1080/11035890401264363

Sergeev, A., Ryabchuk, D. *et al.* 2014. Reconstruction of the palaeorelief of the Littorina Sea coastal zone within

St. Petersburg based on a study of the archaeol243ogical site Okhta 1. *Germania: Anzeiger der Römisch–Germanischen Komission des Deutschen Archäologischen Instituts*, **92**, 33–60.

Smerlor, M. 2020. Geology for society in year 2058: some down-to-earth perspectives. *Geological Society, London, Special Publications*, **499**, https://doi.org/10.1144/SP499-2019-40

Spiridonov, M.A., Rybalko, A.E., Butylin, V.P., Spiridonova, E.A., Zhamoida, V.A. and Moskalenko, P.E. 1988. Modern data, facts and views on the geological evolution of the Gulf of Finland. *Geological Survey of Finland Special Papers*, **6**, 95–100.

Spiridonov, M., Ryabchuk, D., Kotilainen, A., Vallius, H., Nesterova, E. and Zhamoida, V. 2007. The Quaternary deposits of the eastern Gulf of Finland. *Geological Survey of Finland Special Papers*, **45**, 7–19.

Subetto, D.A. 2009. *Lake Sediments: Palaeolimnological Reconstructions*. Herzen Russian State Pedagogical University, St Petersburg, Russia.

Vallius, H. 2016. Sediment geochemistry studies in the Gulf of Finland and the Baltic Sea: a retrospective view. *Baltica*, **29**, 57–64, https://doi.org/10.5200/baltica.2016.29.06

Vassiljev, J. and Saarse, L. 2013. Timing of the Baltic ice lake in the eastern Baltic. *Bulletin of the Geological Society of Finland*, **85**, 9–18, https://doi.org/10.17741/bgsf/85.1.001

Vassiljev, J., Saarse, L. and Rosentau, A. 2011. Palaeoreconstruction of the Baltic Ice Lake in the Eastern Baltic. *In*: Harff, J., Björck, S. and Hoth, P. (eds) *The Baltic Sea Basin*. Springer, Berlin, Heidelberg, 189–202, https://doi.org/10.1007/978-3-642-17220-5_9

Virtasalo, J.J., Ryabchuk, D., Kotilainen, A.T., Zhamoida, V., Grigoriev, A., Sivkov, V. and Dorokhova, E. 2014. Middle Holocene to present sedimentary environment in the easternmost Gulf of Finland (Baltic Sea) and the birth of the Neva River. *Marine Geology*, **350**, 84–96, https://doi.org/10.1016/j.margeo.2014.02.003

Winterhalter, B. 1992. Late-Quaternary stratigraphy of Baltic Sea basins - a review. *Bulletin of the Geological Society of Finland*, **64**, 189–194, https://doi.org/10.17741/bgsf/64.2.007

Zarina, E.P. 1970. Geochronology and palaeogeography of Late Pleistocene of the North-West of Russia. *In*: *Periodization and Palaeogeography of Late Pleistocene*. VSEGEI, Leningrad, USSR, 27–33 (in Russian).

A first approach to a Quaternary geomorphological map of the German seas

Sonja Breuer* and Kristine Asch

Bundesanstalt für Geowissenschaften und Rohstoffe, Stilleweg 2, 30655 Hannover, Germany

⬦ SB, 0000-0003-4995-6284; KA, 0000-0002-7423-7305

*Correspondence: sonja.breuer@bgr.de

Abstract: Our paper presents the first draft of a geomorphological map of the German North Sea and Baltic Sea. The inspiration for this map comes from the international collaboration of marine researchers within the European EMODnet Geology Project (https://www.emodnet-geology.eu/). The current climate change intensifies the natural processes of change in nature. Within the framework of various nature conservation projects, the importance of marine sediment structures on marine fauna and their reproductive cycles, sedimentation conditions, currents, etc. has been investigated. In order to be able to make statements for the German seas and document changes, the current state must first be recorded.

EMODnet, the European Marine Observation and Data Network, was established in 2007 by the European Commission as a long-term marine data initiative from the Directorate General for Maritime Affairs and Fisheries (DG MARE) underpinning its Marine Knowledge 2020 initiative.

EMODnet consists of more than 100 organizations assembling marine data, products and metadata in an organized way and standardized as far as possible to facilitate the use of these often-fragmented resources for public and private users. The aim is to produce open high-quality harmonized marine data that is interoperable and available free of charge. The EMODnet data infrastructure has been realized in several major phases with the target of being fully deployed by 2020.

EMODnet 'Geology' is one of the seven thematic EMODnet projects running in parallel, the so-called 'Lots' (Bathymetry, Geology, Seabed Habitats, Chemistry, Biology, Physics and Human Activities). The Geology Lot (lead Geological Survey of Finland (GTK), 39 partner organizations by now), started in 2009, comprises 12 work packages (see also Moses and Vallius 2020; Vallius *et al.* 2021). The Federal Institute for Geosciences and Natural Resources (BGR) leads the work package Seafloor Geology to compile and harmonize the pre-Quaternary and Quaternary offshore geology of the European seas. EMODnet Geology is currently in its fourth contractual phase.

Geological survey and research organizations everywhere have been developing their own regional or national method to map, describe and depict their geological information for centuries (Asch 2003). For marine data, this is even more the case, as most of the marine domain is investigated in much less detail than the land areas and there is often only geophysical data to provide information on marine geological units and structures. Knowledge is thus heterogeneous: it can be comparatively detailed in small, well-investigated areas while, for example, the surrounding zones are only patchily investigated. In addition, portrayal rules and classification systems show at least as much variety and discrepancies as those used for mapping the onshore geology.

BGR has been involved in the EMODnet Geology project since its beginning. In phase III, for the first time, the theme 'geomorphology' was included in EMODnet Geology and integrated into the work package Seafloor Geology lead by BGR. The tasks of the work package encompass compiling a pre-Quaternary, a Quaternary and a geomorphological map of the European seas with contributions from existing data from all EMODnet Geology partners. BGR also provides geological and geomorphological information about the German part of the North Sea and the Baltic seas to EMODnet Geology and Emodnet Geology Portal by BGR.

Geomorphology

Geomorphology is the study (ancient Greek, λόγος, lógos) of the forms (ancient Greek, μορφή, morphḗ) of the Earth (ancient Greek, γῆ, gê). The British Society for Geomorphology (https://www.geomorphology.org.uk/) describes geomorphology as 'the study of landforms, their processes, form and sediments at the surface of the Earth (and sometimes on other planets). [...] Landforms are produced by erosion or deposition, as rock and

From: Asch, K., Kitazato, H. and Vallius, H. (eds) 2022. *From Continental Shelf to Slope: Mapping the Oceanic Realm.* Geological Society, London, Special Publications, **505**, 39–56.
First published online November 26, 2021, https://doi.org/10.1144/SP505-2021-24

sediment is worn away by these earth-surface processes, transported, and deposited to different localities, wear deposition, as rock and sediment away […] Geomorphologists map the distribution of landforms so as to understand better their occurrence.'

Micallef *et al.* (2018) define submarine or offshore geomorphology as the 'study of landforms and processes in the submarine domain. Within EMODnet Geology, it is understood that 'marine geomorphology describes the bathymetric features and their origin at the seafloor, i.e. it describes the submarine "landscapes" and their genesis'. Combined with geological information, geomorphological information helps to identify landforms as potential habitats for specific organisms and spawning locations of reef fishes, which are crucial habitats. Also identifying specific marine landforms helps in determining hidden natural resources, e.g. mud volcanoes may indicate the existence of gas hydrates. In archaeology, for example, the determination of palaeocoastlines aids the discovery of ancient settlement areas, such as on the Dogger Bank, most probably a submerged moraine (K. Asch and C. Moses, 2016, 'Off-shore geomorphology', EMODnet Geology Phase III position paper, unpublished).

The theme of Geomorphology was added for the first time to the tasks of work package 4 during the third contractual phase of EMODnet Geology. The major challenge here was and is to create a (digital) marine geomorphology map covering the European seas – a 'pioneer work' as such a map has never been created before. The map is still in the process of being built and updated. Based on the contributions of the EMODnet Geology partners it aims to show the landforms ('seaforms') on the seafloor. These geomorphological features in combination with the Quaternary geology visualize the forms, partly the genesis and the material of the uppermost, youngest geological units. This includes geomorphological features of the continental shelf, slope and rise, oceanic trenches and ridges, and the abyssal plain. For example, the shelf features encompass remnants of the last glaciations, such as moraines, marine and fluvial deposits, such as deltaic fans, and aeolic deposits such as dune fields, but also features such as pockmark fields. Volcanic features are distributed not only on the shelf, but also on the continental slope and on the abyssal plain.

Participation in the EMODnet Geology project and the demand for a European geomorphology map initiated this compilation of a Quaternary and Geomorphological map of the North Sea and Baltic Sea as neither a Quaternary nor a geomorphological map of the German seas existed.

One of the challenges of that task was that numerous publications of mapped areas of the geomorphology, the Quaternary or both exist, but a synopsis has

not yet been undertaken. In addition, publications have been published on the topic, but not marked with the key words 'geomorphology' or 'Quaternary', but for example as 'Implementation of the Natura 2000 Guidelines' or 'Mapping of the Flensburg Fjord'. The geomorphology/Quaternary geology of the German offshore areas has not yet been a focus of marine mapping in Germany, where no mapping agency is responsible for geomorphological mapping. However, the Federal Maritime and Hydrographic Agency (Bundesamt für Seeschifffahrt und Hydrographie, BSH) is mapping and publishing bathymetric maps of the German seas in high detail, such as EasyGSH-DB and the nautical chart. The conversion of the data into geomorphological and geological units has not yet been completed.

In order to build a multinational geomorphological layer of the European seas the application of common standards was essential. This encompassed both the use of already existing standards, in particular from the European Commission INSPIRE data specifications (INSPIRE Thematic Working Group Geology 2013) and the IUGS Stratigraphic Chart (Cohen *et al.* 2013, updated), and the creation of additional prototype vocabularies to be applied Europe-wide so that the Quaternary geology and geomorphology will be interoperable across Economic Exclusive Zone boundaries.

This common controlled vocabulary (Asch *et al.* 2020) consists of detailed lists of the properties, terms and definitions needed to describe the geological and geomorphological units within the three datasets 'pre-Quaternary geology', 'Quaternary geology' and marine 'geomorphology'. For all of the features the INSPIRE Data specifications were used as far as possible for description, so that a later transformation of the EMODnet Geology datasets according the INSPIRE rules would be facilitated. The Seafloor Geology vocabulary is complemented by separate guidelines (Asch *et al.* 2020). Both the guidelines and vocabularies have been and are being amended and updated from project phase to phase. The guidelines give a practical overview of the tasks and define the process of transforming semantically the geology and geomorphology data provided by each participating country.

The INSPIRE geomorphology vocabulary (INSPIRE Thematic Working Group Geology 2013) was created for onshore, and not for offshore forms. Thus, together with the project of the IQUAME 2500 (International Quaternary Map of Europe and Adjacent Areas; https://www.bgr.bund.de/IQUAME_en), a working group was formed to set up a specific offshore vocabulary for marine forms (see above, Asch *et al.* 2020). This list is still in the process of being iteratively enlarged and optimized and consists to date of *c.* 90 terms and definitions (see EMODnet Geology vocabulary for seafloor geology at the EModnet Geology Portal by the BGR https://www.bgr.bund.de/EN/Themen/Sammlungen-Grundlagen/GG_geol_Info/Projekte/laufend/EMODnet4/EMODnet4_en.html?nn=1556482).

The terms are subdivided into:

- general physiographic features (e.g. continental shelf, continental slope);
- landforms, including genetic aspects;
- biogenic features (e.g. mussel beds, sunken forests); and
- structural features (e.g. scouring marks, glacial lineations).

Table 1 shows the application of the created term list (https://www.bgr.bund.de/EN/Themen/Sammlungen-Grundlagen/GG_geol_Info/Projekte/laufend/EMODnet4/EMODnet4_en.html) in addition to the use of the corresponding INSPIRE terms (INSPIRE Thematic Working Group Geology 2013) for the features presented on the maps introduced in this publication. The terms serve for all EMODnet Geology partners to describe the geomorphological features within their Economic Exclusive Zone.

German marine areas

Germany's marine areas are divided between two seas – the North Sea and the Baltic Sea (Fig. 1a). The geomorphologies of these two seas are very different: whereas the geomorphology of the North Sea is extremely dominated by tidal processes, the geomorphology of the Baltic Sea is hardly influenced by tidal processes. The sills within the Baltic Sea divide the depositional area into different zones and are themselves often erosion areas owing to increased water currents (Harff et al. 1995). The Baltic Sea receives a high input of fresh water and sediments through rivers. This difference in sedimentation areas makes the interpretation of geomorphological forms challenging. Nevertheless, an attempt was made to compile and interpret individual geomorphological structures in the two marine areas. These maps show a methodical approach to creating geomorphological maps and should encourage other researchers to explore the geomorphology of the German seas more intensively.

North Sea

The North Sea is a classic shelf sea and additionally a marginal part of the Atlantic Ocean. It is located on the European continental shelf and is connected to the other seas by the English Channel and the Norwegian Sea. An area analysis using the Benthic Terrain Modeller (Wright et al. 2012) showed that the entire German North Sea could be described as a shelf sea. The global seafloor geomorphic map

(Harris et al. 2014) also indicates a shelf sea for the German North Sea.

Since the opening of the northern Atlantic Ocean during the Jurassic and Cretaceous periods and the associated uplift of the British Isles, a shallow sea has existed almost throughout the area of today's North Sea (Torsvik et al. 2002).

Over millions of years, the sea has adapted to the respective eustatic sea-level fluctuations and climatic conditions. Since the beginning of the Quaternary before about 2.6 Ma, the eustatic sea-level sank during every glacial period and rose again during every warm period (Streif 1990). When the glaciations in the North Sea area began is still unknown, but there were several ice sheet advances in the Middle and Late Pleistocene (Lamb et al. 2017).

In addition to Germany, there are seven states bordering the North Sea (Fig. 1a). The German sector of the North Sea (Figs 1b & 2) with the Duck's Beak 'Entenschnabel' is characterized by a strong Early and Middle Pleistocene sedimentation with the deposition of sediments more than 800 m thick (Caston 1979). The sediments are mainly shallow marine and near the coast shallow marine to fluvio-deltaic (Cameron et al. 1992). The sedimentation in the Quaternary is strongly dominated by the repeated glacier advances of the Middle and Late Pleistocene (Carr et al. 2006). Numerous geomorphological features within the sediment strata document the development history of the North Sea.

Today, the Wadden Sea covers a large part of the Dutch, German and Danish North Sea. In addition, it is largely protected as a national park in the area of the German North Sea and it is a UNESCO World Heritage Site (https://www.waddensea-worldheritage.org/)

Bathymetry

One of the most important tools for the analysis and interpretation of geomorphological structures on the seafloor except for in-situ studies is bathymetry. If bathymetry has a sufficiently high resolution, the morphological events can be mapped and interpreted. After classification, they can be assigned to geological processes and thus described geomorphologically. For the North Sea, we have different bathymetries to choose from, which differ strongly in quality and extent.

Unfortunately, bathymetry with the best resolution (10 × 10 m grid size) does not cover the entire German area of the North Sea, so a bathymetry with a coarser cell size was chosen for the offshore area. The high-resolution bathymetry is compiled by the EasyGSH-DB project network and made available to the public (https://mdi-de.baw.de/easygsh/Projekt_ENG.html). In addition to the

Table 1 List of the geomorphological features appearing in the figures. They are compared with the translation into the EMODnet Seafloor Geology Vocabulary (https://www.bgr.bund.de/EN/Themen/Sammlungen-Grundlagen/GG_geol_Info/Projekte/laufend/EMODnet4/EMODnet4_en.html) and the corresponding INSPIRE Data Specifications on Geology (INSPIRE Thematic Working Group Geology 2013)

| Geomorphological features | EMODnet Seafloor Geology Vocabulary | | INSPIRE Data Specifications on Geology | | | |
	Landform/physiographic feature	Geomorphology feature type	Event environment	Event process 1	Event process 2	Event process 3
Palaeovalley	Channel	Alluvial and fluvial features	River channel setting	Erosion	Water erosion	
Heligoland Mud Area	Basin	Depression	Continental shelf setting	Deposition from water		
Current channel	Current channel	Depression	Continental shelf setting	Erosion		
Moraine	Moraine	Glacial, glaciofluvial, -lacustrine and -marine features	Subglacial setting	Deposition by or from moving ice		
Sediment bank	Sediment bank	Glacial, glaciofluvial, -lacustrine and -marine features	Continental shelf setting	Deposition by or from moving ice	Deposition from water	Reworking
Tunnel valley	Tunnel valley	Glacial, glaciofluvial, -lacustrine and -marine features	Subglacial setting			
Back barrier channel system	Fluvial erosional landform	Marine, littoral and coastal wetlands features	Tidal channel setting	Erosion	Deposition from water	Reworking
Ebb tidal delta	Delta	Marine, littoral and coastal wetlands features	Deltaic system setting	Deposition from water		
Ebb tidal delta	Delta lobe	Marine, littoral and coastal wetlands features	Tidal setting	Deposition from water	Reworking	
Estuary	Estuary	Marine, littoral and coastal wetlands features	Estuary setting	Sea-level rise		
Estuary	Estuary	Marine, littoral and coastal wetlands features	Estuary setting	Sea-level rise	Deposition from water	

Intertidal flats	Intertidal flats	Marine, littoral and coastal wetlands features	Mud flat setting	Deposition from water		
Intertidal flats	Intertidal flats	Marine, littoral and coastal wetlands features	Tidal flat setting	Erosion	Reworking	
Lagoon	Lagoon	Marine, littoral and coastal wetlands features	Lagoonal setting	Sea-level rise	Deposition from water	
Main ebb channel	Fluvial erosional landform	Marine, littoral and coastal wetlands features	Tidal channel setting	Erosion	Deposition from water	Reworking
Pockmark field	Pockmark field	Marine, littoral and coastal wetlands features	Continental shelf setting	Deposition from fluid		
Shoreface connected ridges	Sand ridge field	Marine, littoral and coastal wetlands features	Marine setting	Reworking	Sedimentary process	
Ridge crest	Ridge crest	Marine, littoral and coastal wetlands features	Marine setting	Reworking	Sedimentary process	
Rocky tidal flat	Bench	Marine, littoral and coastal wetlands features	Tidal setting	Erosion		
Sand bar	Sand bar	Marine, littoral and coastal wetlands features	Inner neritic setting	Reworking	Deposition by or from moving ice	
Sand bar	Sand bar	Marine, littoral and coastal wetlands features	Inner neritic setting	Reworking	Erosion	
Shallow inlets and bays	Shallow inlets and bays	Marine, littoral and coastal wetlands features	Inner neritic setting	Sea-level rise	Deposition from water	
Shoal	Shallow inlets and bays	Marine, littoral and coastal wetlands features	Inner neritic setting	Sea-level rise	Deposition from water	
Wind tidal flat		Marine, littoral and coastal wetlands features	Tidal flat setting			
Sunken forest	Sea-level rise	Marine, littoral and coastal wetlands features	Marine setting	Sea-level rise		

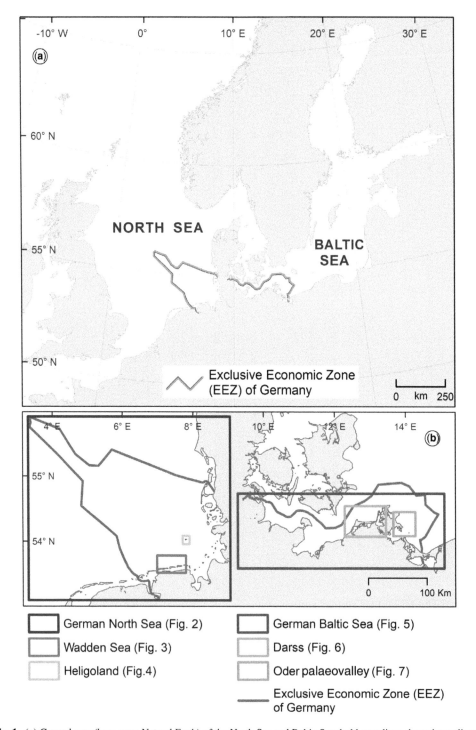

Fig. 1. (**a**) General map (base map: Natural Earth) of the North Sea and Baltic Sea; bold grey lines show the outlines of the German Economic Exclusive Zone (http://www.marineregions.org). (**b**) General map of the German seas; blue rectangles show the position of the detailed maps in the North Sea and green rectangles show the position of the detailed maps in the Baltic Sea; bold grey lines show the outlines of the Economic Exclusive Zone (Flanders Marine Institute 2018).

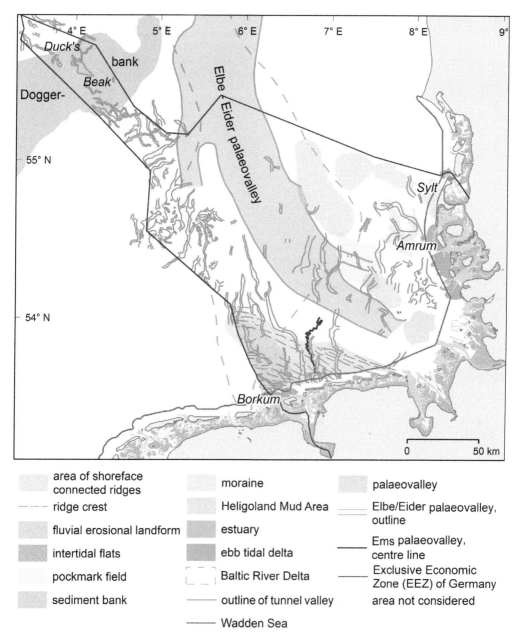

Fig. 2. Compilation of geomorphological structures in the German North Sea; Elbe palaeovalley after Papenmeier and Hass (2019); Ems palaeovalley after Hepp *et al.* (2017); Baltic River Delta after Lamb *et al.* (2017); tunnel valleys after Schwarz (1996), Lutz *et al.* (2009) and Hepp *et al.* (2012); moraine ridges after Streif and Köster (1978), Andresen (1985) and Alappat *et al.* (2010); position of the Dogger Bank after Cotterill *et al.* (2017); intertidal flats after Dijkema (1977); Heligoland Hole after Hebbeln *et al.* (2003).

high-resolution multibeam data, radar data was also used in order to include the extended shallow water areas of the Wadden Sea. The digital terrain model of the EMODnet Bathymetry Portal (http://www. emodnet-bathymetry.eu/), published in September 2018 with a resolution of a quarter of an arc minute, was used for the area of the Duck's Beak and the areas not yet covered by high-resolution bathymetry.

This corresponds to a resolution of about 62 × 116 m in the area of the German North Sea

For both bathymetry datasets, a terrain (elevation) analysis was carried out in order to display and analyse the landform characteristics in detail. For this purpose different raster-based neighbourhood analyses such as slope, aspect and hillshade were performed with the GIS software.

Wadden Sea

The coastal area of the German North Sea is strongly influenced by tidal changes and is therefore defined as a special sea area – the Wadden Sea (https://www.waddensea-secretariat.org/). The Wadden Sea is divided into different areas (Fig. 3), which are distinguished in the map 'Salt marsh data from the 1970s' published by Dijkema (1977). Owing to the high sediment dynamics and strong currents in the Wadden Sea area the zones are not static, but there is a constant displacement of, for example, channels, erosion and accumulation zones (Schäfer

2005). The intertidal flats, biogenic beds and salt marshes were adopted unchanged from the quoted map (Fig. 3). The course of the main ebb channel and other channels was adapted to the current situation using bathymetric maps. Dijkema (1977) did not determine the ebb tidal deltas; however, they are an important geomorphological unit of the Wadden Sea, so they were interpreted and mapped based on EasyGSH-DB bathymetry (Fig. 3).

The large rivers that flow into the North Sea form estuaries, which are also recorded as a separate geomorphological unit. The estuaries have been delineated in accordance with the Water Framework Directive (Fig. 2) on the basis of their salinity (Kampa and Hansen 2004).

Heligoland

On the German offshore island 'Heligoland' (Fig. 4) there is a special form of the Wadden Sea – the Heligoland wave cut platform, which is part of the Heligoland continental shelf. It is formed by the erosion of the cliff of the rocky island. The tidal range in this

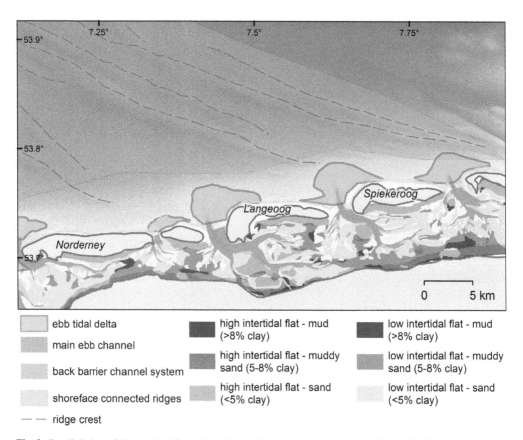

Fig. 3. Detailed view of the intertidal flats of the Wadden Sea classified according to Dijkema (1977); geomorphological structures of the Wadden Sea based on own interpretations of the bathymetric data (EasyGSH-DB).

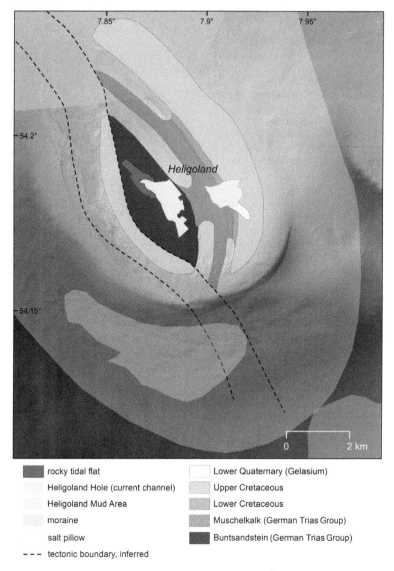

Fig. 4. Map of the Island of Heligoland showing the geology based on the GÜK200 (Barckhausen *et al.* 1973; Kaufmann *et al.* 2014; Reinhold *et al.* 2008); Heligoland Hole and Heligoland Mud Area after Hebbeln *et al.* (2003); position of the moraine after Alappat *et al.* (2010); the bathymetry belongs to the dataset of the EasyGSH-DB. Topographic base: DTK 250 (BKG 2020).

area is *c.* 2.3 m (Klein and Mittelstaedt 2001). The 2.5 m deep contour and aerial photographs were used to define the border of the wave cut platform.

Two other interesting geomorphological features can be found on the seabed south and SE of the island of Heligoland. The almost flat shelf shows two depressions. First, the Heligoland Hole (Fig. 4) with a water depth of about −50 m is a distinct depression in the otherwise very shallow shelf area of the North Sea. It is an almost unfilled depression.

Second, the area SE of Heligoland, the so-called Heligoland Mud Area (Fig. 2), belongs to it. The 'Mud area' does not stand out morphologically from its surroundings, except by the high clay content of the sediments deposited there. Up to 30 m thick Holocene sediments are drilled (Hebbeln *et al.* 2003). The area is comparable with the unfilled Heligoland Hole. Both depressions are caused by halotectonic processes related to the rise of the Heligoland diapir (Hebbeln *et al.* 2003) and represent the rim syncline of the salt dome (Fig. 4).

Quaternary (geo-)morphological features

During the last 2.6 Ma, the area of the North Sea was strongly influenced by the various ice ages and interglacial periods of the Pleistocene. During the Elsterian glaciation, almost the entire North Sea basin was covered by ice. Also during the Saalian and the early Weichselian glaciations significant phases of glacier activity took place which geomorphologically shaped the present seabed and deposited glacial sediments (Graham et al. 2011). The relicts of these glaciations are today's subaquatic moraines, palaeovalleys and filled tunnel valleys.

Moraines

Hardly any terminal moraine ridges have been published from the seabed of the German North Sea. A publication by Streif and Köster (1978) shows moraine sediments on the islands Amrum and Sylt and offshore the west of the islands (Fig. 2). Also the Office of Agriculture and Water Management Husum which belongs to the State Agency for Coastal Protection, National Park and Marine Protection Schleswig–Holstein, published a report in 1985 (Andresen 1985) showing the moraines in this area. Both maps were compared with the current bathymetric map, as the moraine ridges are still visible in the bathymetry. From all of this information, the most accurate location of moraine systems has been compiled (Fig. 2). In addition, the till plateaus are the remains of such moraine systems (Fig. 2), and are described for example by Alappat et al. (2010) offshore Borkum.

The Dogger Bank (Fig. 2) in the central North Sea, which cuts across the German sea area in the outer Duck's Beak, is also regarded as a moraine, although it is often described in the literature as a sandbank. Dogger Bank was either an island or connected to the mainland during various water lowstands during the last glacial period (Cotterill et al. 2017).

Tunnel Valleys

The main morphological evidence for glaciation in the North Sea is restricted to subglacial tunnel valleys (Graham et al. 2011). In the area of the German North Sea, Schwarz (1996) and Lutz et al. (2009) determined the position of the tunnel valleys (Fig. 2) using geophysical methods (2D and 3D seismic data). In Lutz et al.'s dataset (2009), different generations of tunnel valleys can be subdivided which most probably belong to the different phases of the Elsterian glaciation. Some of the data from Lutz et al. (2009) and Schwarz (1996) contain duplications that have been removed to make the map easier to read. For this purpose, the quality of the data,

the context and the database were used as references. D. A. Hepp et al. (2012) provides another tunnel valley that has been included in the dataset. The special aspect of this tunnel valley is that it extends in an east–west direction in contrast to the others (Fig. 2).

Palaeovalleys

The submerged valley of the Palaeo-Elbe forms one of the most prominent structures of the German North Sea (Fig. 2), which is still visible today in the bathymetric data. The valley developed during the Weichselian glacial period, when the sea-level was 130 m lower than it is today. Meltwater, which flowed NW along the ice sheet, fed the Palaeo-Elbe. During the Holocene, rising sea-level flooded the valley (Papenmeier and Hass 2019). D. Hepp et al. (2017) discovered another river structure in the south that can be interpreted as a submerged extension of today's Ems River (Fig. 2).

Bedforms

Looking at the high-resolution bathymetry of the North Sea, numerous bottom features formed by tidal currents and other currents can be recognized. In the area of the southern North Sea so-called tidal ridges have formed, the distribution of which extends into the Dutch part of the North Sea (Fig. 2). These are flat giant ripples caused by the extremely strong tidal current in the English Channel (Schäfer 2005). Two different presentation modes were chosen for the map. Firstly, the ridge crests of the ripples were interpreted and digitized in order to classify the ridge spacing and ridge heights, and secondly, the distribution area of the tidal ridges was delimited (Fig. 3).

The pockmarks are structures that occur repeatedly in the North Sea as well as buried structures in the seismic data (Fig. 2). Owing to the high sediment dynamics of the North Sea, these structures can be levelled or filled with sediment. Krämer et al. (2017) report a pockmark field at the edge of the Elbe palaeovalley, as it formed after a strong storm event in 2015. The total area is 915 km^2 and up to 1200 pockmarks/square kilometre have been recorded.

Paleogene and Neogene – development of the Baltic River System

During the Paleogene, a large river system developed by the uplift of the Fennoscandian Shield, which drained west and flowed along today's Baltic Sea basin, the Baltic River System. In the Miocene the river system reached today's North Sea basin (Overeem et al. 2001) and filled the North Sea Basin with Duck's Beak thick sediment packages, leaving extensive delta systems (Thöle *et al.* 2014). The delta fills are now overlaid by Quaternary deposits (Thöle *et al.* 2014); however, geophysical methods can make these structures visible again. Lamb *et al.* (2017) have shown the position of the Neogene shelf deltas in a map showing the palaeoenvironment of the North Sea 2.58 Ma ago.

Baltic Sea

The Baltic Sea is a marginal sea of the Atlantic Ocean. The relatively flat Baltic Sea is about 190 km wide and 1600 km long (Fig. 1a). The average depth of the Baltic Sea is 55 m (Hupfer *et al.* 2003) and the deepest point is 459 m (Tuuling *et al.* 2011). Several shallow sills (partly only 7 m water depth) in the Danish part limit the inflow of fresh water from the Atlantic Ocean. Therefore, the Baltic Sea is one of the largest brackish water seas in the world (Tuuling *et al.* 2011). In the literature, the formation of the Baltic Sea is attributed to glacial erosion by the various Quaternary glaciations or tectonic processes. The appearance of today's Baltic Sea is obviously a combination of both.

The Baltic basins lie on suture zones. They were affected by crustal thinning and possibly aborted rift formation during the Mesoproterozoic and the late Neoproterozoic (Winterhalter *et al.* 1981). The continental plate collision in the Paleozoic period led to the formation of a fold and thrust belt. In the Mesozoic up to the Cretaceous, there was only a little tectonic activity. During the Cretaceous tectonic inversion occurred in the southwestern part of the basin (Winterhalter *et al.* 1981). During the Neogene period a large river system was formed, the Baltic River System, which drained from east to west through the Fennoscandian and Baltic Shields along the present Baltic Sea and deposited large delta sediments in the area of the present North Sea (Overeem *et al.* 2001). The rivers cut deeply into the land surface and gouged out huge canyons (Tuuling *et al.* 2011). The Baltic River continued to be active until *c.* 1.2 Ma, when the drainage network was destroyed by erosion beneath the Menapian ice sheet (Hall and van Boeckel 2020).

Until the Early Pleistocene, there is no evidence for the existence of the Baltic Sea and thus the present Baltic Basin is a Quaternary phenomenon that is of Holsteinian age (Meyer 2003). During the Pleistocene, various glaciations and glacial lakes filled the area of the today's Baltic Sea basin. The glacial erosion strongly influenced the morphology of the sea bottom, especially when the glacial flow direction coincided with structurally weak zones in the bedrock and considerable deepening and widening of channels and valleys were caused by glacial gouging (Winterhalter *et al.* 1981). Major glacial landforms

have been formed owing to glacial erosion. The primary features are the topographic depressions that hold the Bothnian Bay, the Bothnian Sea and the Baltic Sea, which are separated by submarine sills (Hupfer *et al.* 2003; Hall and van Boeckel 2020). A prominent threshold in the area of the German Baltic Sea is the Darss Sill (Figs 5, 6). The sill, which is also hydrographically important, is currently located at <20 m water depth, with the base of the Saale period and younger glacial succession (moraines; Figs 5 & 6) resting on Late Cretaceous chalk at 40–80 m depth (Lemke *et al.* 1994). The Darss Sill has been lowered by glacial erosion in the Middle and Late Pleistocene. Prior to glacial erosion, the sill probably stood at or above present sea-level (Hall and van Boeckel 2020).

First, the Baltic proglacial lake was formed in the area of today's Baltic Sea by the melting Scandinavian Ice Sheet. The next phase of the Baltic Sea development is the phase of the Yoldia Sea. Characteristic of this is an at least temporary influx of salty water into the Baltic Sea basin. With the isostatic rise of Sweden, the connection between the Baltic Sea basin and the Kattegat became visibly narrower (Lemke 1998). A new phase of Baltic Sea development, the Ancylus Sea phase, began. The first signs of maritime influences appeared in the Mecklenburg Bay around 8000 BC, when marine water masses flowed into the area from the North Sea during the Littrorina Transgression. They represent the transition from limnic–terrestrial to brackish–marine conditions and belong to the development of the Littorina Sea. Based on sedimentological and stratigraphical findings, an extraordinarily rapid rise in sea-level can be demonstrated (Lemke 1998). The changes in the environment of Baltic Sea bottom water is documented by sediment physical, geochemical and microfossil proxies (Harff *et al.* 2011). The Littorina Transgression formed the recent geographic picture of the Baltic Sea. The formation of today's marginal sea of the Atlantic Ocean is attributed to two important processes (Hupfer *et al.* 2003): on the one hand, the melting of Scandinavia's ice masses and the rising sea-level worldwide, and on the other hand, isostatic uplift owing to the lack of ice loading (Harff *et al.* 2005).

Bathymetry

For the Baltic Sea, there are some bathymetric datasets, which differ clearly in quality, extent and resolution. The best-resolved and edited dataset is from the BSH. The data have a resolution of 50×50 m and can be downloaded free of charge from the map viewer 'Geoseaportal' BSH (2019) (https://www.geoseaportal.de/). In addition to this dataset, the seabed relief map published by the Leibniz

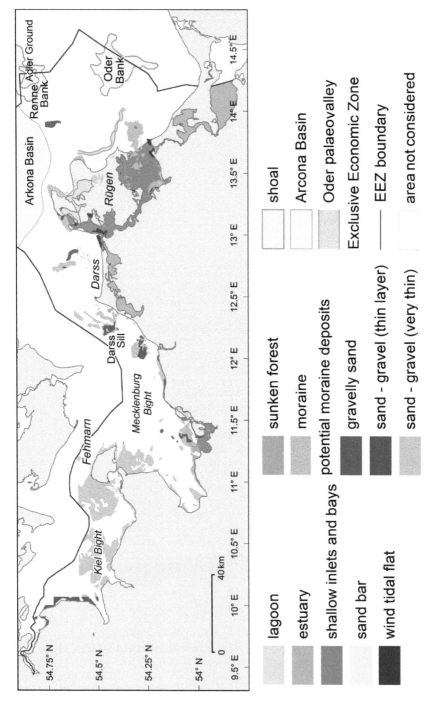

Fig. 5. Compilation of geomorphological structures in the German Baltic Sea: potential moraine deposits after GÜK250: classification of lagoon, estuary and bays after Herrmann *et al.* (2015); moraines and wind tidal flats after Herrmann *et al.* (2015); shoals after Zeiler *et al.* (2008). Topographic base: EEA (2018), EEA coastline for analysis 1:100 000.

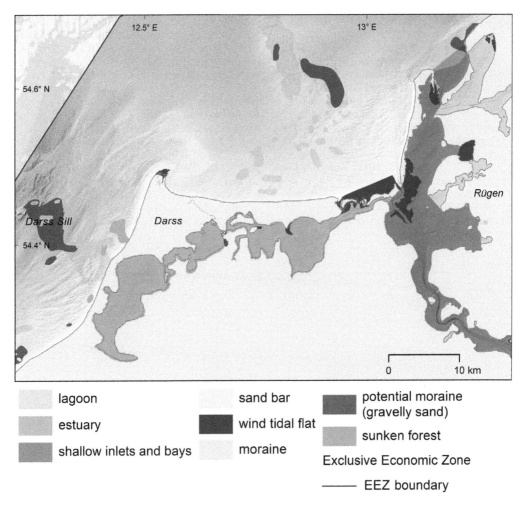

Fig. 6. Detailed view of the geomorphological features between the Darss Peninsula and the Island of Hiddensee (German Baltic Sea); base map is the bathymetry after Bundesamt für Seeschifffahrt und Hydrographie (BSH bathy); potential moraine deposits after GÜK250; classification of lagoon, estuary, bays, moraines and wind tidal flats after Herrmann *et al.* (2015); sunken forest from own interpretation. Topographic base: EEA (2018), EEA coastline for analysis 1:100 000.

Institute for Baltic Sea Research and BSH was used. These can be found as geological maps under 'Topical Charts and Atlases' on the BSH website: (https://www.bsh.de/EN/PUBLICATIONS/publications_node.html). The institutes have cartographically processed the bathymetry dataset so that the subaquatic structures can be represented in the best possible way. For the marine areas outside the German Baltic Sea, the EMODnet Bathymetry dataset from September 2018 was used (see 'North Sea Bathymetry').

A terrain analysis was also applied to the bathymetric data for the Baltic Sea, which significantly improved the interpretation of the data.

Seafloor classification

Since there is still no geomorphological or morphological classification for the Baltic Sea, the draft of the geomorphological map started with a classical spatial analysis. Where are basins and depressions located? Where are sills, banks and grounds to be found? In addition, how are they separated from each other?

This first classification was based on sea navigation maps published by the BSH. The subaquatic structures are limited to certain depth contours (−5, −10 and −20 m). The digitized contours of these structures were compared with the bathymetric map and, with comprehensible agreement, included in the map as a structure (Fig. 5).

The included features are interpreted as shoals (Zeiler *et al.* 2008) which are of glacigenic origin like the Oder Bank, Rønne Bank and Adler Ground (Fig. 5). Within the southwestern Baltic Sea, the Arkona Basin (Fig. 5) is the deepest basin with a water depth of up to 45 m (Zeiler *et al.* 2008).

For the Baltic Sea region of Mecklenburg–Western Pomerania we were able to identify geomorphological features based on data collected by the authorities (Herrmann *et al.* 2015). The authorities are implementing the Fauna–Flora-Habitat Directive of 1992 issued by the EU Commission. This directive also lists marine habitat types for which the Member States have special protection obligations

(Council of the European Union 1992). For this purpose, the nature of the seabed must be sufficiently well known. Numerous surveys have therefore been initiated to obtain sufficient data for the implementation of this directive in the Natura 2000 project (https://www.bmu.de/themen/natur-biologische-vielfalt-arten/naturschutz-biologische-vielfalt/gebietsschutz-und-vernetzung/natura-2000/). Herrmann *et al.* (2015) describes the subaquatic structures of the seabed. In general, six different geomorphological characteristics are distinguished in the publication. The subdivision of the 'Bodden' landscape into lagoon, estuary and bay is based on the salinity of the seawater (Fig. 5). The distribution of subaquatic moraines, sandbanks and wind tidal flats was mapped, but without guarantee of

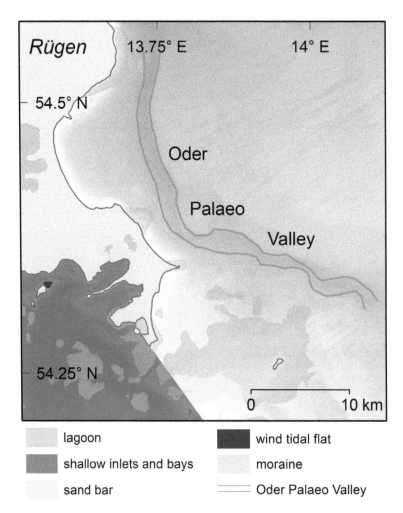

Fig. 7. Detailed view of the geomorphological features east of the Rügen Island; base map is the bathymetry after BSH (BSH bathy); geomorphological features after Herrmann *et al.* (2015).

completeness (Herrmann *et al.* 2015). In order to verify the occurrence of moraines or sandbanks, a survey of the General Geological Map of the Federal Republic of Germany (GÜK250), scale 1:250 000 (Bundesanstalt für Geowissenschaften und Rohstoffe 2019), was carried out. This query was intended to select all areas containing sand and gravel. These areas correspond well with the areas mapped by Herrmann *et al.* (2015). The mapped areas are much more accurate than the query in terms of their extent and distribution and have been incorporated into the geomorphological map (Fig. 5).

For the Baltic Sea region of Schleswig–Holstein no data from the mapping for the implementation of the Habitats Directive have been received; therefore excerpts from the GÜK250 (Bundesanstalt für Geowissenschaften und Rohstoffe 2019) for this area were taken.

Palaeovalleys

The bathymetric dataset of the German Baltic Sea shows several channels and valleys (Fig. 6). However, these structures cannot be assigned to a specific geological event without further scientific investigation. They may be of glacial origin or related to post-glacial flooding of the basin. Therefore, these features have not been included in the map at this time, except for a channel or valley located east of the island of Rügen and very clearly visible in the bathymetric data, the Oder palaeovalley (Fig. 7). The Oder palaeovalley is mentioned in the publication of Lampe *et al.* (2007), but the exact location and course of the valley is not shown. Therefore, the course of the valley was reinterpreted based on the bathymetric data. The valley can only be perfectly mapped in a limited area. In the other areas thick sediment layers or other geomorphological structures superimpose the morphology.

Sunken forests

The distribution of tree and wood residues in the GÜK250 in the Baltic Sea was also analysed using a query (Fig. 6). In the Darss Peninsula, several polygons were written out which correspond to publications (Lampe and Lorenz 2010; Herrmann *et al.* 2015) on this area. These publications describe a large area of tree trunks and wooden remains belonging to a sunken forest flooded by the rapid sea-level rise (20 m) during the Littorina Transgression.

Discussion and conclusions

Despite the existence many wonderful publications of seminal science, some of which are reported here, the selection of data and its compilation took up a large part of the project phase. Unfortunately, it was not possible to address all of the published data and maps, but an attempt was made to substantiate the most important geomorphological structures with the relevant publications. In order to publish a definitive geomorphological map of the German seas, significantly more time would have to be invested and the German marine research centres would have to be more involved.

Nevertheless, the geomorphological map presented here is the first of its kind and will be revised and, above all, expanded in further follow-up projects. It should also be considered whether the focus should not rather be on a purely digital application so that the geomorphological structures can be presented in different information layers. This makes sense because geomorphological structures of different sizes often overlap or because different time periods are to be worked out in a large geomorphological structure.

The information that can be derived from the geomorphological map is a valuable support for all marine activities in the scientific exploration of the different marine earth systems, such as the seabed and fauna habitats, and the commitment to protect the marine environment while developing the marine economy. Examples include projects to explore sites for wind farms or deep-sea cables or the search for areas for sand and gravel extraction. These measures address the economy, social needs and habitat protection.

Acknowledgements We acknowledge with thanks the support of the EMODnet Geology project. In addition, we would like to thank the wonderful colleagues of the EMODnet Geology community for their support and inspired work on the prototype of the European offshore geomorphology vocabulary and Henry Vallius, the EMODnet Geology coordinator, for his encouragement and support. Special thanks go to Alexander Müller of BGR for his excellent GIS-cartographic work and patient optimization of the maps.

Author contributions SB: writing – original draft (lead); **KA**: writing – original draft (equal).

Funding We acknowledge with thanks the support of the EMODnet Geology project under the EASME/ EMFF/2018/1.3.1.8 – lot 1/SI2.811048_EMODnet – Geology of the European Union.

Data availability The data that support the findings of this study are available from the corresponding author upon reasonable request.

References

Alappat, L., Vink, A., Tsukamoto, S. and Frechen, M. 2010. Establishing the Late Pleistocene–Holocene sedimentation boundary in the southern North Sea using OSL dating of shallow continental shelf sediments. *Proceedings of the Geologists' Association*, **121**, 43–54, https://doi.org/10.1016/j.pgeola.2009.12.006

Andresen 1985. Fachplan Küstenschutz Sylt. Landesbetrieb für Küstenschutz, Nationalpark und Meeresschutz Schleswig–Holstein (LKN.SH).

Asch, K. 2003. The 1: 5 000 000 International Geological Map of Europe and Adjacent Areas: Development and Implementation of a GIS-enabled Concept. Geologisches Jahrbuch, SA 3, Stuttgart: E. Schweizerbart'sche Verlagsbuchhandlung.

Asch, K., Müller, A., Breuer, S. and Gdaniec, P. 2020. EMODnet Phase IV, WP 4 Vocabulary: Pre-Quaternary, Quaternary, Geomorphology. BGR (Hannover) Bundesanstalt für Geowissenschaften und Rohstoffe.

Cameron, D.J., Crosby, A., Balson, P., Lott, K., Bulat, G., J, J. and Harrison, D. 1992. *The Geology of the Southern North Sea*. HMSO for the British Geological Survey.

Carr, S.J., Holmes, R., van der Meer, J.J.M. and Rose, J. 2006. The last glacial maximum in the North Sea basin micromorphological evidence of extensive glaciation. *Journal of Quaternary Science*, **21**, 131–153, https://doi.org/10.1002/jqs.950

Caston, V.N.D. 1979. The quaternary sediments of the North Sea. *In*: Banner, F.T., Collins, M.B. and Massie, K.S. (eds) *The North-west European Shelf Seas: The Sea Bed and the Sea in Motion Part I; Geology and Sedimentology*. Elsevier Oceanography Series. Elsevier, 195–270.

Cohen, K.M., Finney, S.C., Gibbard, P.L. and Fan, J.-X. 2013, updated. The ICS international chronostratigraphic chart. *Episodes*, **36**, 199–204 https://doi.org/10.18814/epiiugs/2013/v36i3/002

Cotterill, C., Phillips, E., James, L., Forsberg, C., Inge Tjelta, T., Carter, G. and Dove, D. 2017. The evolution of the Dogger Bank, North Sea: A complex history of terrestrial, glacial and marine environmental change. *Quaternary Science Reviews*, **171**, 136–153, https://doi.org/10.1016/j.quascirev.2017.07.006.

Dijkema, K.S. 1977. Salt marsh data from the 1970s. Common Wadden Sea Secretariat.

Council of the European Union. 1992. Council Directive 92/43/EEC of 21 May 1992 on the Conservation of Natural Habitats and of Wild Fauna and Flora.

Graham, A.G.C., Stoker, M.S., Lonergan, L., Bradwell, T. and Stewart, M.A. 2011. The Pleistocene glaciations of the North Sea basin. *In*: Ehlers, J., Gibbard, P.L. and Hughes, P.D. (eds) *Quaternary Glaciations Extent and Chronology, Developments in Quaternary Sciences*. Elsevier, 261–278.

Hall, A. and van Boeckel, M. 2020. Origin of the Baltic Sea basin by Pleistocene glacial erosion. *GFF*, **142**, 237–252, https://doi.org/10.1080/11035897.2020.1781246

Harff, J., Lemke, W., Tauber, F. and Emelyanov, E.M. 1995. Geologische Kartierung der Ostsee. *Geowissenschaften*, **13**, 442–447.

Harff, J., Lampe, R., Lemke, W., Lübke, H., Lüth, F., Meyer, M. and Tauber, F. 2005. The Baltic Sea – a model ocean to study interrelations of geosphere, ecosphere, and anthroposphere in the coastal zone. *Journal of Coastal Research*, **213**, 441–446, https://doi.org/10.2112/1551-5036(2005)21[579:AITMIC]2.0.CO;2

Harff, J., Endler, R. *et al*. 2011. Late quaternary climate variations reflected in Baltic Sea sediments. *In*: Harff, J., Björck, S. and Hoth, P. (eds) *The Baltic Sea Basin*. Springer, 99–132.

Harris, P.T., Macmillan-Lawler, M., Rupp, J. and Baker, E.K. 2014. Geomorphology of the oceans. *Marine Geology*, **352**, 4–24, https://doi.org/10.1016/j.margeo.2014.01.011

Hebbeln, D., Scheurle, C. and Lamy, F. 2003. Depositional history of the Helgoland mud area, German Bight, North Sea. *Geo-Marine Letters*, **23**, 81–90, https://doi.org/10.1007/s00367-003-0127-0

Hepp, D., Warnke, U., Hebbeln, D. and Mörz, T. 2017. Tributaries of the Elbe palaeovalley: features of a hidden palaeolandscape in the German Bight, North Sea. *In*: *Under Sea: Archaeology and Palaeolandscapes of the Continental Shelf*. 211–222, Springer, https://doi.org/10.1007/978-3-319-53160-1_14.

Hepp, D.A., Hebbeln, D., Kreiter, S., Keil, H., Bathmann, C., Ehlers, J. and Mörz, T. 2012. An east–west-trending quaternary tunnel valley in the south-eastern North Sea and its seismic–sedimentological interpretation. *Journal of Quaternary Science*, **27**, 844–853, https://doi.org/10.1002/jqs.2599

Herrmann, C., von Weber, M., Zscheile, K. and Gosselck, F. 2015. Nationalpark unter Wasser – Marine Lebensräume in Ostsee und Bodden. *Meer und Museum*, **25**, 72–88, https://www.researchgate.net/publication/275955484_Nationalpark_unter_Wasser_-_Marine_Lebensraume_in_Ostsee_und_Bodden/citations.

Hupfer, P., Harff, J., Sterr, H. and Stigge, H.-J. 2003. Wasserstandsentwicklung in der südlichen Ostsee während des Holozäns. *In*: Hupfer, P., Harff, J., Sterr, H. and Stigge, H.-J. (eds) *Die Küste 66 – Die Wasserstände an der Ostseeküste – Entwicklungen – Sturmfluten – Klimawandel*. Kuratorium für Forschung im Küsteningenieurwesen.

INSPIRE Thematic Working Group Geology. 2013. D2.8.II.4 INSPIRE Data Specification on Geology – Technical Guidelines. European Commission Joint Research Centre (Ispra), p. 232ff. https://inspire.ec.europa.eu/documents/Data_Specifications/INSPIRE_DataSpecification_GE_v3.0.pdf

Kampa, E. and Hansen, W. 2004. *Guidance on Heavily Modified and Artificial Water Bodies*. Springer, Berlin.

Kaufmann, D., Heim, S., Jähne, F., Steuer, S., Bebiolka, A., Wolf, M. and Kuhlmann, G. 2014. GSN – Generalisiertes, erweitertes Strukturmodell des zentralen deutschen Nordsee-Sektors – Konzept zur Erstellung einer konsistenten Datengrundlage für weiterführende Modellierungen im Bereich des zentralen deutschen Nordsee-Sektors; zweite überarbeitete Auflage. Bericht, Hannover (Bundesanst. Geowiss. Rohstoffe).

Klein, E.H. and Mittelstaedt 2001. Gezeitenströme und Tidekurven im Nahfeld von Helgoland. Berichte des Bundesamtes für Seeschifffahrt und Hydrographie (BSH).

Krämer, K., Holler, P. *et al*. 2017. Abrupt emergence of a large pockmark field in the German Bight, southeastern North Sea. *Scientific Reports*, **7**, 5150, https://doi.org/10.1038/s41598-017-05536-1

Lamb, R.M., Harding, R., Huuse, M., Stewart, M. and Brocklehurst, S.H. 2017. The early Quaternary North Sea basin. *Journal of the Geological Society*, **175**, 275–290, https://doi.org/10.1144/jgs2017-057.

Lampe, R. and Lorenz, S. 2010. *Eiszeitlandschaften in Mecklenburg-Vorpommern*. Geozon Science Media.

Lampe, R., Meyer, H., Ziekur, R., Janke, W. and Endtmann, E. 2007. Holocene evolution of the irregularly sinking southern Baltic Sea coast and the interactions of sea-level rise, accumulation space and sediment supply. *Bericht der Römisch-Germanischen Kommission*, **88**, 14–46.

Lemke, W. 1998. Sedimentation und paläogeographische Entwicklung im westlichen Ostseeraum (Mecklenburger Bucht bis Arkonabecken) vom Ende der Weichselvereisung bis zur Litorinatransgression. *Meereswiss Ber, Warnemünde*, **31**, digitale Neuauflage (2015).

Lemke, W., Kuijpers, A., Hoffmann, G., Milkert, D. and Atzler, R. 1994. The Darss Sill, hydrographic threshold in the southwestern Baltic: late quaternary geology and recent sediment dynamics. *Continental Shelf Research*, **14**, 847–870, https://doi.org/10.1016/0278-4343(94)90076-0

Lutz, R.K., Gaedicke, C., Reinhardt, L. and Winsemann, J. 2009. Pleistocene tunnel valleys in the German North Sea: spatial distribution and morphology. *Zeitschrift der Deutschen Gesellschaft für Geowissenschaften*, **160**, 225–235, https://doi.org/10.1127/1860-1804/2009/0160-0225

Meyer, M. 2003. Modelling prognostic coastline scenarios for the southern Baltic Sea. *Baltica*, **16**, 21–30.

Micallef, A., Krastel, S. and Savini, A. 2018. Introduction. *In*: Micallef, A., Krastel, S. and Savini, A. (eds) *Submarine Geomorphology*. Springer, Cham, 1–9.

Moses, C.A. and Vallius, H. 2020. Mapping the geology and topography of the European Seas (European Marine Observation and Data Network, EMODnet). *Quarterly Journal of Engineering Geology and Hydrogeology*, **54**, https://doi.org/10.1144/qjegh20 20-131

Overeem, I., Weltje, G.J., Bishop-Kay, C. and Kroonenberg, S.B. 2001. The late Cenozoic Eridanos Delta system in the Southern North Sea Basin: a climate signal in sediment supply? *Basin Research*, **13**, 293–312, https://doi.org/10.1046/j.1365-2117.2001.00151.x

Papenmeier, S. and Hass, C. 2019. The paleo Elbe River: 10.000 yrs after flooding. In *GEOHAB2019, GEOHAB 2019, 13–17 May 2019, Saints Petersburg*.

Schäfer, A. 2005. *Klastische Sedimente – Fazies und Sequenzstratigraphie*. Springer Spektrum.

Schwarz, C. 1996. Boreholes 89/3, 89/4 and 89/9 from the German North Sea shelf – sedimentological and magnetostratigraphical results and lithostratigraphical correlation. *Geologisches Jahrbuch A*, **146**, 33–137.

Streif, H. 1990. Quaternary sea-level changes in the North Sea, an analysis of amplitudes and velocities. *In*: Brosche, P. and Sündermann, J. (eds) *Earth's Rotation Eons to Days*. Springer, Berlin, 201–214.

Streif, H. and Köster, R. 1978. The geology of the German North Sea coast. *Die Küste*, **32**, 30–50.

Thöle, H., Gaedicke, C., Kuhlmann, G. and Reinhardt, L. 2014. Late Cenozoic sedimentary evolution of the German North Sea – a seismic stratigraphic approach. *Newsletters on Stratigraphy*, **47**, 299–329, https://doi.org/10.1127/0078-0421/2014/0049

Torsvik, T.H., Carlos, D., Mosar, J., Cocks, L.R.M. and Malme, T. 2002. Global reconstructions and North Atlantic palaeogeography 400 Ma to recent. *In*: Eide, E.A. (ed.) *BATLAS–Mid Norway Plate Reconstructions Atlas with Global Atlantic Perspectives*. Geological Survey of Norway, 18–39.

Tuuling, I., Bauert, H., Willman, S. and Budd, G. 2011. The Baltic Sea – geology and geotourism highlights.

Vallius, H.T.V., Kotilainen, A.T., Asch, K.C., Fiorentino, A., Judge, M., Stewart, H.A. and Pjetursson, B. 2021. Discovering Europe's seabed geology: the EMODnet concept of uniform collection and harmonization of marine data. *Geological Society, London, Special Publications*, **50**, https://doi.org/10.1144/SP505-2019-208.

Winterhalter, B., Flodén, T., Ignatius, H., Axberg, S. and Niemistö, L. 1981. Geology of the Baltic Sea. *In*: Voipio, A. (ed.) *The Baltic Sea, Elsevier Oceanography Series*. Elsevier, 1–121.

Wright, D.J., Pendleton, M. *et al.* 2012. ArcGIS Benthic Terrain Modeler (BTM), v. 3.0, Environmental Systems Research Institute, NOAA Coastal Services Center, Massachusetts Office of Coastal Zone Management. http://esriurl.com/5754

Zeiler, M., Schwarzer, K., Bartholomä, A. and Rickleps, K. 2008. Seabed morphology and sediment dynamics. *Die Küste*, **74**, 31–44, Kuratorium für Forschung im Küsteningenieurwesen.

Maps

Barckhausen, J., Lang, H.D., Mengeling, H., Meyer, K.D., Elwert, D. and Voss, H.H. 1973. Geologische Übersichtskarte der Bundesrepublik Deutschland 1:200 000 (GÜK200) – CC 2310 Helgoland; Bundesanstalt für Geowissenschaften und Rohstoffe, BGR.

BKG. 2020. DTK 250 ©GeoBasis-DE. Bundesamt für Geodäsie und Kartographie (Frankfurt/M.). http://www.bkg.bund.de

Bundesanstalt für Geowissenschaften und Rohstoffe. 2019. Geologische Übersichtskarte der Bundesrepublik Deutschland 1:250.000 (GÜK250) (WMS). https://geoviewer.bgr.de/mapapps4/resources/apps/geoviewer/index.html?lang=de. Ausschnitt Helgoland

EEA. 2019. Coastline for analysis, scale 1:100 000. Copyright European Environment Agency (Copenhagen). https://semantic.eea.europa.eu/factsheet.action?uri=http://www.eea.europa.eu/data-and-maps/data/eea-coastline-for-analysis-2

Flanders Marine Institute. 2018. Maritime Boundaries Geodatabase: Maritime Boundaries and Exclusive Economic Zones (200NM), version 10. http://www.marineregions.org/; https://doi.org/10.14284/312

Reinhold, K., Krull, P. and Kockel, F. 2008. Salzstrukturen Norddeutschlands – 1:500 000. BGR, Hannover/Berlin.

Websites

BSH. 2019. Shelf Geo Explorer – Seabed sediments 1:10 000. https://www.geoseaportal.de/wss/service/SGE_Seabed_Sediments_1_to_10000/guest

BSH bathy. https://www.bsh.de/EN/PUBLICATIONS/ publications_node.html

EasyGSH. http://mdi-de.baw.de/easygsh/index.html

EMODnet Bathymetry Portal. http://www.emodnet-bathy metry.eu/

EMODnet Geology Portal. https://www.emodnet-geology.eu/

EMODnet Geology Portal at BGR. https://www.bgr.bund. de/EN/Themen/Sammlungen-Grundlagen/GG_geo l_Info/Projekte/laufend/EMODnet4/EMODnet4_en. html?nn=1556482

IQUAME 2500 project. https://www.bgr.bund.de/ IQUAME_en

Revealing the secrets of Norway's seafloor – geological mapping within the MAREANO programme and in coastal areas

Reidulv Bøe*, Lilja Rún Bjarnadóttir, Sigrid Elvenes, Margaret Dolan, Valérie Bellec, Terje Thorsnes, Aave Lepland and Oddvar Longva

Geological Survey of Norway (NGU), Postal Box 6315 Torgarden, 7491 Trondheim, Norway

SE, 0000-0002-3343-613X; TT, 0000-0002-4040-2122; AL, 0000-0002-8713-7469
*Correspondence: reidulv.boe@ngu.no

Abstract: Results from geological mapping within the MAREANO (Marine Areal Database for Norwegian Coasts and Sea Areas) programme and mapping projects in the coastal zone reveal a rich and diverse seafloor in Norwegian territories. The geomorphology and sediment distribution patterns reflect a complex geological history, as well as various modern-day hydrodynamic processes. By early 2019, MAREANO has mapped more than 200 000 km^2 (c. 10%) of Norwegian offshore areas, spanning environmental gradients from shallow water to more than 3000 m depth, with ocean currents in places exceeding 1 m s^{-1} and water temperatures below $-1°C$. Inshore, along the 100 000 km-long Norwegian coastline, the Geological Survey of Norway (NGU) has conducted a series of seabed mapping projects in collaboration with local communities, industry and other stakeholders, resulting in detailed seabed and thematic maps of seabed properties covering c. 10 000 km^2 (11% of the areas). Bathymetric and geological maps produced by MAREANO and coastal mapping projects provide the foundation for benthic habitat mapping when combined with biological and oceanographic data. Results from the mapping conducted over the past decade have significantly increased our understanding of Norway's seabed and contributed to the knowledge base for sustainable management. Here we summarize the main results of these mapping efforts.

The multidisciplinary Norwegian seabed mapping programme MAREANO (Thorsnes et al. 2008; MAREANO 2019) is a collaboration between the Geological Survey of Norway (NGU), the Institute of Marine Research (IMR) and the Norwegian Mapping Authority (Norwegian Hydrographic Service (NHS)). The programme is financed by the Ministry of Trade, Industry and Fisheries, and the Ministry of Climate and Environment. These ministries, along with the ministries of Petroleum and Energy, Local Government and Modernisation, and Transport and Communications, form the MAREANO Steering Board.

Between 2006 and early 2019, more than 200 000 km^2 of seabed have been mapped (Fig. 1), corresponding to around 10% of the Norwegian offshore area. The areas mapped span broad environmental gradients with water depths extending to more than 3000 m, ocean currents exceeding 1 m s^{-1} and seawater temperatures below $-1°C$. Dramatic landscapes (Fig. 2) have been observed, with canyons up to 1 km deep formed by fluid-flow processes and sliding, and locally almost subvertical margins. Continental slopes vary in width from 30 km to more than 100 km, with gradients locally reaching 60°. Shelf plains and banks (30–300 m

water depth) and cross-shelf troughs (200–500 m water depth) occur over wide areas, and a rich faunal diversity has been observed (e.g. Bellec et al. 2008, 2009, 2010, 2016, 2017a, b, 2019; Chand et al. 2008, 2009, 2012; Thorsnes et al. 2008, 2009, 2016a, b, 2017; Bøe et al. 2009, 2012, 2015, 2016; Buhl-Mortensen et al. 2009a, b, 2012, 2015; Dolan et al. 2009, 2012b; Elvenes et al. 2012, 2013, 2016; Rise et al. 2013, 2015, 2016a, b; Elvenes 2014; King et al. 2014; Bjarnadóttir et al. 2016, 2017; Diesing and Thorsnes 2018).

Results from MAREANO (MAREANO 2019) have contributed significantly to the revision of Norway's management plan for the Barents Sea–Lofoten areas, as well as the management plan for the Norwegian Sea (Fig. 2). These plans are used by Norwegian authorities in their management of the northern seas, particularly in relation to fisheries and petroleum activities.

Spatial management of the Norwegian nearshore areas is the responsibility of coastal municipalities. Municipal jurisdiction extends to 1 nautical mile offshore of a baseline joining the outermost islets and skerries. Coastal marine areas cover c. 90 000 km^2, comprising a wide range of environments from rocky, exposed shallows to fjords up to 1300 m

From: Asch, K., Kitazato, H. and Vallius, H. (eds) 2022. *From Continental Shelf to Slope: Mapping the Oceanic Realm.* Geological Society, London, Special Publications, **505**, 57–69.
First published online July 9, 2020, https://doi.org/10.1144/SP505-2019-82

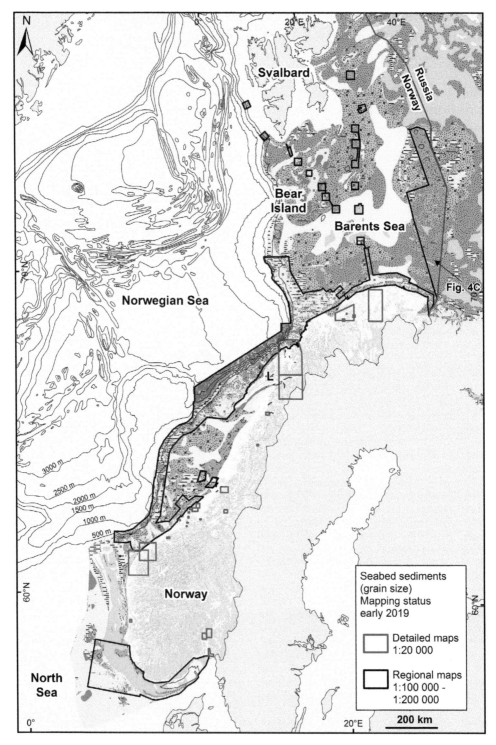

Fig. 1. Areal coverage of seabed sediments (grain size) maps at scale 1:1 000 000–1:4 000 000 by April 2019. Areas mapped in higher detail in Norwegian areas are shown with black (MAREANO and previous projects in the North Sea) and red outlines (coastal mapping projects). L, Lofoten.

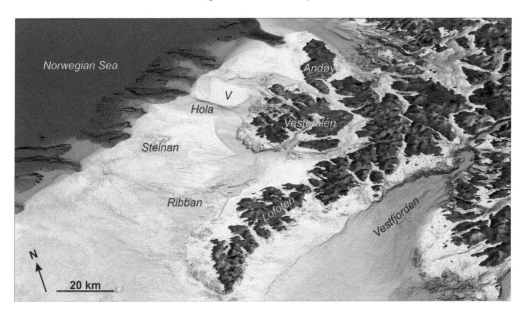

Fig. 2. Example of nearshore–offshore seabed morphology mapped by MAREANO, Lofoten–Vesterålen, north Norway. V, Vesterålsgrunnen. See Figure 1 for the location of Lofoten.

deep. Through a series of mapping projects in cooperation with local authorities and other stakeholders, NGU has published geology-focused seabed maps and derived thematic maps ('marine base maps': Elvenes *et al.* 2019; NGU 2019*a*) covering *c.* 10 000 km² of the coastal areas (Fig. 1; see below). These offer invaluable knowledge to marine spatial planners and the many users of the Norwegian coastal zone.

In this paper, we describe the mapping process and the geological maps produced by NGU. The geological maps (e.g. Fig. 3), along with bathymetry, biological data and oceanographic modelling results, form the basis for further mapping and modelling of benthic habitats (including biotopes, nature types, vulnerable habitats, etc.) both offshore and in the coastal zone (e.g. Buhl-Mortensen *et al.* 2009*a*, *b*, 2012, 2015; Dolan *et al.* 2009, 2012*a*; Bekkby *et al.* 2012; Elvenes *et al.* 2013; Gonzalez-Mirelis and Buhl-Mortensen 2015). Mapping of the environmental chemistry of the seabed sediments (e.g. Pb and PAH) is included in the MAREANO programme and in multiple coastal mapping projects (e.g. Elvenes *et al.* 2018; Jensen *et al.* 2018; Knies and Elvenes 2018).

All maps from MAREANO and NGU's coastal mapping projects are published online and are freely available for viewing, downloading and WMS use (MAREANO 2019; NGU 2019*a*). Additionally, a series of composite, printable PDF MAREANO maps is published online (e.g. Bjarnadóttir *et al.* 2017; NGU 2019*a*).

Methods

MAREANO

Mapping of a new area commences with multibeam echo-sounder surveys by NHS or external contractors according to defined standards (NHS 2018). Other bathymetry data may be available from the petroleum industry, research institutions or the Olex database (mainly single-beam echo-sounder data) (e.g. Elvenes *et al.* 2012).

Multibeam echo sounders are used for detailed mapping of the bathymetry. Furthermore, co-registered backscatter data provide additional information on the composition and structure of the seafloor through the amplitude of the returned signal from the seafloor (e.g. Lurton and Lamarche 2015). Bathymetry data are processed by NHS and subcontractors (data correction and cleaning), and, after quality control, NHS produces terrain models at horizontal resolutions appropriate to the sounding density (in the range 2–50 m). Backscatter datasets are processed from raw data by NGU using industry-standard software to produce mosaics with a pixel resolution of 1–50 m, depending on the data density and quality. Since 2010, water column data have also been acquired as part of the MAREANO multibeam echo-sounder surveys. Subcontractors acquire sub-bottom profiler data (only a few surveys prior to 2018), yielding additional information about the structure and composition of the uppermost *c.* 100 m of the seafloor.

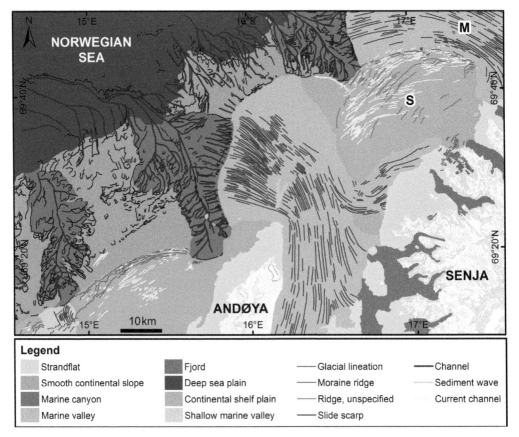

Fig. 3. Landscapes and landforms in an area mapped by the MAREANO programme outside north Norway. The map is made for viewing at a scale of 1:100 000. S, Sveinsgrunnen; M, Malangsdjupet. See Figure 2 for the location of Andøya.

IMR and NGU plan and arrange common sampling cruises. Given MAREANO's wide mapping focus, the station planning must take broad-scale environmental variability into consideration, including both geological and biological diversity, as well as identifying suitable locations for retrieving samples for chemical analysis. High-resolution bathymetry and backscatter data, supplemented with available oceanographic model data, are fundamental to this station planning process.

While early MAREANO station planning was essentially expert driven (but guided by a simple unsupervised classification of the physical environment), MAREANO has now phased in more objective and automated methods (Thorsnes *et al.* 2015). The sampling effort is matched as far as possible to the scale of the map product(s) and available budgets. Approximately 10 stations per 1000 km² have been visited for the collection of video data, allowing visual observation of seabed sediments and megafauna. A proportion of these stations (generally two

per 1000 km²) are so-called full stations where a range of sampling gear (see below) is used in support of multidisciplinary mapping (geology, biology, chemistry). These general averages in terms of station density have been adapted from area to area in recent years, depending on the complexity of the seabed and/or the length of the video lines. In 2018, the sampling density for video and geological grab samples was increased to 20 stations per 1000 km², in connection with a reduction in the length of video lines from 700 to 200 m. This change allows more locations to be documented which generally increases the environmental space observed within a similar timeframe.

MAREANO employs the towed video platforms Campod and Chimaera for seabed video surveying deployed from relatively large, stable research vessels with dynamic positioning (e.g. R/V *G.O. Sars*). These platforms are equipped with a low-light charge-coupled device (CCD) (forward-looking) and high-definition (HD) video cameras, in addition to lights, scale indicator and a geopositioning transponder accurate to *c.* 2% of the water depth. The HD camera has a manual zoom and focus, and is mounted on a pan-and-tilt device. The video platforms are used both temporarily parked on the seabed for detailed studies and in a transect mode towed by the ship along a predefined survey line. The height above the seabed is maintained by a winch operator using visual observations from the forward-looking camera.

A variety of other physical sampling gear is used by MAREANO at full stations including grab, box corer, multicorer and gravity corer, all providing material and information from the seafloor and uppermost metres of the seafloor for geological and ecological (infauna and epifauna) studies. The multicorer is used for environmental sampling of undistorted core material. Epibenthic sledge and beam trawl are employed to sample fauna on and above the seabed. Oceanographic properties are measured with CTD, ADCP and rosette sampler.

During MAREANO sampling cruises, sub-bottom profiler data (e.g. TOPAS topographical parametric sonar) are acquired during transit between all stations and along selected transects covering features of special interest (e.g. sand waves or pockmarks observed in multibeam bathymetry). Sub-bottom penetration depends on grain size and compaction of the seabed, and may be up to 100 m with a vertical resolution of 0.5–1 m in fine-grained sediments. These data support the geological interpretation and are particularly useful for the production of NGU's sediment genesis map (see below). Sub-bottom profiler data are now also regularly acquired during the MAREANO multibeam mapping cruises.

Coastal mapping

Coastal mapping projects have so far been limited to areas with existing multibeam data available from NHS or other sources (e.g. the Norwegian Defence Research Establishment (FFI)). Additional multibeam data are occasionally acquired during NGU cruises. Pre-cruise station planning for video surveying and physical sampling in the coastal zone is expert-driven, based on multibeam bathymetry and backscatter, and the number and distribution of stations vary depending on seabed complexity and the availability of existing observations. Commonly, 100–200 stations are visited per 1000 km². Automated methods for station planning, like those now adopted by MAREANO, need to be improved before they integrate the complexities of the coastal environment and the geologist's need for ground truthing, particularly when multibeam backscatter datasets from multiple sources are applied, as these can be challenging to harmonize.

In the coastal zone, where surveys are generally conducted from NGU's 17 m-long research vessel R/V *Seisma*, video data are recorded by means of a towed platform equipped with one low-light CCD camera and one HD camera, as well as lights, scale indicator and geopositioning transponder. Camera settings are locked during operation, and the platform is kept at 0.5–1 m above the seabed while the ship is moving at low speed. Video lines are generally 50–300 m long. Their final lengths are adapted en route depending on the heterogeneity of the observed seabed. The main sampling gears consist of grab and multicorer or Niemistö-corer.

Sub-bottom profiler (TOPAS) data are also acquired on coastal mapping cruises on R/V *Seisma* during transits between stations, as well as along selected transects, and data are used to support the geological interpretations.

Results

In MAREANO, geological seabed maps based on high-quality multibeam echo-sounder data (5 m grids) and seabed ground-truth data include seabed sediments (grain size), seabed sediments (genesis), sedimentary environment, and landscapes and landforms. These are all mapped for use at the scale 1:100 000 (digitizing scale *c.* 1:50 000) (Table 1) and coarser. Marine base maps for the coastal zone are generally made for use at the scale 1:20 000 (digitizing scale *c.* 1:10 000). In both cases, map scales may vary depending on the purpose, multibeam data quality and available ground-truth data.

Landscapes and landforms

The marine landscape mapping performed by MAREANO delimits broad-scale morphological elements (Thorsnes *et al.* 2009). Landscape classification is based on the national nature description and typification system NiN (Nature types in Norway: Artsdatabanken 2019). NiN defines landscapes as large geographical areas with a visually homogeneous character. Through a semi-automated GIS method (Elvenes 2014), bathymetry data and derived terrain attributes (e.g. slope. curvature, relative relief, relative vertical position) are used to categorize all areas of the seabed as one of the following classes: strandflat; smooth continental slope; marine canyon; marine valley; shallow-marine valley; fjord; deep sea plain; continental slope plain; continental shelf plain; and hilly/mountainous marine landscape (Fig. 3).

Strandflat is the crystalline platform which characterizes large parts of the Norwegian coast and in many areas contrasts sharply with the sedimentary rocks of the continental shelf. Fjords and marine valleys are the results of concentrated glacial erosion

Table 1. *Map products and mapping scales – MAREANO and coastal mapping projects*

	MAREANO		Coastal mapping projects	
Map products:	Included:	Map scale/raster resolution:	Included:	Map scale/raster resolution:
Seabed sediments (grain size)	Yes	1:100 000–1:3 000 000	Yes	1:10 000–1:50 000
Seabed sediments (genesis)	Yes	1:100 000	No	
Sedimentary environment	Yes	1:100 000	No	
Accumulation areas	No		Yes	1:10 000–1:50 000
Anchoring conditions	No		Yes	1:10 000–1:50 000
Digability	No		Yes	1:10 000–1:50 000
Slope >30°	No		Yes	1:10 000–1:50 000
Slope (raster)	No		Where permitted	1–5 m
Seabed terrain (raster)	Yes	2–25 m	Where permitted	1–5 m
Backscatter (raster)	Yes	1–10 m	Where permitted	1–5 m
Landforms	Yes	1:100 000	No	
Marine landscapes	Yes	1:100 000–1:1 000 000	No	
Mapping strategy	Large and coordinated mapping efforts, long-term planning		Smaller projects in cooperation with local authorities and stakeholders	
Video transects				
Length	Standardized: 700 m pre-2018, 200 m from 2018		Adjusted to local conditions, often 200–300 m	
Number	Standardized but varying from area to area, typically 5–20 transects per 1000 km²		Adjusted to local conditions, often *c.* 100 transects per 1000 km²	
Video platform	Towed and stationary on seabed		Towed	
Samples	Sediment grab and other sampling equipment, standardized number of samples		Mainly sediment grab; number of samples adjusted to local conditions	

during repeated glaciations, with fjords incising the mainland. Continental shelf plain is the residual low-relief landscape between marine valleys on the continental shelf.

For areas with multibeam data coverage, MAREANO's marine landscape maps are based on bathymetry data with a horizontal resolution of 50 m and are at a scale of 1:100 000. In other areas, maps are based on best available resolution data such as Olex or IBCAO bathymetry, which are generally of lower quality (larger uncertainty) than multibeam data. Landscape delineation in areas without multibeam data coverage is, therefore, conducted at coarser map scales – typically 1:500 000–1:1 000 000. Marine landscape maps do not form part of the coastal marine base maps, although some mapped areas are covered by the MAREANO classification.

Marine landforms (Figs 3 & 4) are mapped in MAREANO based on detailed bathymetry and sub-bottom profiler data, supported by video observations of the seabed. Landforms are interpreted and digitized manually in GIS as polygons or lines,

depending on the type of landform and size. Examples of landforms mapped are drumlin, moraine, esker, meltwater channel, crevasse-fill ridge, glacial lineation, glaciotectonic hole, glaciotectonic hill, sediment wave field, channel, canyon, slide scarp, slide front, slide fan, submarine slide and pockmark area. A collaboration has been developed between MAREANO, the British MAREMAP programme and the Irish INFOMAR programme to develop a common framework for morphological and geomorphological mapping (Dove *et al.* 2016).

Thousands of cold-water coral reefs occur on the Norwegian continental shelf and in the coastal zone (e.g. Mortensen *et al.* 2001; Bøe *et al.* 2016; Thorsnes *et al.* 2016a, 2017; Jarna *et al.* 2017). From 2018, offshore coral carbonate mounds are mapped by a methodology that combines image segmentation and spatial prediction based on multibeam bathymetry. The results of Diesing and Thorsnes (2018) show that, for a limited study area, the image-object mean planar curvature is the most important predictor, and their approach allows the presence and absence of carbonate mounds to be mapped

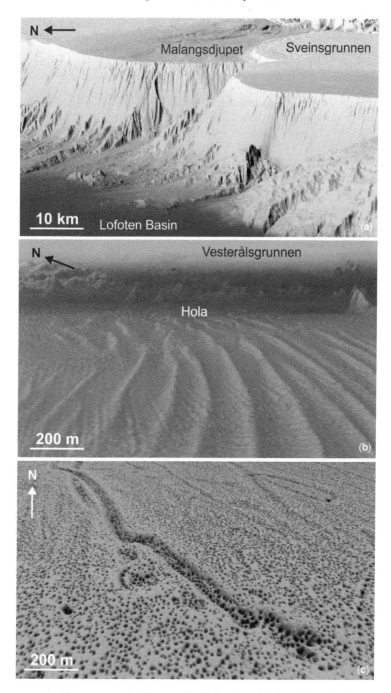

Fig. 4. Landscapes and landforms mapped by MAREANO. (**a**) Continental shelf and slope with canyons and slides outside Vesterålen, north Norway. In this area, water depths increase from around 50 m on the Sveinsgrunnen Bank to 200 m in the Malangsdjupet Trough and to more than 2000 m in the Lofoten Basin. See Figure 3 for the location of Sveinsgrunnen and Malangsdjupet. (**b**) Sand waves (up to 5 m high) and coral reefs (up to 17 m high) in the Hola Trough outside Vesterålen, north Norway. Water depths are 70–90 m on the Vesterålsgrunnen Bank and 200–270 m in the Hola Trough. See Figure 2 for the locations of Hola and Vesterålsgrunnen. (**c**) Iceberg plough marks and pockmarks in the Barents Sea. In this area, with water depths of around 270 m, pockmarks are 20–50 m across and 2–5 m deep, while plough marks are 60–70 m wide and 6–7 m deep. See Figure 1 for the location.

with high accuracy. This method is currently being scaled up for application to wider areas. Prior to the development of Diesing and Thorsnes' (2018) method in 2018, coral reefs and associated sediments were manually digitized and classified as 'bioclastic sediments' (Bellec *et al.* 2014). In the seabed sediments (grain size) map (see the following subsection), coral carbonate mounds are indicated as 'mud, sand and gravel of biological origin', while, in the seabed sediments (genesis) map, they are classified as 'bioclastic sediment'.

Seabed sediments (grain size)

The seabed sediments (grain size) map (Fig. 5) reflects the sediment or bottom type in the uppermost *c.* 10 cm of the seabed, categorized as one of 35 defined classes (NGU 2019*b*). Most of the classes comprise a mixture of grain sizes (e.g. 'gravelly sand' or 'mud and sand with gravel, cobbles and boulders'), which is a signature of predominantly glacially influenced environments.

For expert-driven interpretation and compilation of the grain-size map, all available data (i.e. multi-beam bathymetry and backscatter, videos, seabed samples taken with grab, box corer, and multicorer, as well as sub-bottom profiler data) are used by the geologist, and published literature is consulted where available. The classification of sediment type is determined by the final scale of the map and the degree of detail in the data used for interpretation and map compilation.

Seabed sediments (genesis)

The seabed sediments (genesis)/Quaternary geology map reveals processes on the seafloor during and after the last ice age. The map describes deposits and bottom types in the upper 1–2 m of the seafloor (i.e. not only the surface deposits influenced by the most recent processes).

For mapping of seabed sediments (genesis), the geologist chooses from amongst 32 sediment/bottom type classes (NGU 2019*c*). Examples include

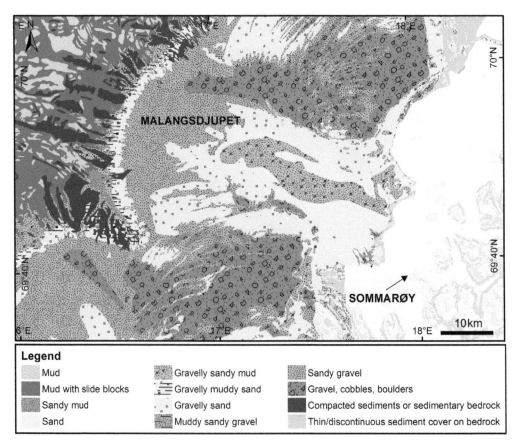

Fig. 5. Seabed sediments (grain size) in an area mapped by the MAREANO programme outside north Norway. The map is made for viewing at a scale of 1:100 000.

suspension deposit, glaciomarine deposit, bedload (traction) deposit, contourite, glaciofluvial deposit, till, mass-movement deposit, debris-flow deposit and exposed bedrock. Vast areas, especially on the continental shelf, are dominated by sediments deposited in glacial environments.

The map is based on the seabed sediments (grain size) map, as well as the landscapes and landforms maps, in addition to further interpretation of multibeam bathymetry, and sub-bottom profiler and seismic data. The classification of sediment type is determined by the final scale of the map and the degree of detail in the data used for interpretation and map compilation. So far, seabed sediments (genesis) maps have only been compiled within MAREANO for the offshore areas.

Sedimentary environment

The sedimentary environment map is based on the seabed sediments (grain size) map and the datasets used for producing that map. A predefined number of classes is used for compilation (NGU 2019d). The main purpose of the map is to visualize areas of erosion and deposition of sediments, and how bottom currents influence the seabed.

Deposition of fine-grained sediments (mud and sandy mud) primarily occurs in deep or sheltered waters. Erosion may remove fines and deposit sand where bottom currents become weaker or where sand is transported back and forth by tidal currents. The shallowest areas are often dominated by erosion, although fine-grained sediments may accumulate in local, topographical depressions. A lag deposit of sandy gravel, cobbles and boulders is often formed where bottom currents (wave, tidal or oceanographic currents) are strong. Grain size generally indicates the strength of the bottom currents; mud suggests weak bottom currents, while coarser sediments or erosion suggest stronger currents.

Marine base maps in the coastal zone

The marine base maps published by NGU since 2003 (Sandberg et al. 2005; Longva et al. 2008; Thorsnes et al. 2013; Elvenes et al. 2019) present geological information relevant to end users outside the geological community in a format that is comprehensible to geologists and non-geologists alike. As described above, sediment type mapping in Norwegian nearshore areas is based on pre-existing multibeam echo-sounder data ground-

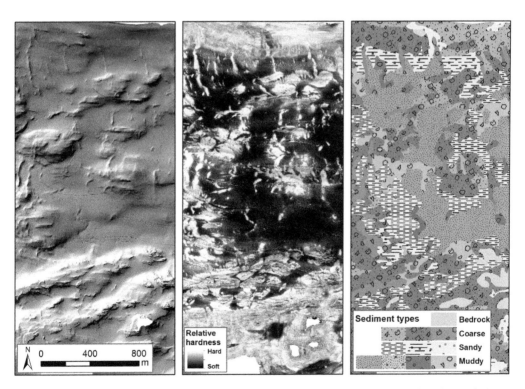

Fig. 6. Example of high-resolution multibeam bathymetry and backscatter data used for compilation of the seabed sediments (grain size) map at a scale of 1:20 000 from a Norwegian fjord.

truthed by video observation and physical samples. Figure 6 shows an example of shaded bathymetry, backscatter and interpreted sediment types from a Norwegian fjord.

Since grain size can be difficult for the non-specialist to interpret with respect to everyday applications, the detailed, full-coverage maps of seabed sediment types (grain size) are supplemented with thematic maps based on expert knowledge of sediment properties and on high-resolution multibeam data. A full stack of marine base maps will contain shaded relief bathymetry and slope data of the highest permitted resolution (given military restrictions), seabed sediments (grain size), anchoring conditions, digability and accumulation basins (examples shown in Fig. 7). Most marine base maps published since 2015 have a scale of 1:20 000. This suite of applied marine base maps conveys useful information on the coastal marine environment to managers, industry, fishermen, recreational users, marine scientists, etc., even in areas where access to high-resolution multibeam bathymetry is restricted by the Norwegian defence authorities.

Harmonized maps for EMODnet Geology

The EMODnet (European Marine Observation and Data Network) Geology portal (EMODnet Geology 2019) aims to provide harmonized information on marine geology in Europe. NGU has delivered harmonized datasets at different scales (both MAREANO and coastal data) on landscapes and landforms, seabed substrates (seabed sediments (grain size)), the Quaternary (seabed sediments (genesis)), the pre-Quaternary, mineral occurrences, sediment accumulation rates, geological events and probabilities, and coastal behaviour.

Summary

Results from MAREANO and coastal mapping projects show that Norway has a rich and diverse seafloor. The geomorphology and sediment-distribution patterns reflect a long geological history and complex modern-day hydrodynamic processes. By 2019, approximately 10% of the Norwegian seafloor had been mapped: *c.* 200 000 km^2

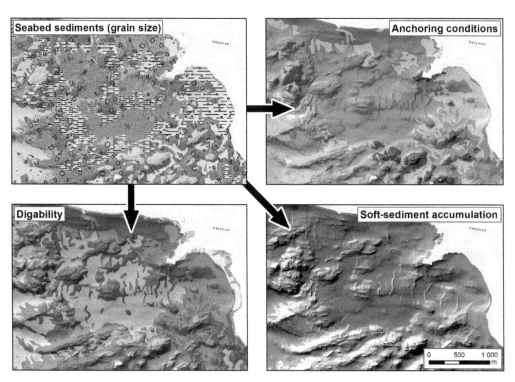

Fig. 7. Marine base maps. Example of use of the seabed sediments (grain size) map (in addition to detailed bathymetry) for compilation of the derived thematic maps for anchoring conditions, digability and accumulation basins. See Figures 5 and 6 for the legends to the seabed sediments (grain size) map. In the thematic maps, the most favourable conditions for anchoring are shown in green; for digging, in light grey; and for soft-sediment accumulation, in blue.

offshore and 10 000 km^2 in the coastal zone. Geological maps by MAREANO include seabed sediments (grain size), seabed sediments (genesis), sedimentary environment, and landscapes and landforms. These are made for use at a scale of c. 1:100 000. Marine base maps for the coastal zone include seabed sediments (grain size) and the derived maps – anchoring conditions, digability and accumulation basins – made for use at a scale of c. 1:20 000. Bathymetric and geological maps produced by MAREANO and coastal mapping projects, along with biological and oceanographic data, have been found to form an invaluable basis for further mapping and modelling of benthic habitats, with grain-size and landscape maps often serving as important predictor variables. In addition, mapping of seabed chemistry (including pollution) is included in the working programmes of both MAREANO and the coastal mapping projects.

Acknowledgements We would like to thank all participants of the MAREANO programme and the coastal mapping projects, onshore and offshore, for their invaluable cooperation. Matthias Forwick and an anonymous reviewer are thanked for constructive comments to a previous version of the manuscript.

Funding This research received no specific grant from any funding agency in the public, commercial, or not-for-profit sectors.

Author contributions RB: conceptualization (equal), data curation (equal), formal analysis (equal), funding acquisition (equal), investigation (equal), methodology (equal), project administration (equal), resources (equal), supervision (equal), validation (equal), visualization (equal), writing – original draft (lead), writing – review & editing (lead); **LRB**: funding acquisition (equal), methodology (equal), project administration (equal), resources (equal), writing – review & editing (supporting); **SE**: formal analysis (equal), investigation (equal), project administration (equal), visualization (equal), writing – review & editing (supporting); **MD**: conceptualization (supporting), methodology (equal), writing – review & editing (supporting); **VB**: formal analysis (equal), investigation (equal), visualization (equal); **TT**: conceptualization (equal), funding acquisition (equal), methodology (equal), project administration (equal), resources (equal); **AL**: data curation (lead), validation (equal), visualization (equal); **OL**: conceptualization (equal), funding acquisition (equal), methodology (equal).

Data availability statement The datasets generated during and/or analysed during the current study are available in the NGU repository, www.mareano.no and www.ngu.no.

References

Artsdatabanken. 2019. *Natur i Norge*. Artsdatabanken, Trondheim, Norway, https://www.artsdatabanken.no/NiN [accessed 30 January 2019].

Bekkby, T., Moy, F.E. *et al.* 2012. The Norwegian program for mapping of marine habitats – providing knowledge and Maps for ICZMP. *In*: Moksness, E., Dahl, E. and Støttrup, J. (eds) *Global Challenges in Integrated Coastal Zone Management*. Wiley-Blackwell, Chichester, UK, 21–30.

Bellec, V., Wilson, M., Bøe, R., Rise, L., Thorsnes, T., Buhl-Mortensen, L. and Buhl-Mortensen, P. 2008. Bottom currents interpreted from iceberg ploughmarks revealed by multibeam data at Tromsøflaket, Barents Sea. *Marine Geology*, **249**, 257–270, https://doi.org/10.1016/j.margeo.2007.11.009

Bellec, V.K., Dolan, M.F.J., Bøe, R., Thorsnes, T., Rise, L., Buhl-Mortensen, L. and Buhl-Mortensen, P. 2009. Sediment distribution and seabed processes in the Troms II area – offshore North Norway. *Norwegian Journal of Geology*, **89**, 29–40.

Bellec, V.K., Bøe, R., Rise, L., Slagstad, D., Longva, O. and Dolan, M.F.J. 2010. Rippled scour depressions on continental shelf bank slopes off Nordland and Troms, North Norway. *Continental Shelf Research*, **30**, 1056–1069, https://doi.org/10.1016/j.csr.2010.02.006

Bellec, V., Thorsnes, T. and Bøe, R. 2014. *Mapping of Bioclastic Sediments – Data, Methods and Confidence*. NGU Report **2015.043**, https://www.ngu.no/upload/Publikasjoner/Rapporter/2015/2015_043.pdf

Bellec, V., Rise, L., Bøe, R. and Dowdeswell, J. 2016. Glacially related gullies on the upper continental slope, SW Barents Sea margin. *Geological Society, London, Memoirs*, **46**, 381–382, https://doi.org/10.1144/M46.31

Bellec, V.K., Bøe, R., Rise, L., Lepland, A. and Thorsnes, T. 2017a. *Seabed Sedimentary Environments and Sediments (Genesis) in the Nordland VI Area off Northern Norway*. NGU Report **2017.046**, https://www.ngu.no/upload/Publikasjoner/Rapporter/2017/2017_046.pdf

Bellec, V.K., Bøe, R., Rise, L., Lepland, A., Thorsnes, T. and Bjarnadóttir, L.R. 2017b. Seabed sediments (grain size) of Nordland VI, offshore north Norway. *Journal of Maps*, **13**, 608–620, https://doi.org/10.1080/17445647.2017.1348307

Bellec, V.K., Bøe, R. *et al.* 2019. Sandbanks, sandwaves and megaripples on Spitsbergenbanken, Barents Sea. *Marine Geology*, **416**, https://doi.org/10.1016/j.margeo.2019.105998

Bjarnadóttir, L.R., Ottesen, D., Dowdeswell, J.A. and Bugge, T. 2016. Unusual iceberg ploughmarks on the Norwegian continental shelf. *Geological Society, London, Memoirs*, **46**, 283–284, https://doi.org/10.1144/M46.126

Bjarnadóttir, L.R., Ottesen, D. *et al.* 2017. *Geologisk havbunnskart, Kart 65000900, Mai 2017. M 1:100 000*. Norges geologiske undersøkelse, Trondheim, Norway, https://www.ngu.no/upload/Kart%20og%20data/Maringeologiske%20kart/GEOLOGISK_HAVBUNNSKART_65000900.pdf

Bøe, R., Bellec, V.K., Dolan, M.F.J., Buhl-Mortensen, P.B., Buhl-Mortensen, L. and Rise, L. 2009. Giant sand waves in the Hola glacial trough off Vesterålen,

North Norway. *Marine Geology*, **267**, 36–54, https://doi.org/10.1016/j.margeo.2009.09.008

Bøe, R., Bellec, V.K., Rise, L., Buhl-Mortensen, L., Chand, S. and Thorsnes, T. 2012. Catastrophic fluid escape venting-tunnels and related features associated with large submarine slides on the continental rise off Vesterålen–Troms, North Norway. *Marine and Petroleum Geology*, **38**, 95–103, https://doi.org/10.1016/j.marpetgeo.2012.08.008

Bøe, R., Skarðhamar, J. *et al.* 2015. Sandwaves and sand transport on the Barents Sea continental slope offshore northern Norway. *Marine and Petroleum Geology*, **60**, 34–53, https://doi.org/10.1016/j.marpetgeo.2014.10.011

Bøe, R., Bellec, V., Dolan, M., Buhl-Mortensen, P., Rise, L. and Buhl-Mortensen, L. 2016. Cold-water coral reefs in the Hola glacial trough off Vesterålen, North Norway. *Geological Society, London, Memoirs*, **46**, 309–310, https://doi.org/10.1144/M46.8

Buhl-Mortensen, P.B., Buhl-Mortensen, L., Dolan, M., Dannheim, J. and Kröger, K. 2009*a*. Megafaunal diversity associated with marine landscapes of northern Norway: a preliminary assessment. *Norwegian Journal of Geology*, **89**, 163–171.

Buhl-Mortensen, P., Dolan, M.F.J. and Buhl-Mortensen, L. 2009*b*. Prediction of benthic biotopes on a Norwegian offshore bank using a combination of multivariate analysis and GIS classification. *ICES Journal of Marine Science*, **66**, 2026–2032, https://doi.org/10.1093/icesjms/fsp200

Buhl-Mortensen, L., Bøe, R., Dolan, M.F.J., Buhl-Mortensen, P., Thorsnes, T., Elvenes, S. and Hodnesdal, H. 2012. Banks, troughs and canyons on the continental margin off Lofoten, Vesterålen, and Troms, Norway. *In:* Harris, P.T. and Baker, E.K. (eds) *Seafloor Geomorphology as Benthic Habitat*. Elsevier, Amsterdam, 703–715, https://doi.org/10.1016/B978-0-12-385140-6.00051-7

Buhl-Mortensen, L., Hodnesdal, H. and Thorsnes, T. (eds). 2015. *The Norwegian Sea Floor, New Knowledge from MAREANO for Ecosystem-Based Management*. MAREANO.

Chand, S., Rise, L. *et al.* 2008. Active venting system offshore Northern Norway. *Eos, Transactions of the American Geophysical Union*, **89**, 261–262, https://doi.org/10.1029/2008EO290001

Chand, S., Rise, L., Ottesen, O., Dolan, M.F.J., Bellec, V. and Bøe, R. 2009. Pockmark-like depressions near the Goliat hydrocarbon field, Barents Sea: morphology and genesis. *Marine and Petroleum Geology*, **26**, 1035–1042, https://doi.org/10.1016/j.marpetgeo.2008.09.002

Chand, S., Thorsnes, T. *et al.* 2012. Multiple episodes of fluid flow in the SW Barents Sea (Loppa High) evidenced by gas flares, pockmarks and gas hydrate. *Earth and Planetary Science Letters*, **331–332**, 305–314, https://doi.org/10.1016/j.epsl.2012.03.021

Diesing, M. and Thorsnes, T. 2018. Mapping of cold-water coral carbonate mounds based on geomorphometric features: an object-based approach. *Geosciences*, **8**, 34, https://doi.org/10.3390/geosciences8020034

Dolan, M.F.J., Mortensen, P.B., Thorsnes, T., Buhl-Mortensen, L., Bellec, V. and Bøe, R. 2009. Developing seabed nature-type maps offshore Norway: initial results from the MAREANO programme. *Norwegian Journal of Geology*, **89**, 17–28.

Dolan, M.F.J., Elvenes, S. *et al.* 2012*a*. *Marine grunnkart i Sør-Troms: Rapport om biotopmodellering*. NGU Report **2012.070**, https://www.ngu.no/upload/Publikasjoner/Rapporter/2012/2012_070.pdf

Dolan, M.F.J., Thorsnes, T., Leth, J., Alhamdani, Z., Guinan, J. and Van Lancker, V. 2012*b*. *Terrain Characterization from Bathymetry Data at Various Resolutions in European Waters – Experiences and Recommendations*. NGU Report **2012.045**, https://www.ngu.no/upload/Publikasjoner/Rapporter/2012/2012_045.pdf

Dove, D., Carter, G. *et al.* 2016. *Seabed Geomorphology: A Two-Part Classification System*. British geological Survey, Marine Geoscience Programme, Open Report **OR/16/001**, http://nora.nerc.ac.uk/id/eprint/514946/1/Seabed_Geomorpholgy_classification_BGS_Open_Report.pdf

Elvenes, S. 2014. *Landscape Mapping in MAREANO*. NGU Report **2013.035**, https://www.ngu.no/upload/Publikasjoner/Rapporter/2013/2013_035.pdf

Elvenes, S., Buhl-Mortensen, P. and Dolan, M.F.J. 2012. *Evaluation of Alternative Bathymetry Data Sources for MAREANO: A Comparison of Olex Bathymetry and Multibeam Data for Substrate and Biotope Mapping*. NGU Report **2012.030**, https://www.ngu.no/upload/Publikasjoner/Rapporter/2012/2012_030.pdf

Elvenes, S., Dolan, M.F.J., Buhl-Mortensen, P. and Bellec, V.K. 2013. An evaluation of compiled single-beam bathymetry data as a basis for regional sediment and biotope mapping. *ICES Journal of Marine Science*, **71**, 867–881, https://doi.org/10.1093/icesjms/fst154

Elvenes, S., Bøe, R. and Rise, L. 2016. Post-glacial sand drifts burying De Geer moraines on the continental shelf off North Norway. *Geological Society, London, Memoirs*, **46**, 261–262, https://doi.org/10.1144/M46.25

Elvenes, S., Knies, J. and Rasmussen, T. 2018. *Forurensningsstatus i havbunnssedimenter i Ofotfjorden, Tysfjorden og Tjeldsundet*. NGU Report **2017.047**, https://www.ngu.no/upload/Publikasjoner/Rapporter/2017/2017_047.pdf

Elvenes, S., Bøe, R., Lepland, A. and Dolan, M.F.J. 2019. Marine base maps, Søre Sunnmøre, Norway. *Journal of Maps*, **15**, 686–696, https://doi.org/10.1080/17445647.2019.1659865

EMODnet Geology. 2019. *European Marine Observation and Data Network*. EMODnet Geology, http://www.emodnet-geology.eu/ [accessed 30 January 2019].

Gonzalez-Mirelis, G. and Buhl-Mortensen, P. 2015. Modelling benthic habitats and biotopes off the coast of Norway to support spatial management. *Ecological Informatics*, **30**, 284–292, https://doi.org/10.1016/j.ecoinf.2015.06.005

Jarna, A., Elvenes, S. and Bøe, R. 2017. *Kartlegging av korallforekomster i Romsdalsfjorden, Harøyfjorden og rundt Gossa ved hjelp av dybdedata fra multistråleekkolodd*. NGU Report **2017.033**, https://www.ngu.no/upload/Publikasjoner/Rapporter/2017/2017_033.pdf

Jensen, H.K.B., Knies, J. and Bellec, V. 2018. *Miljøgeokjemiske data og dateringsresultater fra MAREANO Øst – MAREANO*. NGU Report **2018.018**, https://www.ngu.no/upload/Publikasjoner/Rapporter/2018/2018_018.pdf

King, E.L., Bøe, R, Bellec, V.K., Rise, L., Skarðhamar, J., Ferré, B. and Dolan, M. 2014. Contour current driven continental slope-situated sandwaves with effects from secondary current processes on the Barents Sea margin offshore Norway. *Marine Geology*, **353**, 108–127, https://doi.org/10.1016/j.margeo.2014. 04.003

Knies, J. and Elvenes, S. 2018. *Sedimentasjonsmiljø og historisk utvikling i forurensningsstatus i sjøområdene i Ofot-regionen*. NGU Report **22018.007**, https://www.ngu.no/upload/Publikasjoner/Rapporter/2018/2018_007.pdf

Longva, O., Arvesen, B., Ulsrud, E., Hestvik, O.B., Martinsen, J. and Roaldsnes, T. 2008. *Nye marine grunnkart i fiskeri- og havbruksnæringen – sluttrapport*. NGU Report **2008.034**, https://www.ngu.no/upload/Publikasjoner/Rapporter/2008/2008_034.pdf

Lurton, X. and Lamarche, G. (eds). 2015. *Backscatter Measurements by Seafloor-Mapping Sonars. Guidelines and Recommendations*. GeoHab Report, http://geohab.org/wp-content/uploads/2018/09/BWSG-REPORT-MAY2015.pdf

MAREANO. 2019. *MAREANO Collecting Marine Knowledge*. MAREANO, http://mareano.no/en [accessed 26 August 2019].

Mortensen, P.B., Hovland, M.T., Fosså, J.H. and Furevik, D.M. 2001. Distribution, abundance and size of Lophelia pertusa coral reefs in mid-Norway in relation to seabed characteristics. *Journal of the Marine Biological Association*, **81**, 581–597, https://doi.org/10.1017/S002531540100426X

NGU. 2019a. *Geological Survey of Norway*. http://www.ngu.no [accessed 26 August 2019].

NGU. 2019b. *Classification of Sediments Based on Grain Size Composition (Folk, 1954, Modified)*. Norges geologiske undersøkelse, Trondheim, Norway, https://www.ngu.no/Mareano/Grainsize.html [accessed 27 March 2019].

NGU. 2019c. *Seabed Sediments – Genesis*. Norges geologiske undersøkelse, Trondheim, Norway, https://www.ngu.no/Mareano/SedGenesis.html [accessed 27 March 2019].

NGU. 2019d. *Sedimentary Environment*. Norges geologiske undersøkelse, Trondheim, Norway, https://www.ngu.no/Mareano/SedEnvironment.html [accessed 27 March 2019].

NHS. 2018. *Technical Specifications*. Norwegian Hydrographic Service (NHS), Stavanger, Norway, https://www.mareano.no/resources/files/om_mareano/arbeidsmater/standarder/Appendix-B-Technical-Specifications-1.pdf [accessed 25 March 2019].

Rise, L., Bøe, R. *et al.* 2013. The Lofoten–Vesterålen continental margin, North Norway: canyons and mass-movement activity. *Marine and Petroleum Geology*, **45**, 134–149, https://doi.org/10.1016/j.marpetgeo.2013.04.021

Rise, L., Bellec, V.K., Chand, S. and Bøe, R. 2015. Pockmarks in the southwestern Barents Sea and Finnmark fjords. *Norwegian Journal of Geology*, **94**, 263–282.

Rise, L., Bellec, V., Ottesen, D., Bøe, R. and Thorsnes, T. 2016a. Hill–hole pairs on the Norwegian continental shelf. *Geological Society, London, Memoirs*, **46**, 203–204, https://doi.org/10.1144/M46.42

Rise, L., Bøe, R., Bellec, V., Thorsnes, T. and Dowdeswell, J. 2016b. Canyons and slope instability on the Lofoten–Vesterålen continental margin, North Norway. *Geological Society, London, Memoirs*, **46**, 407–408, https://doi.org/10.1144/M46.36

Sandberg, J.H., Thorsnes, T., Bekkby, T., Longva, O., Christensen, O., Andresen, K.H.B. and Lepland, A. 2005. Future perspectives for ICZPM in relation to aquaculture. *In*: Howell, B. and Flos, R. (eds) *Aquaculture Europe: Lessons from the Past to Optimise the Future*. EAS Special Publications, **35**, 53–58.

Thorsnes, T., Buhl-Mortensen, L. and Skyseth, T. 2008. Integrated mapping of the seafloor and ecosystems in the Arctic – the MAREANO programme. *Gråsteinen*, **12**, 115–125.

Thorsnes,T., Erikstad, L., Dolan, M.F.J. and Bellec, V.K. 2009. Submarine landscapes along the Lofoten–Vesterålen–Senja margin, northern Norway. *Norwegian Journal of Geology*, **89**, 5–16.

Thorsnes, T., Sandberg, J.H., Longva, O., Røyland, G., Jakobsen, P.-A. and Hestvik, O.B. 2013. *Nye marine grunnkart i fiskeri-og havbruksnæringen – Fase 2*. NGU Report **2013.037**, https://www.ngu.no/upload/Publikasjoner/Rapporter/2013/2013_037.pdf

Thorsnes, T., van Son, T.C. *et al.* 2015. *An Assessment of Scale, Sampling Effort and Confidence for Maps Based on Visual and Acoustic Data in MAREANO*. NGU Report **2015.043**, https://www.ngu.no/upload/Publikasjoner/Rapporter/2015/2015_043.pdf

Thorsnes, T., Bellec, V.K. and Dolan, M.F.J. 2016a. Cold-water coral reefs and glacial landforms from Sula Reef, mid-Norwegian shelf. *Geological Society, London, Memoirs*, **46**, 307–308, https://doi.org/10.1144/M46.74

Thorsnes, T., Rise, L., Bellec, V.K. and Chand, S. 2016b. Shelf-edge slope failure and reef development: Trænadjupet Slide, mid-Norwegian shelf. *Geological Society, London, Memoirs*, **46**, 413–414, https://doi.org/10.1144/M46.75

Thorsnes, T., Bjarnadóttir, L.R. *et al.* 2017. National programmes: Geomorphological Mapping at Multiple Scales for Multiple Purposes. *In*: Micallef, A., Krastel, S. and Savini, A. (eds) *Submarine Geomorphology*. Springer, Cham, Switzerland, 535–552.

Mapping Ireland's coastal, shelf and deep-water environments using illustrative case studies to highlight the impact of seabed mapping on the generation of blue knowledge

Ronan O'Toole[1]*, Maria Judge[1], Fabio Sacchetti[2], Thomas Furey[2], Eoin Mac Craith[1], Kevin Sheehan[2], Sheila Kelly[1], Sean Cullen[1], Fergal McGrath[2] and Xavier Monteys[1]

[1]Geological Survey Ireland, Beggars Bush, Haddington Rd., Dublin 4, Ireland

[2]Marine Institute, Rinville, Oranmore, Co. Galway, Ireland

ROT, 0000-0003-0242-5611; FS, 0000-0002-2098-7071; EMC, 0000-0002-7919-6303; SK, 0000-0002-5774-6212

*Correspondence: ronan.o'toole@gsi.ie

Abstract: Through Ireland's national seabed mapping programme, Integrated Mapping for the Sustainable Development of Ireland's Marine Resource (INFOMAR), the collaboration between Geological Survey Ireland and the Marine Institute continues to comprehensively map Ireland's marine territory in high resolution. Through its work, the programme builds on earlier Irish seabed mapping efforts, including the Irish National Seabed Survey project in producing seabed mapping products that support Ireland's blue economy, European marine policy and international efforts to understand our global oceans. INFOMAR uses a variety of marine technologies to deliver accurate bathymetric maps and useful data products to end users through a free and open source licensing agreement. To reflect the diversity of applications these data products serve, a series of four case studies are presented here focusing on marine geophysical and geological data from locations within Ireland's marine territories. The case studies illustrate how data generated through seabed mapping may be interpreted to directly impact the generation of blue knowledge across a variety of marine environments ranging from shallow coastal and shelf waters to the deep oceanic depths of the continental slope of Ireland's marine area. The impact of Ireland's seabed mapping efforts is further considered in the context of national, European and international initiatives where Ireland's marine knowledge resource is leveraged to deliver positive benefit to the programme's stakeholders.

A history of Irish seabed mapping

Deep-water hydrographic and geophysical survey operations to designate the boundaries of the Irish continental margin in support of Ireland's United Nations Convention on the Law of the Sea (UNCLOS) maritime territorial claims began offshore Ireland in 1996 (Naylor *et al.* 1999), conducted by Ireland's Petroleum Affairs Division on behalf of the Government of Ireland. Findings reinforced the need for a comprehensive assessment of the entire Irish seabed. The Geological Survey Ireland (GSI)-managed Irish National Seabed Survey followed (INSS, 2000–06), an ambitious but successful programme to survey Ireland's entire deep-water territory beyond 200 m water depth (Verbruggen and Cullen 2008). With national interests and development opportunities largely coastal and shelf based, and with one of the most detailed offshore cohesive seabed mapping knowledge resources available globally in 2006, mapping the gaps naturally evolved into a follow-on national seabed survey initiative through a joint venture between GSI and the Marine Institute. The **IN**tegrated mapping **FO**r the sustainable development of Ireland's **MA**rine **R**esource (INFOMAR) programme was initiated to survey the remaining shelf and coastal waters between 2006 and 2026, to deliver a seamless baseline bathymetry dataset to underpin the future management of Ireland's marine resource (Dorschel *et al.* 2010).

To leverage the €80 m financial support required after INSS for mapping the gaps in the coastal, shelf and inshore waters of Ireland, a comprehensive review was commissioned to consider the programme approach, cost and survey priorities. This informed a two-phase programme strategy subsequently developed (INFOMAR 2007) that outlined the approach, outputs and anticipated beneficiaries. Phase 1, which was completed in 2015, focused on mapping 26 priority bays and 3 priority offshore areas that were deemed to be of most economic

From: Asch, K., Kitazato, H. and Vallius, H. (eds) 2022. *From Continental Shelf to Slope: Mapping the Oceanic Realm.* Geological Society, London, Special Publications, **505**, 71–96.

First published online September 9, 2020, https://doi.org/10.1144/SP505-2019-207

significance to the country. Phase 2, which commenced in 2016, is focused on mapping the remaining unsurveyed marine areas and building on the knowledge and expertise generated through the initial project phase.

Following Government of Ireland approval and INFOMAR's commencement in 2006, the joint programme management of GSI and Marine Institute coordinated a seminal cost benefit analysis (CBA) to investigate the economic impact of the seabed mapping initiative across all marine sectors (PwC 2008). Taking a conservative approach, a 4–6 times return on investment was reported, depending on the duration over which the programme was completed. This proved to be a critical assessment in securing future annual programme investment, particularly despite national fiscal challenges from 2008 onwards. Subsequent independent reviews commissioned by the programme carried out by PwC and Risk Solutions have further supported the case for continued seabed mapping, with key recommendations tabled and implemented incrementally year-on-year (PwC 2013; Risk Solutions 2016).

INFOMAR is a key cross-sectoral enabling action in Ireland's integrated marine plan, 'Harnessing Our Ocean Wealth' (Government of Ireland 2012) with an oversight Board and Technical Advisory Committee (TAC) governance structure ensuring its relevance to all key stakeholders nationally. The primary marine bathymetry dataset derived from full coverage high-resolution multibeam echosounder surveying, is critical for the development of Ireland's marine knowledge, economy and policy, as well as the protection of its marine environment. With an ethos of improving efficiency and embracing innovation, domestically the programme supports the needs of Irish society, industry and government, while internationally it contributes to numerous EU Directive-related reporting, regulatory and monitoring obligations (PwC 2013).

Knowledge as a marine resource

Ireland's seabed mapping efforts initially began with the aim of developing a marine baseline dataset to underpin national security as well as future economic, environmental, infrastructural and policy decisions for Ireland as set out in the INFOMAR Proposal and Strategy. With more than 20 years of seabed mapping undertaken to date, this endeavour is being steadily achieved with over 700 000 km^2 of the seafloor within the Irish designated area mapped to date in high resolution (Fig. 1).

Shallow water mapping reveals uncharted rocks and unknown shipwrecks, delivering safe navigation data within the busiest zones for marine traffic. Data describing the geomorphology of the seabed enable

accurate oceanographic modelling to assess coastal flooding and erosion risk, as well as state-of-the-art ecosystem investigations such as assessing the aquaculture-carrying capacity of inshore waters. Shelf and offshore mapping provides a foundation for aggregate resource and habitat assessments, informing permitting and development decisions, while enabling protection of key fish spawning and nursery grounds (Sutton 2008). Collectively, these coastal and offshore marine data enrich the efforts of Ireland's research community who analyse the bathymetry data for a multitude of applications.

As of 2019, Ireland's database of marine data has grown in excess of 120 terabytes (TB) and continues to expand. The database comprises a range of geophysical data measurements including multibeam echosounder (MBES) bathymetry and backscatter, shallow seismic profiles, gravity, magnetics, side-scan sonar and oceanographic water column profiles. It also houses information on the many physical ground-truthing samples and interpreted observations including, for example, shipwreck discoveries of which there are currently 426 listed in the INFOMAR database.

One of the key drivers behind the successful uptake of Ireland's seafloor mapping data has been the Irish government's Open Data Initiative (DPER 2017). Supported by the Union (EU) Open Data Directive (EU 2019), this major government initiative ensures that INFOMAR data are freely available to the public. Online access is favoured as a direct route to the range of high-quality data and data products produced. The revised programme website, relaunched in November 2018 has been developed with the aim of providing a contemporary feel for the end user, while strengthening the value of programme outputs. Provision of straightforward access to the data is a key objective. This has been accomplished through the production of web map services (WMS) and availability of embedded data viewers. A simple web-viewer available on the website homepage allows for the visualization and exploration of Ireland's marine territory in detail and is capable of displaying multiple layers of seafloor information. The programme's official data download portal is the Interactive Web Data Delivery System (IWDDS), which is accessible through the INFOMAR website and provides free and open data to programme stakeholders.

Key products, designed around stakeholder requirements include Geographical Information System (GIS)-compatible datasets: bathymetry, backscatter and shaded relief as geo-referenced images; sediment samples and shipwrecks as point files; sediment classification and survey coverage as polygon files; and survey track lines as polyline files. In addition, raw data are made available on request for those working with non-standard applications and

The Real Map of Ireland

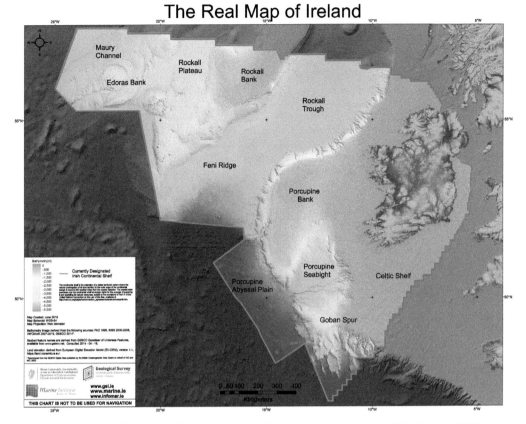

Fig. 1. 'The Real Map of Ireland' – Irish Designated Area, Coordinate Reference System: Web Mercator (EPSG code: 3857).

software. For users who are unfamiliar with GIS processes or other technical software, the data are available in formats that can be easily displayed on widely available free software and viewers such as Google Earth.

Integration of regional marine data from the world's oceans enhances our understanding of the Earth's coasts, seas and oceans as a globally connected system. To support the integration of marine knowledge for cross-border, European and international collaborations, Ireland's seabed mapping datasets are analysed and interpreted by programme staff to produce standardized products and metadata compatible with international initiatives. The European Marine Observation and Data Network (EMODnet) is a large-scale pan-European marine data initiative (Kaskela et al. 2019). Funded by the European Commission, it aims to implement the EU's Marine Knowledge 2020 strategy (EU 2010). The INFOMAR seabed-mapping programme has contributed data to EMODnet's Bathymetry, Geology and Seabed Habitats lots. Through these projects,

pan-European data products that include Irish data are freely available for global dissemination and usage.

INFOMAR programme data are further distributed through international open-access data portals: for example, the National Oceanic and Atmospheric Administration (NOAA) archives, where the NOAA Bathymetric Data Viewer enables users to view and download raw and processed seabed mapping data from Ireland's past and present seabed mapping programmes. Through this well-known facility, Irish seabed data reach a broad global network of potential end users. These data resources are also integrated in the Nippon Foundation's Generalised Bathymetric Chart of the Ocean (GEBCO) compilation and are one of the largest data contributors to the Seabed 2030 initiative (Mayer et al. 2018). Additionally, through participation in international partnerships such as AORA (Atlantic Ocean Research Alliance), ASMIWG (Atlantic Seabed Mapping International Working Group) and CHERISH (Climate, Heritage and Environments of Reefs, Islands and

Headlands) Ireland's seabed mapping results are further distributed to a broad international community of multidisciplinary data end users.

Irish seabed mapping case studies: enhancing our knowledge of Ireland's marine environment

Gauging the impact of Ireland's marine data dissemination strategy and wider activity at national, European and international levels is a subject that will be considered further in this paper; however, to illustrate the impact of how baseline seabed mapping datasets and products can be utilized to enhance our understanding of the marine environment, generating new knowledge and insights, a series of four case studies is presented in this paper.

The case studies focus on a variety of marine environments found within Ireland's designated marine territory (Fig. 1), traversing from the coastal, shallow waters of the inshore marine areas, progressing across the continental shelf and onwards to the continental slope and deep ocean floor. These case studies consist of four individual interpretations of Irish marine data that illustrate how analysis of Ireland's marine data resources can enhance the understanding of Ireland's marine environment, geology and submerged landscapes.

The first case study focuses on the Hook Head peninsula and Waterford Estuary. This study illustrates how high-resolution MBES datasets can be used to explore a potential submerged landscape. Offshore extrapolation of onshore geological features and interpretation of overlaying bathymetric characteristics reveals information about submerged landscapes and the Last Glacial Maximum (LGM). The second case study is located further out to sea, SE of Ireland, on the Irish continental shelf and represents a detailed overview of the programme's mapping rationale, approach and procedures when surveying and analysing a specific survey block. Geological and geophysical data including multibeam sonar data are recorded to reveal substrate geology and environmental information. As the site is known for its significance to the Irish fishing sector as a *Nephrops* fishery, the impact of the blue knowledge generated through seafloor, substrate and habitat mapping is also described, facilitating enhanced stock assessments. The third case study presented focuses on a dynamic shelf environment offshore North Donegal. The study details interpretation of baseline multibeam sonar data that infer a high-energy dynamic environment from mobile sedimentary seabed features. A resurveying of the area is used in correlation with existing hydrographic models to examine the driving forces of sediment mobility in this region. The study is made all the more

relevant by the possibility of future development of renewable energy infrastructure at the site and provides a wealth of knowledge on the types of considerations that will be instrumental for future offshore development plans. The fourth case study details an overview of deep-water geological and geophysical data. These data were used in conjunction with INSS and INFOMAR bathymetry data to produce the first bedrock map of Ireland's offshore Exclusive Economic Zone (EEZ) as part of work undertaken for the EMODnet Geology project. Now complete, the map details a chronology of Ireland's geological history dating back to the Grampian. Geological knowledge of this kind can be applied in support of multiple applications including: habitat mapping; marine spatial planning; environmental conservation; and resource mapping.

Taken together, these case studies are effective in demonstrating the value of seabed mapping in furthering our understanding of the broad range of marine environments that make up Ireland's marine resource. Finally, this paper considers the impact of Ireland's seabed mapping efforts as they relate to national, European and international initiatives, education, research and industry and reflects on Ireland's future activities and prospects in the sphere of marine and ocean science at the beginning of the United Nations Decade of Ocean Science for Sustainable Development (2020–30) movement as Ireland continues in its aim to map, observe and predict its coasts, seas and oceans.

Case study 1: a coastal submerged landscape; Hook Head – a seabed within a seabed

The extent of exposure of Ireland's continental shelf due to a lower sea-level during and following the last glaciation is still being understood, but bathymetry data from Ireland's seabed mapping data resource present compelling evidence for potential submerged landscapes in the country's coastal and nearshore environment. Submerged landscapes are features on or below the seafloor that can be reasonably deduced to have been subaerially exposed in the past, before sea-level rose to where it is today. One place where INFOMAR bathymetry data help build a strong case for the presence of submerged landscapes is Waterford Estuary on the south coast of Ireland. In this study, high-resolution shaded relief bathymetry from Waterford Estuary and its surrounding area has been used to infer the presence of these submerged seafloor features in conjunction with contemporary geological knowledge from the area in order to consider the relationship between terrestrial observations and the potential submerged landscape features identifiable in the shaded relief bathymetry data. By visualizing bathymetry data at high resolution and adjusting parameters such as hill shading and water depth colour-scale, these data can be used to identify and represent images of underwater outcrop and submerged, ancient landscape features.

High-resolution MBES bathymetry data featured in this study were acquired during ongoing INFO-MAR inshore survey operations. Acoustic soundings data were processed using a Teledyne-CARIS HIPS & SIPS™ hydrographic software package. Tidal and navigation corrections, sound velocity and noise cleaning processes were applied to the data in order to generate a high resolution bathymetric grid of the study area in the World Geodetic System 1984 (WGS84) Coordinate Reference System (CRS) and vertically referenced to Lowest Astronomical Tide (LAT) using the United Kingdom Hydrographic Office (UKHO) Vertical Offshore Reference Frame (VORF) model (Ziebart et al. 2007). The resulting bathymetric grids were exported in a standard ESRI™ format and brought into ARC-GIS™ software for further analysis and juxtaposition with GSI's 1:100 000-scale geological map series. This methodology allowed for a broad examination of the study area's seafloor morphology in the context of potential submerged landscapes and the area's geological history as determined from terrestrial-based studies.

Hook Head, County Wexford is a prominent peninsula in the SE of Ireland located at the eastern bank of Waterford Estuary (Fig. 2). It is comprised of rocks ranging from Cambrian in age to Devonian and continues upwards into a Lower Carboniferous limestone assemblage (GSI 2018). The geological record of Hook Head and its modern, adjacent seabed details an evolution through continental collision, both ancient and modern sea-level as well as glaciation. Around the base of its scenic lighthouse, Carboniferous coral fossils are exposed in abundance in subaerial conditions (Sleeman et al. 1974). One can walk over rocks that were once submerged calcareous deposits on a tropical seafloor akin to the present-day Bahamas. This once coral-rich seabed, however, now lithified and uplifted, forms the modern limestone seabed as we move from the headland, underwater, down below contemporary sea-level. This rock, detailed using high-resolution nearshore bathymetry (Fig. 3), is a substrate for modern sea life. Modern geological processes and sea-level change have facilitated the phenomenon where the modern seabed sits directly atop the ancient Carboniferous seabed, each separated in time by more than 300 Ma.

On land, the solid geology of Hook Head is well documented (Tietzsch-Tyler and Sleeman 1994). The peninsula is long and narrow, consisting of

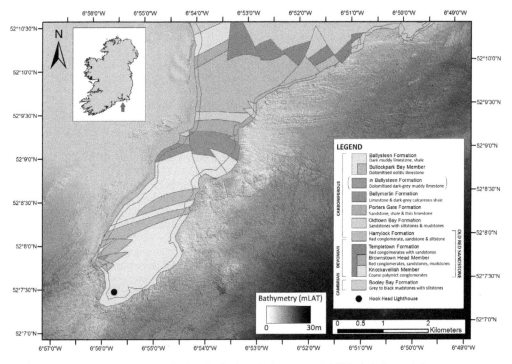

Fig. 2. Geology of Hook Head peninsula, Co. Wexford overlaying grey-scale hillshade bathymetric layer.

Fig. 3. INFOMAR bathymetry around the tip of Hook Head, draped over aerial photography (© Ordnance Survey Ireland/Government of Ireland 2020/OSi_NMA_041). The data show the typical strata and weathering patterns often associated with seafloor limestone, supporting the inference that the limestone around the base of the lighthouse extends offshore.

geological formations that, as we step in a southward direction off the ancient Cambrian basement of the mainland, span from the Devonian through to the Carboniferous. The story of sea-level and environmental change here begins with the ancient Cambrian rock, formed from the accumulation of deep-water slope sediments on a continental margin of the long-gone Iapetus Ocean around 500 Ma ago (Tietzsch-Tyler and Sleeman 1994). Following this, continental collision resulted in the building of an ancient mountain chain – the Caledonian Orogeny – during which the Iapetus Ocean was closed and a large continental landmass formed (Tietzsch-Tyler and Sleeman 1994). The Old Red Sandstones of the Devonian, comprising the peninsula's landward end (Fig. 2), were originally laid down in an alluvial setting as these Caledonian mountains were eroded (Woodcock and Strachan 2002). The Lower Carboniferous then saw a gradually northward-advancing sea, with the sedimentary rocks overlying the Old Red Sandstones formed in marine conditions again. At this point, continental drift had brought Ireland's ancestral basement rocks close to the equator and so these shallow seas were tropical and teeming with prehistoric life, such as corals, brachiopods, crinoids and bryozoans (Meere *et al.* 2013). These are the fossils underfoot as one stands on the shoreline below Hook Head lighthouse. As we depart the

southernmost tip of the peninsula and proceed underwater, detailed seafloor bathymetry illustrates the probable continuation of these stratigraphic units offshore (Fig. 3).

Deposits that were once laid down in a warm tropical environment now form the hard substrate below the present-day Celtic Sea. This same substrate was only recently inundated, however, on a geological timescale. During the last ice age, the area was glaciated, with evidence for the advance of a grounded Irish Sea glacier (Ó Cofaigh and Evans 2001). Following the LGM, 27 000 ka BP (Clark *et al.* 2012), the ice sheets receded rapidly, albeit unevenly, due to climatic warming (Chiverrell *et al.* 2013). However, sea-level remained low during a period of isostatic rebound that temporarily outpaced the inevitable sea-level rise due to meltwater (Edwards and Brooks 2008).

Waterford Estuary extended further out to sea and the coastline in the area was further south. This is clearly visible in the bathymetry data in Figure 4. Interpretation of the shaded relief bathymetry (Fig. 4) allows the erosion pattern of the bedrock to be visualized as a possible palaeochannel extension. Beyond the harbour mouth, despite recent marine sedimentation, the expression of the palaeochannel is visible in the bathymetric imagery in the mouth of Waterford Estuary (between Dunmore

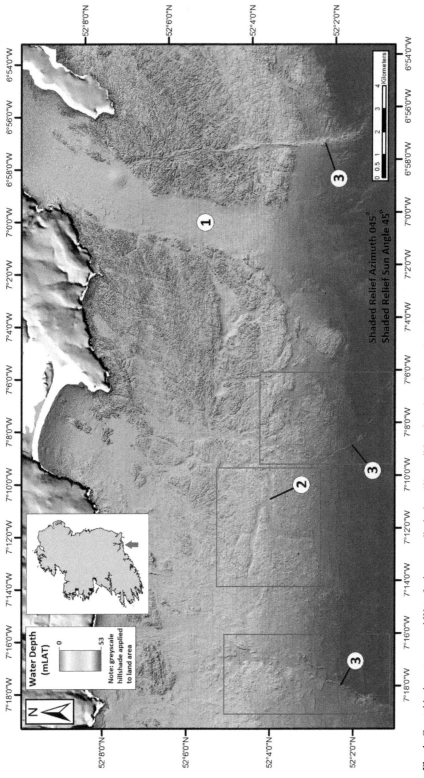

Fig. 4. Coastal bathymetry around Waterford estuary displaying: (1) possible palaeochannel extension; (2) submerged meandering channel feature; and (3) ridge features interpreted as possible eskers.

East and Hook Head) as a readily identifiable sea-floor feature, a bathymetric low, between areas of bedrock. Studies have traced this palaeochannel further offshore and indicate that the channel contained much higher water flows than the present day and so may have been formed by discharges of meltwater, flowing with great energy from the ice and out over the exposed shelf (Gallagher 2002).

Along with the palaeochannel, the ridge features interpreted from the bathymetry marked in Figure 4 have been suggested to be eskers, laid down during glaciation in tunnels of meltwater below the ice (Tóth *et al.* 2016). As the ice retreated, these may have formed prominent subaerial landscape features for a time before the sea transgressed northwards. So, in addition to the Carboniferous transgression, examination of the seabed features in the study area suggests a story of more recent sea-level change too.

High-resolution bathymetry acquired in the region, highlights additional channel features cutting through the bedrock (Fig. 4). Also interpreted from the seafloor imagery is a meandering gap in the bedrock with a ribbon-like geometry. This feature may be related to terrestrial drainage systems, much like the rivers we see on land today. This is inferred as a relic of the terrestrial landscape that post-dates the ice age and pre-dates the significant sea-level rise. Sea-level rise lagged behind the receding ice (Edwards and Brooks 2008).

In summary, high-quality bathymetry data reveal the exposed bedrock offshore Hook Head (Fig. 2), the lithology and structure of which can be extrapolated from detailed geological mapping on land. Here the stratigraphy reveals Cambrian deep-water sedimentation giving way to terrestrial conditions in the Devonian. These are sequentially followed by fossiliferous lithologies laid down in Carboniferous shallow tropical seas. More recent potential Esker deposits result from the last ice age, now drowned by recent sea-level rise. Further offshore, the deepening seabed disappears under a blanket of recent Holocene marine sedimentation. By combining terrestrial geological observations and studies with state-of-the art bathymetric seafloor imagery in the manner described, a picture of ancient and more recent coastal change and evolution can be generated.

In mapping the spatial distribution of probable submerged landscape features to a high degree of accuracy and resolution, data products from the INFOMAR programme support research into past climate change. The dating and further interpretation and study of these features may help constrain the position of Ireland's palaeocoastline at different times in the geological record, thus helping to recreate past sea-level curves. Such information is important for informing models of future sea-level rise and the coastline's response to a changing climate. In addition, using modern bathymetric imagery to enhance the understanding of our ancient and fascinating heritage beyond the coastline and sharing that knowledge with the public promotes Ocean Literacy, which presently aims to raise awareness of the ocean's impact on our lives (NOAA 2013) at a vital time for the protection of the marine environment.

Case study 2: characterizing underwater channels on the Celtic Sea shelf, Southern Ireland

As part of the INFOMAR programme's 2018 seabed mapping campaign, the RV *Celtic Voyager* mapped seabed adjacent to the Ireland/UK EEZ in the Celtic Sea. Figure 5a shows the chart outlining the designated survey area prior to commencement of survey operations. Mapping was conducted over 74 charter days and 3 separate surveys. It is an important area for a number of commercial fish species. Figure 5b shows the final multibeam bathymetry survey coverage for the campaign with prominent channels annotated.

As with all INFOMAR seabed mapping campaigns conducted on the RV *Celtic Voyager*, geophysical datasets acquired included multibeam bathymetry, backscatter and sub-bottom profiler data. Shaded relief and substrate slope angle products are produced from the multibeam bathymetry data and ground-truth data from sediment grabs augmented the backscatter data. For these three surveys in 2018 the RV *Celtic Voyager* mapped a total area of 5650 km^2 in water depths ranging from 78 to 124 m as shown in Figure 5b.

Three large-scale channels are observed within this region of Ireland's continental shelf (Fig. 5b). One of the channels, known colloquially as 'The Trench', had not been previously well defined. As a result of the survey campaign, 'The Trench' is now mapped in high resolution and charted with two additional channels close by. All three channels are detailed here with data acquisition methodology and scientific results. They are significant geomorphological features and form a part of the important *Nephrops* fishery of the Celtic Sea Mud patch. The Trench comprises 177 km^2 of the 14 469 km^2 Celtic Sea Mud patch area currently surveyed for *Nephrops* stock assessment. It is a fishery yielding landings in the region of approximately 5000 tonnes annually over the last decade (Doyle *et al.* 2019) and, in 2018, Irish landings were worth around €56 m (White *et al.* 2019) at first sale.

High-resolution multibeam bathymetry and backscatter (Fig. 6) data were acquired and presented along with the derived shaded relief images illustrated in Figure 7, while substrate slope angle

(a)

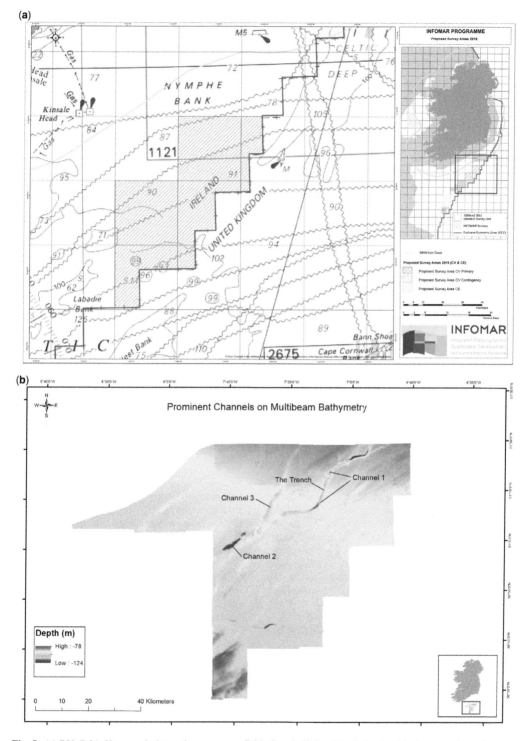

Fig. 5. (a) RV *Celtic Voyager*-designated survey area, Celtic Sea shelf, Southern Ireland and bathymetry chart of mapped area (**b**).

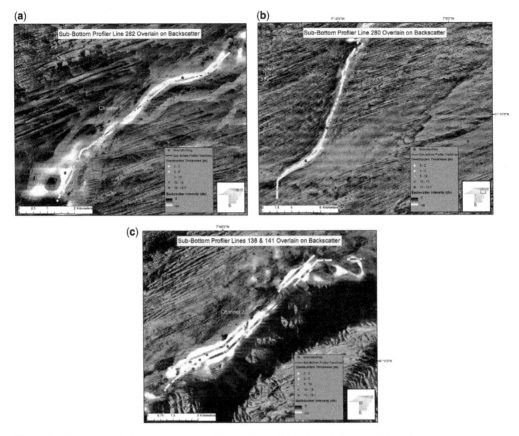

Fig. 6. Sub-bottom profiler lines overlain on backscatter for (**a** & **b**) Channel 1 and (**c**) Channel 2.

products documented in Figure 8 were derived from the underlying bathymetry. Sub-bottom profiler data (Fig. 9) from channels 1 and 2 have been analysed and interpreted resulting in overburden thickness plots, which are displayed over the backscatter and shaded relief imagery respectively in Figures 6 and 7. Ground-truth data and sub-bottom track lines of interpreted profiles are overlain on both the backscatter and shaded relief data. For this study, a combination of bathymetry, backscatter, sub-bottom profiler and ground-truthing data are analysed and interpreted in order to illustrate the impact of integrating multiple data sources for the characterization of a commercially and environmentally significant area of Ireland's marine territory.

The standard survey line pattern selected for this operation represented east–west reciprocal lines with a line spacing of approximately 400 m. A Kongsberg EM2040 high resolution MBES mounted on a retractable pole was used for swathe acoustic acquisition. Backscatter acquired by multibeam sonars contains important information about the seafloor and its physical properties (Lurton *et al.* 2015).

This information provided valuable data for seafloor classification and important auxiliary information for a bathymetric survey. A hull-mounted pinger source 2 × 2 transducer array sub-bottom profiler operating at 3.5 kHz was used for sub-bottom data acquisition. The sweep time was varied appropriately with water depth to maximize ping rate and resolution. The pinger source is chosen as most effective in investigations of the top 20 or 30 m sub-seabed and where sediments are fine to medium grained as readings indicated for this area.

Ground-truthing stations for validating the multibeam interpretations were acquired using a Day grab sampler. The Day grab was deployed from the starboard side of the vessel and gave consistently full samples with no empty returns during the operation. Grab-sample locations were selected based on expert interpretation of the multibeam backscatter data and geographical spread when possible; however, opportunistic samples were also acquired at Sound Velocity Profile (SVP) stations occasionally. A total of ten grab samples were acquired within or proximal to the channels. Samples were photographed and

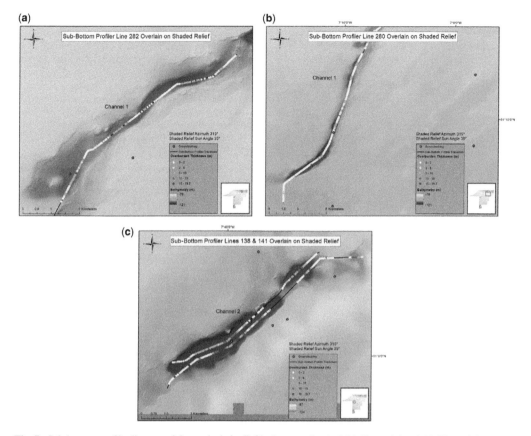

Fig. 7. Sub-bottom profiler lines overlain on shaded relief bathymetry for (**a** & **b**) Channel 1 and (**c**) Channel 2.

described. All samples will undergo particle size analysis, the results of which will be used to create substrate maps which become available through the INFOMAR website as they are generated and finalized. Substrate maps are important for the purposes of fisheries management. They help plan the locations of sampling stations for stock assessment surveys and also can be used to correlate catch data with sediment type in order to estimate the abundance of the stock. Currently, the map for the Celtic

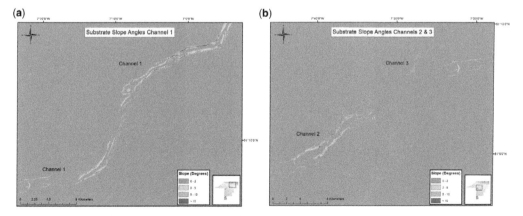

Fig. 8. Celtic Sea channel slope analysis for (**a**) Channel 1 and (**b**) Channels 2 and 3.

Sea is a broadscale European Nature Information System (EUNIS) habitat map made up from interpolated benthic samples and known sediment types based on data from Vessel Monitoring System (VMS) on targeted fisheries (e.g. *Nephrops*). This map is currently being updated using interpreted high-resolution acoustic data from newly acquired MBES data and opportunistic samples. The new map will be classified to Folk and will be refined into a broad benthic habitat map using bathymetry and other environmental data as part of the programme's ongoing commitment to produce high-quality data products from its mapping outputs.

Bathymetry grids, shaded relief and bathymetry geotiff images were created in Teledyne CARIS HIPS & SIPS™ software. Backscatter mosaics were created in QPS FMGT™ software. Geotiffs and grids were imported into ESRI ArcGIS™ software where substrate slope maps can be created from the respective bathymetry grids using the 'Surface Slope' function in ArcToolbox™.

Figure 5b displays the resulting multibeam bathymetry data with the most prominent channels in the area annotated. Channel 1 stretches from the NE to the SW in a sinuous shape spanning a distance of nearly 50 km. Its margins are well defined for the most part. Channel seafloor depth varies from approximately 100 to 120 m and its width varies from 400 m to 2.5 km. The northern limit of this channel extends beyond the boundary of our data. Channel 2 is orientated along a NE to SW axis. It is approximately 18 km in length with well-defined channel margins. An elevated area of substrate separates it from Channels 1 and 3. Seafloor depth ranges

from 103 to 124 m in Channel 2. Channel 3 runs from the north of the area in a south-southwesterly direction where it intersects with the elevated area between Channels 1 and 2. Bathymetry varies from 87 to 106 m within this channel and it deepens from north to south. Channel 3 spans approximately 30 km. All three channels terminate in the same area. A number of other large-scale channels are also observed in the multibeam bathymetry data.

Sub-bottom profiler data for four selected channel infill survey lines were played back through CodaOctopus GeoSurvey™, which is an advanced software package for processing and interpretation of sub-bottom data. Seabed tracking, a bandpass filter with low-cut 1000 Hz and high-cut 4700 Hz, heave correction and a suitable display gain were applied to the data. The bedrock horizon was digitized on each profile using the tagging function in CodaOctopus GeoSurvey™. Two-way travel time to the bedrock horizon was used to calculate overburden thickness. Text files were exported containing position and overburden thickness. The files were imported into ArcGIS, overburden thickness was plotted and then overlain on multibeam backscatter (Fig. 6) and shaded relief data (Fig. 7). Figure 9a and b show raw data for sub-bottom profiler lines 280 and 282 (Channel 1) and Figure 9c and d show raw data for sub-bottom profiler lines 138 and 141 (Channel 2). These four lines were selected for interpretation and analysis. All lines were acquired in the centre of the channels and parallel to channel axes.

Figure 7a and b show the interpreted sub-bottom profiler data for lines 282 and 280 respectively,

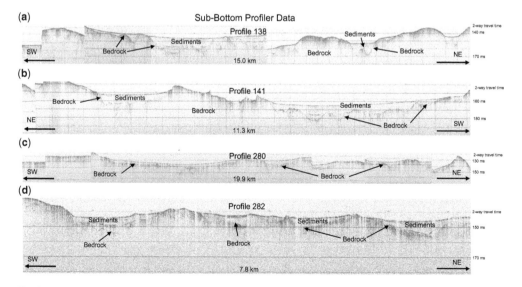

Fig. 9. Sub-bottom profiler data for lines (**a**) 0138, (**b**) 0141, (**c**) 0280 and (**d**) 0282.

overlain on multibeam shaded relief data. Each node on the images represents the overburden thickness at those locations. Bedrock is evident throughout the entire length of both survey lines, mostly at depth but sporadically as seabed outcrop. The maximum overburden thickness on profile line 280 is 7.4 m and 9.2 m on line 282. Greatest overburden thickness correlates with topographic lows on the bedrock horizon, where sediments infill these depressions. Figure 7c shows sub-bottom profiler lines 138 and 141 overlain on multibeam shaded relief data. Outcropping bedrock is signified by an absence of interpreted nodes. The survey track lines are shown in black. Bedrock is evident throughout the entire length of both survey lines; mostly as sub-crop but outcropping bedrock is common. The maximum overburden thickness on line 138 is 17.8 m and 19.7 m on line 141. There is a correlation between overburden thickness and bathymetry on Figure 7c with greatest thickness occurring under the greatest water depths. Deepest water depths also coincide with topographic lows in the bedrock horizon.

Digital imagery (Fig. 9) of the sub-bottom profiler lines infers that bedrock forms the base unit of each survey line. Sub-bottom line 138 (Fig. 9a) has a clearly defined bedrock horizon. Between the top of bedrock and base of the Quaternary is an unconformity. Bedrock outcrops in two distinct sections on the northern half of the line. The bedrock is unconformably overlain by unconsolidated sediments. This unit contains a number of internal reflectors. It is the topmost unit except where bedrock outcrops. Sub-bottom line 141 (Fig. 9b) shows the bedrock unconformity surface to be rugged in character. The digital imagery also shows that the bedrock is unconformably overlain by the unconsolidated sedimentary unit. This unit forms two large infills separated by outcropping bedrock. The maximum thickness of the soft sediment unit is almost 20 m. Several internal reflectors are present within this unit. Sub-bottom line 280 (Fig. 9c) indicates the top of bedrock surface on this profile is smoother in character than that of the previous two analysed profiles. Bedrock only outcrops near the northern end of the profile and it is unconformably overlain by an unconsolidated sedimentary unit elsewhere. This sedimentary unit attains a maximum thickness of over 7 m. Sporadic internal reflectors are present in the unit. The sub-bottom line 281 profile (Fig. 9d) shows that bedrock outcrops at the southern end and in several locations along the profile. The bedrock horizon is rather smooth, similar to the adjacent profile 280. The unconsolidated sedimentary unit attains a maximum thickness of over 9 m.

The channel substrates appear as relatively low-intensity backscatter returns (Fig. 6). These low-intensity backscatter returns coincide with smooth bathymetry data, suggesting fine-grained sediments. Small localized areas of relatively high backscatter returns are also evident within the channels and along channel margins. Correlating backscatter data with the bathymetry suggests that the high backscatter returns along channel margins and within parts of the channels comprise bedrock.

The substrate slope map of Channel 1 and surrounding areas is presented in Figure 8a. A slope scale of 0 to >10° is used with corresponding green to red colour coding. The dark red colour indicates a substrate slope of at least 10°. The substrate in the broad area surrounding the channels is characterized by having a very gentle slope. These gently sloping areas are shaded green. Channel margins are very well defined on the substrate slope map, showing up as yellow and red shading. The substrate slope angles along the channel margins, while well defined, are for the most part moderately sloping, with localized steep slopes. Channel margins with slopes of less than 10° are typical but localized slopes greater than 20° are found within the area. The northern margin of Channel 1 contains the steepest slope angles and is best defined. Maximum slope angles of over 20° are observed. Figure 8b is the substrate slope map for Channels 2 and 3 and their surrounds. Channel 2 is better defined than Channel 3 in terms of slope angles along channel margins. Channel 3 margins are almost all less than 10° but slope angles greater than 10° are evident along Channel 2 margins. Maximum slope angles of over 15° are found along Channel 2.

In summary, INFOMAR hydrographic and geophysical regional mapping surveys carried out in an area of the Celtic Sea, known to be an important fisheries ground for *Nephrops* (Marine Institute 2009), observed large-scale channel seafloor features on the multibeam and sub-bottom profiler data. Analysis of sub-bottom profiler data acquired within Channels 1 and 2 shows unconsolidated sediments with a maximum thickness of over 19 m and sporadic bedrock outcrop. The unconsolidated sediments unconformably overlie bedrock. Multibeam bathymetry data from Channels 1 and 2 indicate that water depths exceed 120 m. Channel 3 is shallower, attaining a maximum depth of 106 m. Relief from channel tops to channel seafloor exceeds 30 m in places. Multibeam backscatter data show that the majority of channel substrate sediments exhibit a relatively low-intensity backscatter. Preliminary inspection of grab samples from Channels 1 and 2 shows that mud and sandy mud sediment compositions are dominant. Substrate slope analysis indicates that the channel seafloors have predominantly gentle slopes. Channel 1 margins show widespread slopes of 10° or more and occasionally slopes of over 20° are found. Channel 2 margins are mostly less than 5° but slopes of over 10° are also noted. Slopes observed in Channel 3 are very gentle, mostly less than 5° and its margins are less well defined than Channels 1 and 2.

In conclusion, *Nephrops* are a common commercial species in the Celtic Sea, occurring in geographically distinct sandy/muddy substrates where the sediment is suitable for them to construct their burrows. VMS data from fishing vessels have historically been used to determine the geographical extent of the fishery, but in recent years the knowledge generated through the interpretation of MBES backscatter and associated sediment data is being used to redefine this extent. This has improved the efficiency and validity of the stock assessment and helped target new areas for underwater video tows, which are an essential part of the stock assessment. The analysis presented here demonstrates how the composition of the seafloor and its associated geomorphological and sedimentological properties may be characterized and interpreted over regional scales using state-of-the-art seabed mapping technology and software processes resulting in an enhanced understanding of the features and structure of Ireland's continental shelf and the generation of knowledge to support the blue economy.

Case study 3: mapping the mobile seabed of Ireland's north coast; a source of risk for locating potential renewable energy sites

Around the world, large sediment waves are still poorly understood seabed features despite being relatively common in many shelf seas (Knaapen and Hulscher 2002; Morelissen *et al.* 2003; Thiébot *et al.* 2015). Early attempts in the 1970s and 1980s to describe sediment movement on European continental shelves relied largely on the interpretation of side-scan sonar data (Kenyon and Stride 1970) and the analysis of bedform asymmetry to determine transport potential and direction (Belderson *et al.* 1982). More recently, thanks to the advent of new technologies such as MBES sensors coupled with improvements in GPS positioning, highly detailed investigations on sediment transport have been conducted on continental shelves globally, in particular thanks to large-scale national mapping initiatives such as the INSS and the INFOMAR programmes (Feldens *et al.* 2012; Denny *et al.* 2013).

The INFOMAR programme places a strong emphasis on the acquisition of high-resolution bathymetry data derived from MBES mapping systems to high degrees of horizontal and vertical accuracy so that the resulting information meets international hydrographic standards to support safe navigation of shipping (IHO 2008). The benefit of adhering to these standards for measuring water depth is that the resulting bathymetric data products

can be gridded to produce comprehensive visualizations of seafloor features that can be used for both qualitative and quantitative analysis (Guinan *et al.* 2009). Furthermore, these mapping endeavours provide the ideal baseline data for any sediment dynamic study when coupled with repeated surveys over the same area at different instances in time.

The understanding of how sediment moves over a continental shelf has critical relevance due to our continuous interaction with and exploitation of the marine environment (EU 2014). The evaluation of the overall sediment volume and its physical characteristics is important for managing economic resources such as aggregates (Alder *et al.* 2010). Equally important is our ability to monitor temporal and spatial sediment movement since this has documented implications for many sectors such as shipping, dredging (Knaapen and Hulscher 2002; Dorst *et al.* 2013) and management of coastal areas under normal and extreme hydrodynamic conditions (Staneva *et al.* 2009). These implications extend further in the context of a developing renewable-energy sector where mobile sediments may pose a risk to renewable energy infrastructure through their interactions on the seabed (Thiébot *et al.* 2015).

Comparing repeat surveys can be used to measure sediment movement from kilometre- to centimetre-scale and to reveal changes in surficial sediment composition through MBES backscatter data (e.g. Németh *et al.* 2002; Ma *et al.* 2014). Additionally the above-mentioned measurements can be used to validate hydrodynamic simulations of water movement, often utilized to provide information about near-seafloor flow velocities (Sheng and Yang 2010; Young *et al.* 2011; Feldens *et al.* 2012).

In Ireland, very large sediment waves, capable of reaching heights of 30 m and able to migrate tens of metres per year have been documented, in particular in the Irish Sea (Evans 2018) and around the Inishowen peninsula (Fig. 10). This case study uses bathymetry acquired during the INSS in 2004 and INFOMAR in 2013 to focus on an area between the north Irish coast and Scotland, where underwater morphologies vary widely from shallow platforms near the coastline in *c.* 20 m water depths to deep troughs up to *c.* 100 m as depicted in Figure 10. Here strong hydrodynamic conditions have the potential to facilitate vast renewable-energy development, but the presence of mobile sediments make this challenging unless sediment transport mechanisms are well understood.

Baseline bathymetric data collected by both mapping initiatives display a range of bedforms including large sediment waves, barchan dunes and gravel waves, all with a range of amplitudes and crest morphologies (Evans *et al.* 2015; Fig. 10). Exposed bedrock, known to be Palaeoproterozoic granitic gneiss, is also present in the NE of the area

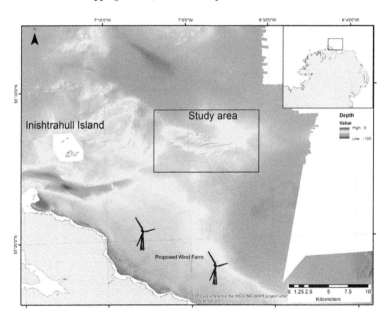

Fig. 10. Overview of study area selection. The sediment waves are clearly visible on the regional bathymetric grids. To the west of the area, sea current energies south of Inishtrahull are some of the highest in Irish waters (>2 m s^{-1}, Rourke *et al.* 2010). Potential renewable energy sites are indicated (Rourke *et al.* 2010).

and forms part of the Inishtrahull Island rock complex (Fig. 10) (Muir *et al.* 1994).

Nine years after the original INSS survey, a joint collaboration between INFOMAR and Ulster University resurveyed these sites in light of future marine plans for the development of renewable energy infrastructure. The resulting high accuracy time-lapse bathymetric data were used to measure horizontal and vertical changes in bedform dimensions. The analysis of multibeam backscatter and sediment data enabled development of a better understanding of sediment distribution, sediment wave composition and allowed inferences on the forces necessary to initiate and sustain sediment transport. All this information was then used in correlation with existing hydrodynamic models to examine the driving forces of sediment mobility in this region. The results of this study indicate that the investigated area has highly mobile sediments with distinct migration directions controlled by local hydrodynamic conditions (Fig. 11).

Initial findings indicate that sediment transport is not linear across the site with crest displacement following a clockwise, rotational movement. Despite the features being highly mobile, surface difference models also suggest that there has not been considerable loss of sediment from the bedform over the nine-year lapse, adding weight to the assumption that while a bidirectional current is in effect across the bedform, hydrodynamic reworking of the sediment remains mostly confined within the bedform

boundaries. The use of multiple repeat surveys has also highlighted oscillation of sand waves at a spatial scale longer than their wavelengths. This suggests the need for shorter time intervals between successive surveys and improved spatial data resolution for both hydrodynamic conditions and sediment distribution to improve the validity of inferences made regarding sediment transport. In summary, the study provides useful knowledge that will have to be taken into account for any future offshore development plans in the area and the results will also hold relevance for other areas identified for the potential development of offshore renewable energy in Ireland (DCENR 2014).

In conclusion, comparison of repeat bathymetric surveys from the INSS and INFOMAR programmes over a nine-year window highlights the advantages of acquiring a high-precision baseline seafloor dataset where subsequent repeat survey activity can be used to measure changes in seafloor properties with a high degree of confidence. The study shows how geomorphological analysis of MBES-generated data provides further insight into the behaviour of sedimentary bedforms within the study area. Analysis of high-precision MBES data coupled with contemporary hydrodynamic models also allows for inferences to be drawn in terms of sediment dynamics, bedform boundaries and the likely requirements for repeat surveys necessary for a comprehensive monitoring campaign within the study area. This knowledge is particularly beneficial in the context

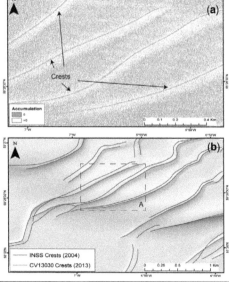

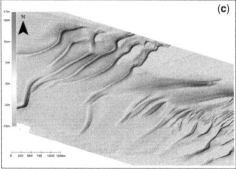

Fig. 11. Example of 'flow accumulation' data where zero values represent crest (**a**). Digitized crests from the 2004 (blue) and 2013 (red) surveys shows the displacement of sediment waves crests in a nine-years' time lapse (**b**) (Evans 2018). Surface difference terrain model derived comparing the 2004 v. the 2013 bathymetry grids (**c**). Faded red indicates area where sediment has accumulated while faded blue are areas with sediment deficit. Grey indicates areas with minimal sediment mobility.

of evaluating prospective renewable energy sites for suitability and risk, where the ability to identify and monitor specific sedimentary seafloor features from seafloor mapping data may help with future site selection or guide design solutions for infrastructure associated with offshore renewable energy devices.

Case study 4: mapping Ireland's offshore geology

The offshore geology map of Ireland has been compiled for the EMODnet Geology project

(Judge 2015) specifically under the remit of work package 4 (WP4): Sea-floor Geology and Geomorphology. The Federal Institute for Geosciences and Natural Resources, Germany (Bundesanstalt für Geowissenschaften und Rohstoffe, BGR) coordinated WP4, creating and sharing work-package guidelines and technical documentation to steer project partners with respect to the preparation of harmonized, standardized data that adhere to Infrastructure for the Spatial Information in Europe (INSPIRE) standards. The ensuing 1:250 000-scale map of Ireland's offshore geology represents the first attempt to characterize the pre-Quaternary stratigraphy offshore Ireland (Fig. 12).

For the purpose of this mapping exercise, pre-Quaternary is defined as the bedrock present directly beneath any Quaternary cover. The area mapped, as defined by EMODnet Geology, represents Ireland's EEZ with an additional 75 km buffer zone. The map has been produced by consolidating available Irish topographical and geological information into one map. The *1:5 Million International Geological Map of Europe and Adjacent Areas* (IGME 5000) map and data were used as an initial baseline dataset (Asch 2005). The IGME 5000 map was completed by the IGME 5000 project comprising 40 European and adjacent countries. The project produced a geological database that includes information on predicted geology offshore Ireland.

INSS and INFOMAR high-resolution MBES data define the bathymetry of the specified area at a resolution of 111 m. These data were used in tandem with the IGME 5000 predictive mapping of Ireland's offshore geology to constrain many of the obvious morphological seabed features and larger geomorphological provenances. Ireland's sedimentary basins and troughs, highs and the Porcupine seabight illustrated in Figure 1 are realized by the INSS dataset. Large intrusive features of magmatic origin including seamounts and dykes are also identified in the bathymetric data. While these kinds of features exhibit a geomorphological expression in the bathymetry, much of the pre-Quaternary bedrock geology of offshore Ireland lies buried beneath overlying marine sediments. Correlation of these geomorphological features identified within the bathymetry data with available geological data, geophysical studies and the baseline IGME 5000 map formed a key part of the interpretative process leading to the development of Ireland's offshore geology map in Figure 12.

The geology of Ireland, as illustrated on the GSI's 1:100 000 bedrock map of Ireland (GSI 2012) defines the boundaries of terrestrial lithological units. Geological units mapped along the coastal zone that show obvious expression offshore on high-resolution bathymetry, have been extrapolated out to sea. Due to the request from the European

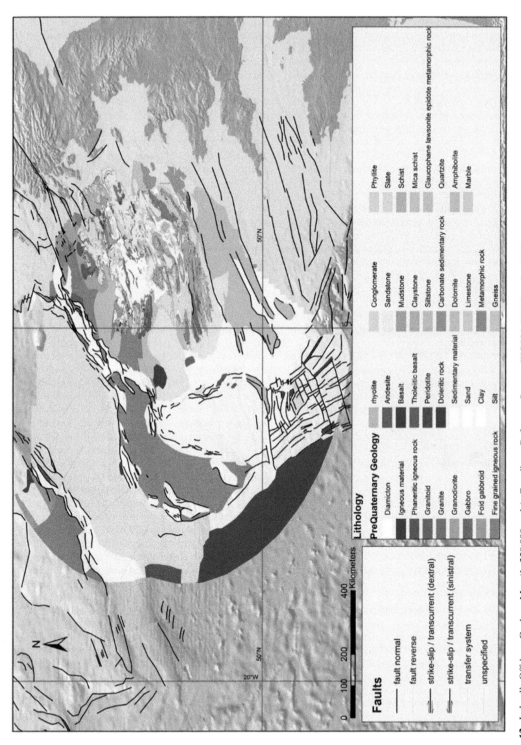

Fig. 12. Ireland's Offshore Geology Map (1 : 250 000 scale), Coordinate Reference System: WGS84 (EPSG code: 4326).

Commission that EMODnet datasets adhere to INSPIRE standards; the nomenclature of onshore geology units mapped with national geological nomenclature and stratigraphic units are transformed in accordance with the INSPIRE vocabulary. These vocabulary define a standardized approach to mapping lithological unites, stratigraphic age, event environments and all relevant information recorded in the data attribution fields.

For inshore waters, outcrops, boundaries and faults evident on the high-resolution bathymetry datasets guide the digitization of coastal geological units. In deeper water, structural datasets are available principally due to extensive research and publication over the past three decades, largely fuelled by petroleum potential. Research and publications on deep water (greater than 200 m) areas offshore Ireland have focused on the crustal structure, tectonostratigraphy, sedimentary development, volcanic province and petroleum potential. Information gleaned from these publications has been incorporated to further constrain the pre-Quaternary outcropping geology of Ireland's offshore EEZ. A comprehensive literature review of previous work has been conducted and elements of this work are incorporated into the offshore geology map and summarized here.

Information detailed in boreholes (Haughton *et al.* 2005), dredge samples (Tyrrell *et al.* 2013), INSS-acquired sub-bottom profiles, seismic profiles (Naylor and Shannon 2005) and deep-towed sidescan sonar (PIP 2004) are used to interpret, identify or track lithological units beneath Quaternary cover. Petroleum Affairs Division (PAD) and Petroleum Infrastructure Programme (PIP) seismic profiles (Morewood *et al.* 2005; O'Reilly *et al.* 2006) as well as GSI gravity and magnetic data are summarized by research endeavours including the Rockall Studies Group (Readman *et al.* 1997; Rockall Studies Group 1998; Hopper *et al.* 2014). Combining the interpretations from these sources allows for reasonable educated assumptions of outcrop types, where other robust ground-truth samples were not available. This material formed the basis of geological edits performed on the baseline IGME 5000 dataset that ultimately resulted in the 1:250 000 map of Ireland's offshore geology (Fig. 9).

Offshore potential field data, refraction and reflection profiles, and well data (Brock *et al.* 1991; Shannon 1991; Naylor *et al.* 1999, 2002; Reston *et al.* 2001; Stoker *et al.* 2005; O'Reilly *et al.* 2006; Shannon *et al.* 2007) provide a comprehensive account of the deep geology of Ireland and its continental margin. This contextual information describes multiple episodes of tectonism through Phanerozoic time (Naylor and Shannon 2011), including failed rifting and subsequent break-up that ultimately resulted in the opening of the Atlantic. These

interpreted geological and geophysical data describe a chronological narrative that lends context to the major geomorphological features we observe offshore Ireland. Detailed Irish data and observations have most recently been incorporated into the first regional systematic compilation and coordinated interpretation of the NE Atlantic for the NAGTEC project (Hopper *et al.* 2014). Marine geological and geophysical datasets were used to build regional models and publish a comprehensive tectonostratigraphic atlas. The NAGTEC Atlas describes in detail the most recent interpretation of the evolution of the NE Atlantic and its conjugate margin pairs. It represents a comprehensive reappraisal of historical studies in the NE Atlantic from the earliest plate tectonic studies. Understanding of how the NE Atlantic region and its continental margins hold unique information is important for many aspects of Earth science, from global geodynamics, palaeoceanography and environmental change (Péron-Pinvidic *et al.* 2017).

The overall seabed surface morphology offshore Ireland as illustrated by the PAD, INSS and INFOMAR bathymetric datasets, summarized in Figure 1, vastly improved knowledge and understanding for the detail of Ireland's continental shelf and Atlantic margin by allowing for a detailed visualization of large-scale seabed features and structures (Dorschel *et al.* 2010). At the westernmost edge of European continent, this region has been affected by multiple orogenic episodes throughout geological history (Naylor and Shannon 2011). The broad continental shelf (greater than 350 km) surrounding Ireland is wide by mean-world standards. Gravity and magnetic data demonstrate that the Moho beneath Ireland lies 30 km below the terrestrial surface (Brock *et al.* 1991). The continental platform surrounding Ireland connects the Irish landmass with Europe to the east and slopes gently westward from terrestrial Ireland to the edge of the shelf break. Along the shelf water depths are in excess of 300 m and here the seabed is generally devoid of major bathymetric features; it has a curved and linear shelf edge (Naylor and Shannon 1982). The shelf edge is defined by steep cliffs that are incised by large canyon systems. Large steep canyons drop off from continental shelf depth of mean 350–4500 m over a mean distance of 30 km, for all but the Porcupine Basin where the Porcupine Seabight etches a more gradual incision into the shelf edge descending from shelf edge to abyssal plane over hyperextended crust. Along the Porcupine Basin's axis stretching factors (the factor by which the lithosphere has been thinned) increase southward deduced from subsidence data for Middle to Late Jurassic rifting (Tate *et al.* 1993). Such lithospheric thinning characteristics are normally associated with the highly thinned crust near the continent–ocean transition of rifted margins (Reston *et al.* 2001).

In the Porcupine Basin, basement is overlain by a thick Upper Carboniferous succession (Stoker *et al.* 2005). This sequence is covered by a thick Paleogene sedimentary material. The bathymetric relief of the Porcupine Seabight cascades from the shelf edge more gently than the rest of the shelf and into the Porcupine Abyssal Plain from *c.* 350 m water depth in the north to greater than 4000 m water depth in the south. An incised channel clearly evident in the bathymetry drains sediment to the base of the Abyssal Plain, which in turn is characterized by oceanic crustal basalts coated with marine sediment.

The Porcupine High, a prominent oblong feature orientated ENE–WSW, is inferred to be comprised of granitic orthogneiss on the northern flanks and metasedimentary rocks on the top of the high, comprising low-grade metamorphic green psammites of Grampian age, similar to outcrops in NW Mayo (Tyrrell *et al.* 2013). Located south of the Porcupine at the southernmost region of Ireland's continental shelf, the Goban Spur is a bathymetric plateau that slopes gently westwards away from the Cretaceous Chalk-dominated Celtic shelf (Naylor and Shannon 2011). The Goban Spur plateau has a thick cover to Upper Paleozoic, probably dominantly Devonian rocks (Naylor and Shannon 2005) with occasions of Mesozoic and Cenozoic cover (Naylor *et al.* 2002). Small perched basins at the Goban province are expressed by complex fault series of NNE–SSW and NNW–SSE controlled basins (Naylor and Shannon 2005).

To the NW of Ireland's offshore region the NE–SW-trending Rockall Trough is the dominant bathymetric feature. The general trend of this trough is believed to have a deep pre-Caledonian (600 Ma) structural origin (Hutton and Alsop 1996; Naylor and Shannon 2005). The Rockall margin was produced by rotational opening of the Rockall Basin in Triassic and Late Jurassic rift events (Thomson and McWilliam 2001). A set of small, elongate, probably Mesozoic basins are located in the footwalls of the main Cenozoic Rockall Basin (Naylor *et al.* 1999). These have a thin Mesozoic and Cenozoic cover (Naylor and Shannon 2005). The Rockall High comprises a distinct province of Precambrian rocks (Naylor and Shannon 2005). It is thought to belong to the Islay terrane, with the boundary against Lewisian rocks lying to the north, in UK waters (Hitchen *et al.* 1997).

The continental shelf of Ireland bears the geological imprints and structures detailing a history of Variscan, Caledonian and older orogenic events. Reactivation of some of these structures has influenced the orientation of major geomorphological features offshore, many of which can be identified

within Ireland's baseline seabed mapping datasets (Dorschel *et al.* 2010). While detailed mapping of all lithological units present offshore is still technically impossible, studies and data summarized here have lent a broad overview to the varying geological compositions recorded. These build a colourful record of events that comprise the geology of Ireland's EEZ as represented on Ireland's offshore geology map in Figure 9.

Finally, The EMODnet Geology project, which has in this case provided the impetus to further amalgamate and interpret the information gleaned from the studies and knowledge resources described, stands as a powerful example of the benefits of scientific co-operation and partnership in furthering our understanding of the ocean realm. In particular, the combination of contemporary geological knowledge from multiple sources with a broad geomorphological analysis of Ireland's shelf and deep-water bathymetry (Judge 2015) has resulted in the production of a geology map for Ireland's offshore area (Fig. 12). This product has been harmonized for compatibility with the EMODnet Geology project's online data discovery viewers and services (Kaskela *et al.* 2019) that enhance the reach and impact of this product by making the knowledge freely available to a wider international body of stakeholders. The combination of geological data with bathymetry data allows users to build up a comprehensive picture of the seabed and its subsurface, providing a vital component for seafloor habitat maps and offering essential tools in marine spatial planning, coastline protection, offshore installation design, environmental conservation, risk management and resource mapping (Kaskela *et al.* 2019).

Discussion: considering the impact of Irish seabed mapping

The collection of four practical interpretative case studies presented in this paper highlights ways in which information developed through Ireland's seabed mapping efforts can be interpreted to enhance scientific knowledge and understanding of coastal, shelf and deep-water marine environments. Their effect in illustrating how the generation of this blue knowledge creates a positive impact in terms of sustainable development will be discussed further in this section. However, in addition to the acquisition, management and delivery of high-resolution baseline datasets supporting cross-sectoral applications, access to Ireland's marine knowledge through the INFOMAR programme continues to underpin key areas of Ireland's blue growth at national, European and international levels (Indecon International Economic Consultants 2017).

The national impact of the INFOMAR programme in this regard can be gauged through its inclusion in Government of Ireland policy, specifically, Harnessing our Ocean Wealth (HOOW) – An Integrated Marine Plan for Ireland (Government of Ireland 2012), the Department of Communications, Climate Action and Environment (DCCAE) Climate Action Plan and Statement of Strategy 2019–21 (DCCAE 2019*a, b*) and the Draft National Marine Planning Framework (NMPF) for public consultation (DHPLG 2019). HOOW represents a vision of sustainable growth for Ireland's blue economy underpinned by coherent policy, planning and regulation, managed in an integrated manner. A key enabler of the Plan is the 'Research, Technology and Innovation' component, where the completion of the INFOMAR seabed mapping programme is listed as a necessary action (Action 23) to provide data, products and services as critical inputs to maritime spatial planning and enablers of infrastructural development, research, education and value-added products (Government of Ireland 2012). The HOOW Review of Progress 2018 (Government of Ireland 2019) notes that Ireland is already well on target to achieve and even exceed the economic targets set out in the Plan and states that European Commission reporting in May 2019 highlighted significant growth across the blue economy of most EU member states, noting the most significant expansion observed was in Ireland and Malta according to Eurostat figures. The HOOW Review of Progress showed that in 2018 Ireland's ocean economy had a turnover of €6.2 billion, giving a total Gross Value Added (GVA) (direct and indirect) figure of €4.19 billion, representing 2% of GDP. Additionally at a national level, the DCCAE's Climate Action Plan 2019 (DCCAE 2019*a*) also stipulates as an action, the need to support the ocean energy research development demonstration pathway for emerging marine technologies (wave, tidal, floating wind) and associated test infrastructure. The completion of mapping for all Irish offshore waters through the INFOMAR programme to support site selection for offshore energy is listed as a key step necessary for the delivery of this action (Action 26). Furthermore, the NMPF (DHPLG 2019) cites INFOMAR as a key reference under overarching marine planning policies, specifically in relation to seafloor integrity as it relates to Good Environmental Status (GES) as per the Marine Strategy Framework Directive (MSFD) (2008/56, EC).

In the European and international contexts the INFOMAR programme has created positive impact through its involvement with European Territorial Cooperation (INTERREG)-funded initiatives including the Joint Irish Bathymetric Survey (JIBS), INIS-Hydro (Ireland, Northern Ireland and Scotland Hydrographic Survey) and CHERISH projects. Additionally, the programme maintains ongoing Irish collaboration with key cross-border and UK-based organizations and agencies including the UKHO, many of whom sit on the programme's Technical Advisory Committee (TAC). Involvement in the Directorate-General for Maritime Affairs (DG MARE) European Commission-funded EMODnet Bathymetry Geology and Habitats projects has increased the international reach and impact of Ireland's national seabed mapping datasets through their inclusion in a harmonized international EU-wide data portal (Kaskela *et al.* 2019) expanding the discoverability and impact of Ireland's marine knowledge many-fold. Information generated through Ireland's seabed mapping efforts represents an important resource for maintaining Ireland's alignment and commitment to several pieces of European legislation including: the Marine Strategy Framework Directive (2008/56, EC); the Habitats Directive (92/43, EEC); the Maritime Spatial Planning Directive (2014/89, EU); the Water Framework Directive (2000/60, EC); the INSPIRE Directive (2007/2, EC); the Open Data and Public Sector information Directive (2019/1024, EU) where data and products produced by the programme are directly relevant to the stipulations of the legislation. The 1992 OSPAR convention is also included as part of the legislative rationale for the programme identified in PwC's most recent external evaluation of the programme (PwC 2013). The DG MARE-funded Blue Growth Report into Scenarios and Drivers for Sustainable Growth from the Oceans, Seas and Coasts cites INFOMAR as a positive example of a member state's Blue Growth initiative (Ecorys 2012). Ireland's obligations to the UN safety of life at sea (SOLAS) convention (UN 1980) are addressed through INFOMAR's data acquisition policies whereby seabed mapping data are acquired according to International Hydrographic Office (IHO) standards for data quality and accuracy (IHO 2008). These data have been provided to the relevant authorities so that nautical charts used for safe navigation incorporate updated information. Ireland's seabed mapping data have also been used to directly support Ireland's marine territorial claim through the UN Convention for the Law of the Sea (UNCLOS) based on geological boundaries (Nelson 2006) whereby a claim for jurisdiction over marine territory has been made by the Irish nation covering an area ten times the size of Ireland's landmass (Government of Ireland 2005). Knowledge resources inferred from Ireland's high-resolution seabed mapping data have been further incorporated in international initiatives and collaborations such as the development of the NAG-TEC Atlas (Péron-Pinvidic *et al.* 2017) European Geological Surveys Research Area (GeoERA) Seabed Mineral Deposits in European Seas: Metallogeny and Geological Potential for Strategic and

Critical Raw Materials (MINDeSEA) project (Gonzalez and MINDeSEA 2019) and the Nippon Foundation's GEBCO Seabed 2030 (Mayer *et al.* 2018) project.

The pioneering open and free data philosophy of the Irish Government (DPER 2017) has led to further international collaborations, with Ireland's seabed mapping data also hosted on the NOAA archive, where users can access raw data files in addition to the standard outputs and metadata already available through the GSI-hosted Interactive Web Data Delivery Service (IWDDS). Prospective data users from anywhere in the world may also access Ireland's seabed data layers and products through a series of web map services (WMS) compatible with standard GIS software packages via the INFOMAR website. Data accessed freely from these resources can be used by the marine and scientific communities under a creative commons licensing agreement, which can be reviewed via the IWDDS.

The INFOMAR programme has supported a wide variety of academia- and industry-led research in Ireland's marine sector by having input into a series of research short calls and supporting subsequent calls funded by the GSI's Geoscience Research initiative (GSI 2016). Currently over 190 publications can be accessed directly through the INFOMAR website and the programme has been cited numerous times within the scientific literature. Furthermore, Irish seabed mapping data underpin multiple scientific and oceanographic research cruises in Irish waters where high-resolution maps of seafloor features are used to drive discovery and research collaborations as in the recent Atlantic Ocean Research Alliance (AORA) Transocean Survey North Atlantic (TRASNA) and Sensitive Ecosystem Assessment and ROV Exploration of Reef (SEA-ROVER) expeditions carried out in waters within Ireland's designated area (Fig. 1).

Demonstrated applications of Ireland's marine data resource supporting the Irish maritime industry include: the safeguarding of navigation and shipping; de-risking the development of offshore renewable energy; informing fisheries and aquaculture management; supporting the development of offshore infrastructure, energy security and resource management; underpinning environmental modelling and monitoring; and enhancing our understanding of coastal behaviour and heritage (PwC 2008, 2013; White *et al.* 2019). Irish seabed mapping data will in most cases represent the best available data to inform and underpin decision making for Maritime Spatial Planning (MSP) as set out in the National Marine Planning Framework Baseline Report where the INFOMAR programme is referred to as a useful source of information (DHPLG 2018). Access to high-resolution seabed information will allow MSP decision makers to manage inherent

scale issues as they relate to specific cross-sectoral maritime activities and associated data requirements.

The educational impact of Ireland's seabed mapping efforts includes ongoing training and skills development of graduates and scientific crews in multidisciplinary techniques for seabed mapping, which are applicable worldwide (Government of Ireland 2012). In 2018 the programme was included on the national educational curriculum via a web portal managed by Scoilnet for both junior and senior cycles and, as of 2020, INFOMAR will pilot an MSc module in GIS and remote sensing in partnership with National University of Ireland Maynooth (NUIM).

External appraisals of Ireland's seabed mapping efforts have been periodically undertaken throughout the life cycle of the INFOMAR programme. The 2008 PwC Marine Mapping Survey Options Appraisal Report included a CBA yielding a ratio of between four and six times to one return on investment in favour of the programme depending on the timeframe. A follow-on 2013 PwC re-evaluation of the programme highlighted a strong rationale for the continued public funding of the INFOMAR programme noting that it supports the attainment of key national and European policy objectives and regulatory obligations. In 2016 a post-project evaluation of INFOMAR Phase 1 (2006–15) carried out by Risk Solutions acknowledged that while the programme is widely respected and that the programme largely achieved its Phase 1 objectives, a radical cut to the programme budget arising from Ireland's financial crises affected its ability to deliver upon all of its initial aspirations. The report recommended that recent funding cuts should be addressed to allow the programme to deliver to its full potential. A review of the Irish Geosciences sector by Indecon reported that the INFOMAR programme directly contributed €24 million to Ireland's ocean economy in 2016 (Indecon International Economic Consultants 2017).

The availability of metrics from the Irish experience to appraise the positive impact of a national-scale/EEZ seabed mapping effort over more than two decades of operations will undoubtedly prove beneficial to the international community and a lasting legacy for the blue knowledge developed through these efforts. Other nations weighing up the challenges and associated benefits arising from national baseline seabed mapping programmes will have a useful reference for estimating the impact of this activity on society, the economy and sustainability, potentially underpinning the case for future international seabed mapping initiatives. The opportunity to share, exchange and transfer Ireland's blue knowledge between other nations and programmes also aligns well with the broader international agenda, especially in the case of the UN Sustainable Development Goals (SDGs) and the complementary

UN Decade of Ocean Science for Sustainable Development (2020–30). In particular, the case studies presented in this paper can be used to illustrate the impact of blue knowledge in supporting some key UN SDGs adopted by all UN member states in 2015.

In summary, Case study 4 highlights how government-based partnership through Ireland's involvement in the pan-European EMODnet Geology project, has resulted in a harmonized geological data product with multiple applications, supporting UN SDG 17 on Partnership. Case study 3 shows how the study of seabed features through multiple repeat surveys (time-series data) offers crucial risk insights for the development of the offshore renewable-energy sector supporting UN SDG 13 on climate action. Case study 2 shows how the interpretation of integrated seabed mapping products can positively impact the ability of Ireland's marine scientists to understand and monitor a key fishing ground through improved stock assessment, aligning with UN SDG 14 on life below sea, while Case study 1 shows how the analysis of Ireland's advanced seabed mapping data allows for the extrapolation of Ireland's unique geological history into the ocean environment, promoting Ocean Literacy, which is a stated goal of UNESCO's UN Decade of Ocean Science for Sustainable Development initiative. This case study directly supports two current principles of Ocean Literacy with modern visualizations of the seafloor underscoring that the ocean is largely unexplored (Principle 7), while the complex geological record and marine fossils inferred from land-based studies onto the submerged landscapes off Hook Head, Co. Wexford, remind us that the ocean made the Earth habitable (Principle 4) (NOAA 2013).

Conclusions

The INFOMAR programme is scheduled to run until the end of 2026. This period will see intensive offshore and inshore seabed surveying campaigns to complete the mapping of Ireland's uncharted marine territories. Analogous to this activity will be a dedicated effort to ensure that data acquired by the programme are processed and distributed to project stakeholders in a way that fulfils the potential of the programme, preserves the legacy of Ireland's seabed mapping efforts and reaches new end users, supporting research, innovation, knowledge-based decision making, climate action strategy and sustainable development. An increasing awareness for the necessity to manage our marine environment in a sustainable way while developing the blue economy will make Ireland's seabed mapping datasets an invaluable baseline from which the nation's future marine knowledge can be spatially and temporally referenced. The need to respond to a changing climate through 'Climate Action', through a philosophy of 'Map, Observe, Predict', will drive future research into the dynamics of our coastal, shelf and deep-water environments. Ireland's high-quality marine datasets will underpin future resurvey and ground-truthing campaigns, supporting modelling of environmental processes. Coastal erosion and sediment transportation will be mapped multitemporally using time-series data acquired at key sites identified from baseline marine datasets. The knowledge resource built up through Ireland's seabed mapping activity will increasingly drive informed policy and optimal decision making in relation to Ireland's maritime activities. Finally, the experience and insight developed through more than two decades of pioneering seabed mapping will be made available to the international community through outward engagement and participation in EU and international initiatives such as the UN Decade of the Ocean 2020–30 aimed at expanding scientific knowledge of our coasts, seas and oceans, tackling climate change and managing the sustainable development of our global marine resource.

Acknowledgements The authors wish to acknowledge the role of colleagues and supporters past and present for their contribution to the development of Ireland's seabed mapping capabilities. From the seagoing crews charting the seafloor to the scientists and programme staff tasked with turning this information into blue knowledge and putting it out into the world, to the members of the public and individuals in government who through their recognition and support for these efforts have ensured that Ireland continues to expand and enhance our knowledge, understanding and sustainable stewardship of our ocean realm. Thanks also to Eimear O'Keeffe for GIS support rendered and Koen Verbruggen for encouraging the development of this article.

Funding This research received no specific grant from any funding agency in the public, commercial, or not-for-profit sectors.

Author contributions ROT: conceptualization (equal), project administration (equal), resources (supporting), supervision (equal), visualization (equal), writing – original draft (lead), writing – review & editing (lead); **MJ**: conceptualization (supporting), formal analysis (equal), supervision (supporting), writing – original draft (supporting), writing – review & editing (equal); **FS**: formal analysis (equal), project administration (equal), visualization (equal), writing – original draft (equal), writing – review & editing (supporting); **TF**: funding acquisition (equal), project administration (equal), resources (equal), writing – original draft (equal), writing – review & editing (supporting); **EMC**: formal analysis (supporting), visualization (supporting), writing – original draft (supporting); **KS**: formal analysis (supporting), visualization (supporting), writing – original draft (supporting); **SK**: writing –

original draft (supporting); **SC**: funding acquisition (equal), project administration (equal), resources (equal), writing – review & editing (supporting); **FMG**: project administration (equal), writing – review & editing (supporting); **XM**: conceptualization (equal), project administration (equal), supervision (supporting), writing – review & editing (supporting).

Data availability statement The datasets generated during and/or analysed during the current study contain Irish Public Sector Data (Geological Survey Ireland & Marine Institute) licensed under a Creative Commons Attribution 4.0 International (CC BY 4.0) licence and are accessible at https://www.infomar.ie/data or upon request from INFOMAR programme management (info@infomar.ie).

References

Alder, J., Cullis-Suzuki, S. *et al.* 2010. Aggregate performance in managing marine ecosystems of 53 maritime countries. *Marine Policy*, **34**, 468–476, https://doi.org/10.1016/j.marpol.2009.10.001

Asch, K. 2005. *IGME 5000: The 1 : 5 Million International Geological Map of Europe and Adjacent Areas*. BGR, Hannover.

Belderson, R.H., Johnson, M.A. and Kenyon, N.H. 1982. Bedforms. *In*: Stride, A.H. (ed.) *Offshore Tidal Sands: Processes and Deposits*. Chapman and Hall, London, 222.

Brock, A., Ryan, P.D. and Shannon, P.M. 1991. The deep geology and geophysics of Ireland and its continental margin. *Journal of the Geological Society, London*, **148**, 129–130, https://doi.org/10.1144/gsjgs.148.1.0129

Chiverrell, R.C., Thrasher, I.M. *et al.* 2013. Bayesian modelling the retreat of the Irish Sea Ice Stream. *Journal of Quaternary Science*, **28**, 200–209, https://doi.org/10.1002/jqs.2616

Clark, C.D., Hughes, A.L.C., Greenwood, S.L., Jordan, C. and Sejrup, H.P. 2012. Pattern and timing of retreat of the last British–Irish Ice Sheet. *Quaternary Science Reviews*, **44**, 112–146, https://doi.org/10.1016/j.quascirev.2010.07.019

Denny, J.F., Schwab, W.C. *et al.* 2013. Holocene sediment distribution on the inner continental shelf of northeastern South Carolina: implications for the regional sediment budget and long-term shoreline response. *Continental Shelf Research*, **56**, 56–70, https://doi.org/10.1016/j.csr.2013.02.004

Department of Communications Climate Action & Environment (DCCAE) 2019*a*. Climate Action Plan 2019.

Department of Communications Climate Action & Environment (DCCAE) 2019*b*. Statement of Strategy 2019–2021.

Department of Communications Energy & Natural Resources (DCENR) 2014. Offshore Renewable Energy Development Plan – A Framework for the Sustainable Development of Ireland's Offshore Renewable Energy Resource.

Department of Housing Planning & Local Government (DHPLG) 2018. National Marine Planning Framework: Baseline Report.

Department of Housing Planning & Local Government (DHPLG) 2019. National Marine Planning Framework: Consultation Draft.

Department of Public Expenditure & Reform (DPER) 2017. Open Data Strategy 2017–2022.

Dorschel, B., Wheeler, A.J., Monteys, X. and Verbruggen, K. 2010. *Atlas of the Deep-Water Seabed: Ireland*. 1st edn. Springer Science and Business Media.

Dorst, L.L., Roos, P.C. and Hulscher, S.J.M.H. 2013. Improving a bathymetric resurvey policy with observed sea floor dynamics. *Journal of Applied Geodesy*, **7**, 51–64, https://doi.org/10.1515/jag-2012-0035

Doyle, J., O'Brien, S., Fitzgerald, R., Vacherot, J.-P., Sugrue, S. and Quinn, M. 2019. *The 'Smalls' Nephrops Grounds (FU22) 2019 UWTV Survey Report and Catch Scenarios for 2020*. Marine Institute.

Ecorys 2012. *Blue Growth – Scenarios and Drivers for Sustainable Growth from the Oceans, Seas and Coasts*. European Commission.

Edwards, R. and Brooks, A. 2008. The island of Ireland: drowning the myth of an Irish land-bridge? *Special Supplement to The Irish Naturalists' Journal*, 19–34.

European Economic Community (EEC) 1992. Council Directive 92/43/EEC of 21 May 1992 on the conservation of natural habitats and of wild fauna and flora. *Official Journal of the European Union*, **206**, 7–50.

European Council (EC) 2000. Directive 2000/60/EC of the European Parliament and of the council of 23 October 2000 establishing a framework for Community action in the field of water policy. *Official Journal of the European Communities*, **22**, 1–73.

European Council (EC) 2007. Directive 2007/2/EC of the European Parliament and of the council of 14 March 2007 establishing an Infrastructure for Spatial Information in the European Community (INSPIRE). *Official Journal of the European Union*, **108**, 1–14.

European Council (EC) 2008. Directive 2008/56/EC establishing a framework for community action in the field of marine environmental policy (Marine Strategy Framework Directive). *Official Journal of the European Union*, **51**, 19–40.

European Union (EU) 2010. Marine Knowledge 2020: marine data and observation for smart and sustainable growth. *Communication from the Commission to the European Parliament and the Council*. 1–13.

European Union (EU) 2014. Directive 2014/89/EU of the European Parliament and of the Council of 23 July 2014 establishing a framework for maritime spatial planning. *Official Journal of the European Union*, **257**, 135–145.

European Union (EU) 2019. Directive (EU) 2019/1024 of the European Parliament and of the Council of 20 June 2019 on open data and the re-use of public sector information. *Official Journal of the European Union*, **172**, 56–83.

Evans, W. 2018. *Hydrodynamic Modelling of Sediment Transport and Bedform Formation on the NW Irish Shelf*. University of Ulster.

Evans, W., Benetti, S., Sacchetti, F., Jackson, D.W.T., Dunlop, P. and Monteys, X. 2015. Bedforms on the northwest Irish Shelf: indication of modern active sediment transport and over printing of paleo-glacial sedimentary deposits. *Journal of Maps*, **11**, 561–574, https://doi.org/10.1080/17445647.2014.956820

Feldens, P., Schwarzer, K., Sakuna, D., Szczuciński, W. and Sompongchaiyakul, P. 2012. Sediment distribution on the inner continental shelf off Khao Lak (Thailand) after the 2004 Indian Ocean tsunami. *Earth, Planets and Space*, **64**, 875–887, https://doi.org/10.5047/eps.2011.09.001

Gallagher, C. 2002. The morphology and palaeohydrology of a submerged glaciofluvial channel emerging from Waterford Harbour onto the nearshore continental shelf of the Celtic Sea. *Irish Geography*, **35**, 111–132, https://doi.org/10.1080/00750770209555800

Geological Survey Ireland (GSI) 2012. *Bedrock 1:100 000*.

Geological Survey Ireland (GSI) 2016. *Geological Survey Ireland Research Roadmap*.

Geological Survey Ireland (GSI) 2018. *Bedrock 1:100 000 Sheet No. 23*.

Gonzalez, F.J. and MINDeSEA 2019. GeoERA-MINDeSEA project: mapping and studying critical elements in the pan-European seabed mineral deposits. *Goldschmidt Abstracts*, 2019, 1174.

Government of Ireland 2005. Submission to the Commission on the Limits of the Continental Shelf pursuant to Article 76, paragraph 8 of the UNCLOS 1982 in respect of the area abutting the Porcupine Abyssal Plain.

Government of Ireland 2012. Harnessing Our Ocean Wealth – An Integrated Marine Plan for Ireland.

Government of Ireland 2019. Harnessing Our Ocean Wealth Review of Progress 2018 – Towards an Integrated Marine Plan for Ireland.

Guinan, J., Grehan, A.J., Dolan, M.F.J. and Brown, C. 2009. Quantifying relationships between video observations of cold-water coral cover and seafloor features in Rockall Trough, west of Ireland. *Marine Ecology Progress Series*, **375**, 125–138, https://doi.org/10.3354/meps07739

Haughton, P., Prage, D. *et al.* 2005. First results from shallow stratigraphic boreholes on the eastern flank of the Rockall Basin, offshore western Ireland. *Geological Society, London, Petroleum Geology Conference Proceedings*, **6**, 1077–1094, https://doi.org/10.1144/0061077

Hitchen, K., Morton, A.C., Mearns, E.W., Whitehouse, M. and Stoker, M.S. 1997. Geological implications from geochemical and isotopic studies of Upper Cretaceous and Lower Tertiary igneous rocks around the northern Rockall Trough. *Journal of the Geological Society, London*, **154**, 517–521, https://doi.org/10.1144/gsjgs.154.3.0517

Hopper, J.R., Funck, T., Stoker, M., Árting, U., Peron-Pinvidic, G., Doornenbal, H. and Garnia, C. 2014. *Tectonostratigraphic Atlas of the North-East Atlantic Region*. Geological Survey of Denmark and Greenland.

Hutton, D.H.W. and Alsop, G.I. 1996. The Caledonian strike-swing and associated lineaments in NW Ireland and adjacent areas: sedimentation, deformation and igneous intrusion patterns. *Journal of the Geological Society, London*, **153**, 345–360, https://doi.org/10.1144/gsjgs.153.3.0345

Indecon International Economic Consultants 2017. *An Economic Review of the Irish Geoscience Sector*.

INFOMAR 2007. *INFOMAR Proposal & Strategy*.

International Hydrographic Organisation (IHO) 2008. *IHO Standards for Hydrographic Surveys S-44*. 5th edn. International Hydrographic Bureau, Monaco.

Judge, M. 2015. European Marine Observation and Data Network (EMODnet): making fragmented marine data relevant and accessible. *IEEE Earthzine*, https://earthzine.org/european-marine-observation-and-data-network-emodnet-making-fragmented-marine-data-relevant-and-accessible/

Kaskela, A.M., Kotilainen, A.T. *et al.* 2019. Picking up the pieces – harmonising and collating seabed substrate data for European maritime areas. *Geosciences (Switzerland)*, **9**, 1–18.

Kenyon, N.H. and Stride, A.H. 1970. The tide-swept continental shelf sediments between the Shetland Isles and France. *Sedimentology*, **14**, 159–173, https://doi.org/10.1111/j.1365-3091.1970.tb00190.x

Knaapen, M.A.F. and Hulscher, S.J.M.H. 2002. Regeneration of sand waves after dredging. *Coastal Engineering*, **46**, 277–289, https://doi.org/10.1016/S0378-3839(02)00090-X

Lurton, X., Lamarche, G., Brown, C., Lucieer, V.L., Rice, G., Schimel, A. and Weber, T. 2015. *Backscatter Measurements by Seafloor-Mapping Sonars – Guidelines and Recommendations*. A report by members of the GeoHab Backscatter Working Group, Salvador da Bahia, Brazil.

Ma, X., Yan, J. and Fan, F. 2014. Morphology of submarine barchans and sediment transport in barchans fields off the Dongfang coast in Beibu Gulf. *Geomorphology*, **213**, 213–224, https://doi.org/10.1016/j.geomorph.2014.01.010

Marine Institute 2009. *Atlas of the Commercial Fisheries around Ireland 2009 Review of the Fisheries*. Oranmore.

Mayer, L., Jakobsson, M. *et al.* 2018. The Nippon Foundation–GEBCO seabed 2030 project: the quest to see the world's oceans completely mapped by 2030. *Geosciences (Switzerland)*, **8**, 63.

Meere, P., Maccarthy, I., Reavy, R.J., Allen, A. and Higgs, K. 2013. *Geology of Ireland: A Field Guide*. Collins Press.

Morelissen, R., Hulscher, S.J.M.H., Knaapen, M.A.F., Németh, A.A. and Bijker, R. 2003. Mathematical modelling of sand wave migration and the interaction with pipelines. *Coastal Engineering*, **48**, 197–209, https://doi.org/10.1016/S0378-3839(03)00028-0

Morewood, N.C., Mackenzie, G.D., Shannon, P.M., O'Reilly, B.M., Readmen, P.W. and Makris, J. 2005. The crustal structure and regional development of the Irish Atlantic margin region. *Geological Society, London, Petroleum Geology Conference Series*, **6**, 1023–1034, https://doi.org/10.1144/0061023

Muir, R.J., Fitches, W.R. and Maltman, A.J. 1994. The Rhinns Complex: Proterozoic basement on Islay and Colonsay, Inner Hebrides, Scotland, and on Inishtrahull, NW Ireland. *Transactions of the Royal Society of Edinburgh: Earth Sciences*, **85**, 77–90, https://doi.org/10.1017/S0263593300006313

National Oceanic and Atmospheric Administration (NOAA) 2013. Ocean Literacy – The Essential Principles and Fundamental Concepts of Ocean Sciences for Learners of All Ages.

Naylor, D. and Shannon, P.M. 1982. *Geology of Offshore Ireland and West Britain*. Graham and Trotman.

Naylor, D. and Shannon, P.M. 2005. The structural framework of Irish Altantic Margin. *Geological Society, London, Petroleum Geology Conference Series*, **6**, 1009–1021, https://doi.org/10.1144/0061009

Naylor, D. and Shannon, P.M. 2011. *Petroleum Geology of Ireland*. Dunedin Academic Press, Edinburgh.

Naylor, D., Shannon, P.M. and Murphy, N. 1999. *Irish Rockall Basin Region – A Standard Structural Nomenclature System*. Department of Marine and Natural Resources.

Naylor, D., Shannon, P.M. and Murphy, N. 2002. Porcupine–Goban region – a standard structural nomenclature system. *Petroleum Affairs Division Special Publication*, **1–2**, 65.

Nelson, L.D.M. 2006. Reflections on the 1982 Convention on the Law of the Sea. *The Law of the Sea: Progress and Prospects*, 7–208, https://doi.org/10.1093/acprof:oso/9780199299614.003.0002

Németh, A.A., Hulscher, S.J.M.H. and de Vriend, H.J. 2002. Modelling sand wave migration in shallow shelf seas. *Continental Shelf Research*, **22**, 2795–2806, https://doi.org/10.1016/S0278-4343(02)00127-9

Ó Cofaigh, C. and Evans, D.J.A. 2001. Sedimentary evidence for deforming bed conditions associated with a grounded Irish Sea glacier, southern Ireland. *Journal of Quaternary Science*, **16**, 435–454, https://doi.org/10.1002/jqs.631

O'Reilly, B.M., Hauser, F., Ravaut, C., Shannon, P.M. and Readman, P.W. 2006. Crustal thinning, mantle exhumation and serpentinisation in the Porcupine Basin, offshore Ireland: evidence from wide-angle seismic data. *Journal of the Geological Society, London*, **163**, 775–787, https://doi.org/10.1144/0016-76492005-079

Péron-Pinvidic, G., Hopper, J.R., Stoker, M.S., Gaina, C., Doornenbal, J.C., Funck, T. and Árting, U.E. 2017. The NE Atlantic region: a reappraisal of crustal structure, tectonostratigraphy and magmatic evolution. *Geological Society, London, Special Publications*, **447**, 1–9, https://doi.org/10.1144/SP447.17

Petroleum Infrastructure Programme (PIP) 2004. TOBI Rockall Irish Margin project (R97/14).

PricewaterhouseCoopers (PwC) 2008. *INFOMAR Marine Mapping Study Options Appraisal Report: Final Report*.

PricewaterhouseCoopers (PwC). 2013. INFOMAR External Evaluation.

Readman, P.W., O'Reilly, B.M. and Murphy, T. 1997. Gravity gradients and upper-crustal tectonic fabrics, Ireland. *Journal of the Geological Society, London*, **154**, 817–828, https://doi.org/10.1144/gsjgs.154.5.0817

Reston, T.J., Pennell, J., Stubenrauch, A., Walker, I. and Perez-Gussinye, M. 2001. Detachment faulting, mantle serpentinization, and serpentinite- mud volcanism beneath the Porcupine Basin, southwest of Ireland. *Geology*, **29**, 587–590, https://doi.org/10.1130/0091-7613(2001)029<0587:DFMSAS>2.0.CO;2

Risk Solutions 2016. *INFOMAR Phase 1 Post-Project Evaluation*. Warrington.

Rockall Studies Group 1998. *Report on the compilation, presentation and interpretation of gravity and magnetic data. Rockall Studies Group Project R98/1*. BGS Technical Report **WK/99/12/c**. Version 1.0: 1/11/99.

Rourke, F.O., Boyle, F. and Reynolds, A. 2010. Marine current energy devices: current status and possible future applications in Ireland. *Renewable and Sustainable Energy Reviews*, **14**, 1026–1036, https://doi.org/10.1016/j.rser.2009.11.012

Shannon, P.M. 1991. The development of Irish offshore sedimentary basins. *Journal of the Geological Society, London*, **148**, 181–189, https://doi.org/10.1144/gsjgs.148.1.0181

Shannon, P.M., McDonnell, A. and Bailey, W.R. 2007. The evolution of the Porcupine and Rockall basins, offshore Ireland: the geological template for carbonate mound development. *International Journal of Earth Sciences*, **96**, 21–35, https://doi.org/10.1007/s00531-006-0081-y

Sheng, J. and Yang, B. 2010. A nested-grid ocean circulation model for simulating three-dimensional circulation and hydrography over Canadian Atlantic coastal waters. *Terrestrial Atmospheric and Oceanic Sciences*, **21**, 27–44, https://doi.org/10.3319/TAO.2009.06.08.01(IWNOP)

Sleeman, A.G., Johnston, I.S., Naylor, D. and Sevastopulo, G.D. 1974. The stratigraphy of the Carboniferous rocks of Hook Head, Co. Wexford. *Proceedings of the Royal Irish Academy. Section B: Biological, Geological, and Chemical Science*, **74**, 227–243.

Staneva, J., Stanev, E.V. *et al.* 2009. Hydrodynamics and sediment dynamics in the German Bight. A focus on observations and numerical modelling in the East Frisian Wadden Sea. *Continental Shelf Research*, **29**, 302–319, https://doi.org/10.1016/j.csr.2008.01.006

Stoker, M.S., Praeg, D., Hjelstuen, B.O., Laberg, J.S., Nielsen, T. and Shannon, P.M. 2005. Neogene stratigraphy and the sedimentary and oceanographic development of the NW European Atlantic margin. *Marine and Petroleum Geology*, **22**, 977–1005, https://doi.org/10.1016/j.marpetgeo.2004.11.007

Sutton, G. 2008. *Irish Sea Marine Aggregate Initiative (IMAGIN) Technical Synthesis Report*. Marine Institute.

Tate, M., White, N. and Conroy, J.J. 1993. Lithospheric extension and magmatism in the Porcupine Basin west of Ireland. *Journal of Geophysical Research: Solid Earth*, **98**, 13905–13923, https://doi.org/10.1029/93JB00890

Thiébot, J., Bailly du Bois, P. and Guillou, S. 2015. Numerical modelling of the effect of tidal stream turbines on the hydrodynamics and the sediment transport – application to the Alderney Race (Raz Blanchard), France. *Renewable Energy*, **75**, 356–365, https://doi.org/10.1016/j.renene.2014.10.021

Thomson, A. and Mcwilliam, A. 2001. The structural style and evolution of the Bróna Basin. *Geological Society, London, Special Publications*, **188**, 401–410, https://doi.org/10.1144/GSL.SP.2001.188.01.25

Tietzsch-Tyler, D. and Sleeman, A.G. 1994. *Geology of South Wexford: A Geological Description of South Wexford and Adjoining Parts of Waterford, Kilkenny and Carlow to Accompany the Bedrock Geology 1:100 00 Scale Map Series, Sheet 23, South Wexford*. Geological Survey Ireland.

Tóth, Z., Wheeler, A., Mccarron, S. and Monteys, X. 2016. Esker ridges and seismostratigraphic evidence for a southerly once flow extending onto the present nearshore continental shelf of the Celtic Sea, SE Ireland. Paper presented at the American Geophysical Union Fall Meeting, New Orleans, USA, December 2016, Abstract Number: C53C-0749.

Tyrrell, S., Chew, D.*et al.* 2013. Dredging up the past: new insights into the geology of the Porcupine High, offshore western Ireland. Paper presented at the 56th Irish Geological Research Meeting (IGRM), 1–3 March 2013, University of Ulster, Magee Campus.

United Nations (UN) 1980. International Convention for the Safety of Life at Sea (SOLAS).

Verbruggen, K. and Cullen, S. 2008. Mapping the Irish Seabed: the Irish National Seabed Survey and INFOMAR Projects. *The Journal of Ocean Technology*, **3**, 44–49.

White, J., Aristegui, M. *et al.* 2019. *The Labadie, Jones and Cockburn Banks Nephrops Grounds (FU20-21) 2019 UWTV Survey Report and Catch Scenarios for 2020*, Marine Institute.

Woodcock, N.H. and Strachan, R.A. 2002. *Geological History of Britain and Ireland*. Blackwell Publishing.

Young, E.F., Meredith, M.P., Murphy, E.J. and Carvalho, G.R. 2011. High-resolution modelling of the shelf and open ocean adjacent to South Georgia, Southern Ocean. *Deep Sea Research Part II: Topical Studies in Oceanography*, **58**, 1540–1552, https://doi.org/10.1016/j.dsr2.2009.11.003

Ziebart, M., Iliffe, J.C., Turner, J., Oliveira, J.F. and Adams, R. 2007. VORF – The UK Vertical Offshore Reference Frame: enabling Real-time Hydrographic Surveying. *Proceedings of ION GNSS2007*, Fort Worth, Texas, USA.

Integrated thematic geological mapping of the Atlantic Margin of Iberia

Pedro Terrinha[1,2]*, Teresa Medialdea[3], Luis Batista[1,2], Luis Somoza[3], Vitor Magalhães[1,2], Francisco Javier González[3], João Noiva[1], Ana Lobato[3], Marcos Rosa[1], Egidio Marino[3], Pedro Brito[1,2], Marta Neres[1,2] and Carlos Ribeiro[4]

[1]IPMA, Instituto Português do Mar e da Atmosfera, Rua C do Aeroporto, 1749-077 Lisboa, Portugal

[2]IDL, Instituto Dom Luíz, University of Lisbon, FCUL, Campo Grande Edifício C1, 1749-016 Lisboa, Portugal

[3]IGME, Instituto Geológico y Minero de España, Ríos Rosas, 23, 28003 Madrid, Spain

[4]Department of Geociências, University of Évora; Instituto Ciências da Terra; Rua Romão Ramalho, 59, 7000-671 Évora, Portugal

PT, 0000-0002-6824-6002; TM, 0000-0002-7969-5751; LB, 0000-0002-0830-1612; LS, 0000-0001-5451-2288; VM, 0000-0002-3205-771X; JN, 0000-0002-5625-3432; EM, 0000-0001-9392-0647; MN, 0000-0003-3939-4636; CR, 0000-0001-7492-1425
*Correspondence: pedro.terrinha@ipma.pt

Abstract: This paper synthesizes the geology of the Atlantic Margin off the coast of Iberia and surrounding Abyssal Plains using published thematic mapping freely downloadable from EMODNET-Geology portal at different scales. Selected information was chosen in order to highlight mineral occurrences and natural hazards overlaid on geological and morphological maps. Altogether, this information is published and interpreted here for the first time; nevertheless this exercise can be carried out by anyone interested and allows different visualizations of geological objects. Cross-correlations of geological objects and processes can easily arise. Because all of the information (each piece of data and metadata) in the EMODNET-Geology portal has bibliographic references associated, readers are able to find the original source of information. It is shown that clicking in and out of layers of information (that cannot be found all together in a single scientific paper) allows quick cross-correlation using the EMODNET Geology thematic portal. This allows a free, versatile and quick way of cross-correlating geological objects and processes in vast marine areas and their comparison with onshore geology.

Iberia is a trapezoidal continental block located between stable Eurasia and NW continental Africa. The onshore geology of Iberia has a stratigraphic record spanning from terranes of Proterozoic age involved in the Paleozoic Wilson cycle through the Alpine Wilson cycle, from Triassic through Quaternary times. The Paleozoic cycle ended with the formation of Pangea during Permian times when Iberia was trapped at the westernmost tip of the Tethys Ocean between Laurasia and Gondwana (Nance *et al.* 2012; Terrinha *et al.* 2019*a*; Vergés *et al.* 2019 and references therein).

The Atlantic Margin of Iberia (Fig. 1) encompasses three different tectonic sectors, the West Iberia Margin (WIM), the North Iberia Margin (NIM) and the SW Iberia Margin (SWIM) that formed as a result of rifting and seafloor spreading of the North Atlantic Ocean, Bay of Biscay and Western Tethys Ocean (or Neo-Tethys), respectively. The

NIM and SWIM were strongly involved in continental collision and subduction during the Alpine orogeny and both abut two orogenic belts, the Pyrenees and the Betics, respectively in the north and south of the Iberia Peninsula. The Pyrenees resulted from continental collision between Iberia and stable Eurasia in Late Cretaceous–Paleogene times (Boillot and Capdevila 1977). The Betics formed as a result of a complex tectonic history of oceanic subduction of the Tethys Ocean, slab roll-back and formation of back-arc basins that ended with accretion of the Betic terranes in southern Spain (e.g. Maldonado *et al.* 1999; Schettino and Turco 2011; Vergés *et al.* 2019). The West Galicia Margin that experienced the rifting of the North Atlantic Ocean is described here together with the NIM.

Mesozoic rifting structures are very well exposed in the NW of the WIM, in the Galicia Bank region. However, further south, the nature of the highly

From: Asch, K., Kitazato, H. and Vallius, H. (eds) 2022. *From Continental Shelf to Slope: Mapping the Oceanic Realm.* Geological Society, London, Special Publications, **505**, 97–115.
First published online September 8, 2020, https://doi.org/10.1144/SP505-2019-90

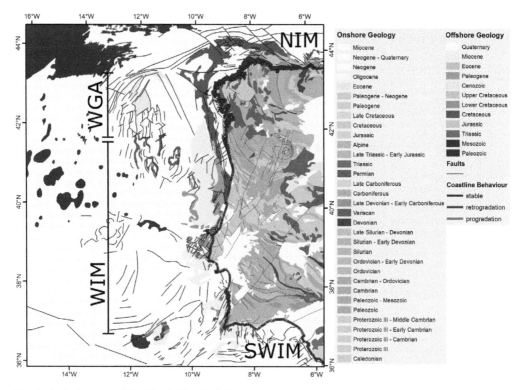

Fig. 1. Geological map of the Atlantic Margin of Iberia. Onshore simplified geological map is also shown (adapted from BGR 1/5 000 000 scale map (Asch 2007)).

extended lithosphere of the Iberia, Tagus and Horse-shoe abyssal plains, covered by *c.* 5 km of syn- and post-rift sediments, is still debated, as well as the Ocean–Continent Transition off the WIM and SWIM. Iberia shows widespread evidence of the Alpine orogeny both in the Paleozoic basement and in the Mesozoic rift basins, as well as major intra-oceanic tectonic structures (Terrinha *et al.* 2019c). The Atlantic Ocean floor bordering West Iberia is of Mesozoic age and displays major intra-oceanic compressional structures, mostly of Cenozoic age.

The NIM, together with the conjugate Armorican margin in SW France, border the Bay of Biscay formed by rifting and seafloor spreading that lasted until Late Cretaceous (Gallastegui *et al.* 2002). The subsequent Alpine evolution was determined by NW–SE convergence between the Iberia and Eurasia plates, which finished at the end of the Miocene (Álvarez-Marrón *et al.* 1997). This motion resulted in continental collision in the Pyrenees, subduction of the Bay of Biscay oceanic lithosphere beneath Iberia and deformation of the NIM (Boillot and Malod 1988; Tugend *et al.* 2014; Cadenas *et al.* 2018). While subduction of the Bay of Biscay finished at the end of the Pyrenean orogeny, present-day active subduction in the SWIM (also known as the

Gibraltar Arc) is still a matter of discussion (Gutscher *et al.* (2002); Zitellini *et al.* (2009).

Large-magnitude earthquakes and tsunamis originating near the Eurasia–Africa plate boundary have struck in the past and thus constitute a major societal concern. The historic 1 November 1755 Great Lisbon Earthquake, with an estimated magnitude of *c.* M8.1–8.9 (Buforn *et al.* 1988; Johnston 1996), is by far the best known example. The earthquake was felt as far away as Finland and the tsunami effects were felt heavily in southern England and Holland. Additionally, tsunami deposits were recorded across the Atlantic (Dourado *et al.* 2018). The 28 February 1969 earthquake, M7.9, located under the Horseshoe Abyssal Plain (Fukao 1973; Stich *et al.* 2005; Custódio *et al.* 2015), is also an important event supporting the present-day tectonically active state of this margin (Duarte *et al.* 2019).

Metallic deposits and aggregates on the seafloor are just some of the many marine resources that require sustainable management. Seafloor mineral deposits may represent the most important yet least explored resource of Critical Raw Materials and base metals on the planet. Their occurrence is related to various geological processes that encompass dynamic processes from the lithospheric origin (hot spot volcanism) through interaction of seawater and oceanic currents with the seafloor. Fe–Mn crusts and nodules and phosphorites are the most important marine mineral deposits owing to their economic and industrial potential for high-tech applications (Hein *et al.* 2013). They are important as potential mineral resources of Co, Te, Ni, Tl, Y, P, rare earth elements, platinum group elements and other metals. Fe–Mn crusts and phosphorites are usually found on seamounts and submarine banks, where vigorous currents have kept the rocks swept clean of sediments for millions of years. In general, crusts with greater thickness and mineral richness, and therefore with greater economic interest, are usually at between 800 and 2500 m water depth. Nodules are frequent in abyssal plains at water depths between 4000 and 6000 m, but they can also be found forming nodule fields in the flanks and flat tops of seamounts and banks (see minerals in https://www.emodnet-geology.eu/map-viewer/?bmagic=y&baslay=baseMapGEUS&layers =emodnet_mineral_occurrences).

Geological data from the marine environment are a valuable asset in addressing potential natural and anthropogenic hazards in the marine environment and developing policies for the protection of vulnerable areas of the coast and the deep ocean. In order to facilitate the access to and free use of marine data and information by public and private users, the European Commission launched the European Marine Observation and Data Network (EMODNET, http://www.emodnet.eu) underpinning its Marine Knowledge 2020 strategy. The EMODNET portal gathers and provides free access to information on the geology, bathymetry, chemistry, physics, seabed habitats, biology, and human activities of the European seas that were otherwise fragmented and difficult to reach and to cross-correlate. In addition to the data collection, the 160 participating institutions also carried out a harmonization effort to facilitate interpretation of the information (Kaskela *et al.* 2019).

The main goals of this paper are first to illustrate the wealth of the geological information available on the Atlantic offshore Iberia and second to show that combining information of different work packages on a single portal allows immediate visualization of georeferenced information on a variety of topics, some of them apparently unconnected.

Data: sources and harmonization

The geological products of EMODNET Geology are organized in the following six packages: Sea-Floor Geology; Events and Probabilities; Seabed Substrate; Submerged Landscapes; Marine Minerals; and Coastal Behaviour. The geographical scope of the project includes the European continental margins (Baltic, Barents, Black, North and Mediterranean seas, the Celtic Sea, the Bay of Biscay, the Iberian Margin and the Macaronesia region), including the Turkish sector, as well as areas of the seafloor of the legal continental shelf beyond 200 nautical miles.

EMODNET-Geology products include Pre-Quaternary and Quaternary geological and geomorphological maps, and maps of seabed substrate and accumulation rate of recent sediments, marine mineral layers including all types of naturally occurring geological raw materials, metals and hydrocarbons, coastline migration and geological events, concerning earthquakes, submarine landslides, volcanoes, tsunamis, fluid emissions and Quaternary tectonics. In addition, maps of submerged landscape features and palaeoenvironmental indicators, including estimated age where known, are also provided. Each layer is complemented by an attribute table that provides, in addition to the location, information of interest such as references, data source, ages and other valuable data. The information in the EMODNET Geology portal relative to the Atlantic offshore Iberia results essentially from a variety of sources: (1) collections of maps, geological models, analytical and numerical data from published papers or published cartography (e.g. from geological surveys), theses and marine campaign reports; (2) data from open access databases (e.g. International Seabed Authority; European projects); and (3) interpretation of data from the authors of this paper of seismic profiles, maps and multibeam bathymetry (Fig. 3).

All of the collected information was organized and classified using attribute tables for each entry. All attributes comply with the INSPIRE directive (https://inspire.ec.europa.eu/). Organizing information according to the INSPIRE directive and the use of a common vocabulary established in the project was the first level of harmonization, especially with regard to stratigraphy, lithology, tectonics and geomorphology.

The second level of harmonization was the matching of datasets and published work, such as cartography across political borders or criteria on coastline behaviour. This harmonization agreement was achieved during work meetings between the involved partners and also fixed bases for future mapping. The final harmonized coastline behaviour classification resulted from the work by Ponte Lira *et al.* (2016) on sandy coasts and information from the EUROSION dataset on hard rock coastlines (Lenôtre *et al.* 2004).

The outputs (maps, documents and additional data) are freely delivered at different scales (including 1,000,000 and 1:250,000 and 1:100,000 or finer scales when data are adequate) through a data portal (http://www.emodnet-geology.eu). This data portal offers direct download of bathymetric data and OGC Web Services that can provide open standard

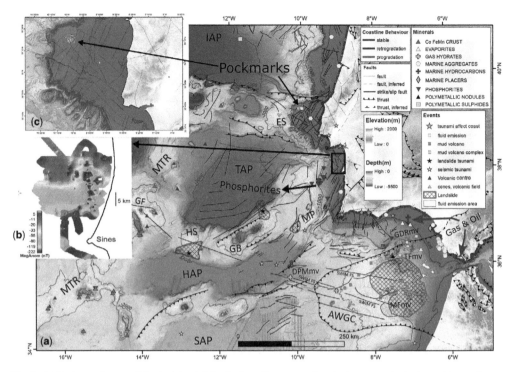

Fig. 2. (**a**) Bathymetric map of Iberia Atlantic Margin (http://www.emodnet-bathymetry.eu). Onshore altimetry is also shown for offshore–onshore correlation of morphology (http://srtm.csi.cgiar.org). Also shown is information on coastline migration, coastline impacted by tsunamis, landslides and mass transport deposits, mineral deposits (Fe–Mn crusts and nodules, phosphorites, methane hydrates), fluid emissions (mud volcanoes and pockmarks of Quaternary age) and volcanoes. AWGC, Accretionary Wedge of the Gulf of Cadiz; DPMmv, Deep Portuguese Margin mud volcano field; ES, Estremadura Spur pop-up; GB, Gorringe Bank; GDRmv, Guadalquivir Ridge mud volcano field; HS, Hirondelle seamount; IAP, Iberia abyssl plain; MF, Morocco mud volcano field; MP, Marquês de Pombal plateau; MTR, Madeira–Tore Rise; SAP, Seine abyssal plain; SWIM FS, SouthWest Iberia Fault system; TAP, Tagus abyssal plain; TFmv, Tasyo mud volcano field. (**b**) High-resolution map of marine magnetic anomalies north of Sines. (**c**) Location of pockmarks on the Estremadura Spur.

interfaces in order to exchange geospatial data between systems. It is important to point out that those OGC Web Services allow all bathymetric data to be combined with data from other portals (Chemistry, Biology, Physics, Bathymetry, Seabed Habitats and Human Activities) developed as part of the EMODNET initiative. These possibilities make the EMODNET outputs a powerful tool for decision-making, research and spatial planning, coastline protection, offshore installation design, environmental conservation, risk management and resource mapping among other activities related to the marine environment

Geology and geomorphology off west and SW Iberia margins

The submarine morphology is strikingly different from that of the continent as the morphological features are not prolongations of the onshore ones. The analysis of Figures 1–3a allows comparison of the seafloor and the onshore morphology and geology. The lack of continuity of stratigraphic units and morphology from the shelf to the mainland across the coastline is obvious. There are various causes for this besides the dominant depositional character of the marine environment on the continental slope and abyssal plains. Another important reason is that at present we are in a high-stand sea-level period that allows the continental shelf to be 45 km wide and usually covered by flat-lying Quaternary sediments. This sediment cover interrupts the outcrop pattern of old terrains, such as Mesozoic or Paleozoic, that underwent orogenic shortening and consequently display linear trends. Moreover, beyond the shelf, the Paleozoic scarcely crops out because the rifted margin is filled with Mesozoic syn-rift and Meso-Cenozoic post-rift sediments. The continuous sedimentation accommodated by post-oceanic

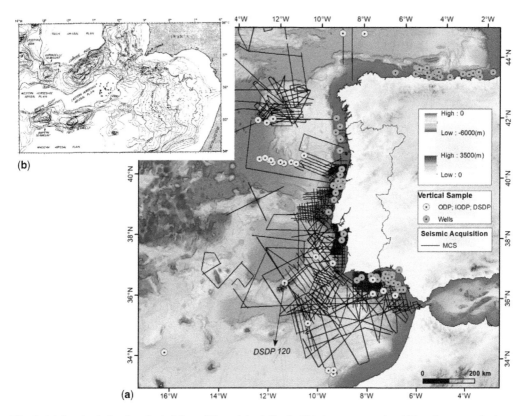

Fig. 3. (**a**) Geophysical and geological data offshore Atlantic Iberia. This is a representative (although not complete) map of data used in many literature scientific papers and reports used as a basis for the EMODNET Geology portal; (**b**) the high-resolution bathymetry of Purdy (1975) is shown for comparison with other figures of this work.

drifting thermal subsidence was interrupted by the Alpine compressive tectonics and basement exhumation. The two most outstanding cases of basement or Mesozoic rock exposure associated with the Alpine compression are the Gorringe Bank mantle outcrop (e.g. Purdy 1975; Sartori *et al.* 1994; Jiménez-Munt *et al.* 2001; Martínez-Loriente *et al.* 2013; Sallarès *et al.* 2013) and the Estremadura Spur pop-up (Neves *et al.* 2009; Neres *et al.* 2014; Terrinha *et al.* 2019*b*, *c*).

The Gorringe Bank has a core of serpentized mantle up-thrust from depths of about 5 km below sea-level to near sea-level (top at 25 m below sea-level) probably during Mid-Miocene times. Mantle rocks were collected, dredged and drilled by DSDP 120 (Fig. 3) (Ryan *et al.* 1973; Auzende *et al.* 1984; Girardeau *et al.* 1998). Serpentinized peridotite ridges that resulted from mantle exhumation during the slow-spreading hyper-extended continental rifting are a distinctive tectonic feature of the West Iberia Margin. These were drilled by the ODP Leg 149, sites 897–901 (Pinheiro *et al.* 1992; Sawyer *et al.* 1994).

The Estremadura Spur (ES in Fig. 2) is a 300 km long pop-up structure that connects the tectonically inverted Lusitanian Basin with the Tore–Madeira Rise (MTR). The ES is punctuated by volcanoes, sills and dykes of Cretaceous age with evident magnetic signature. Although these are coeval with the Cretaceous volcanic and plutonic events that are known onshore, the volume and extent of this offshore sub-volcanic system is still unknown (Fig. 2; Neres *et al.* 2014; Terrinha *et al.* 2018, 2019*a*, *b*).

The *c.* 1100 km-long MTR (Fig. 2) is a magmatic ridge that borders the West Iberia Abyssal Plains and the Seine Abyssal Plain; it is a major geological and morphological feature that has no continental equivalent. The MTR is an over-thickened segment of oceanic lithosphere (Peirce and Barton 1991) that straddles the Africa–Eurasia plate boundary in the Atlantic. It is punctuated by several volcanic seamounts mostly of Cretaceous age (*c.* 100–80 myr) in the Eurasia plate and Paleogene through Quaternary in the Africa plate and at the Gloria Fault plate boundary (Geldmacher *et al.* 2006; Merle *et al.* 2006, 2009, 2018; Terrinha *et al.* 2019*b*).

The accretionary wedge of the Gulf of Cadiz in the SWIM is tectonically active at present with widespread formation of mud volcanoes (Fig. 2), many of which exhalate hydrocarbon-rich fluids (Ivanov *et al.* 2000; Somoza *et al.* 2003; Van Rensbergen *et al.* 2005; León *et al.* 2012; Hensen *et al.* 2015). Mud volcano development is triggered by compressional stress along the European–African boundary. Mud volcanoes can be sourced through westward-directed thrusts or deep-routed dextral strike-slip faults (Medialdea *et al.* 2009; Terrinha *et al.* 2019*b*), arguably part of the present-day diffuse plate boundary between Africa and Eurasia in the Atlantic (Zitellini *et al.* 2009; Hensen *et al.* 2015).

The west and SW Iberia margins: natural hazards and mineral occurrences

The SWIM lies near the Eurasia–Africa plate boundary, also known as the Azores–Gibraltar Fracture Zone, about which many papers have been published (Srivastava *et al.* 1990; Schettino and Turco 2009, 2011). The MTR marks a rheological boundary in this part of the Atlantic lithosphere. To the west of the MTR the plate boundary is clearly delineated by the Gloria Fault and recent advances have been made in understanding the kinematics and nature of the crust (e.g. McKenzie 1978; Neres *et al.* 2016; Batista *et al.* 2017). Here the seismicity is less frequent but still producing very-large-magnitude strike-slip events such as the 1941 M8.4 one (Buforn *et al.* 1988; Johnston 1996). To the east of the MTR seismicity is frequent, dominated by strike-slip and thrusting, and large-magnitude earthquakes have been generated, like the historical Great Lisbon earthquake (1 November, 1755, estimated magnitude M8.1–8.9; Buforn *et al.* 1988; Johnston 1996) and more recently the M7.9 earthquake on 28 February 1969 in the Horseshoe Abyssal Plain (Fukao 1973; Stich *et al.* 2005; Custódio *et al.* 2015).

The origin of seismicity in the SWIM is a matter of debate as well as the location of the Eurasia–Africa plate boundary. Two interfering plate boundaries have been proposed. One is a diffuse plate boundary (Sartori *et al.* 1994) materialized by the lateral strike-slip SWIM fault system that extends for *c.* 600 km from the MTR through NW Morocco (Zitellini *et al.* 2009). The second is the subduction of a remnant of the oceanic Tethyan slab under the Gibraltar Arc (Gutscher *et al.* 2002). Deformation on the accretionary wedge of the Gulf of Cadiz (Duarte *et al.* 2011) supports the present-day activity of both deformation mechanisms, strike-slip and thrusting, respectively. Recent advances on the location of seismicity were based on ocean bottom passive experiments (Grevemeyer *et al.* 2016; Silva *et al.* 2017) that allowed understanding that seismicity is located in

the mantle and at the intersection of the strike-slip SWIM faults and NE–SW striking thrusts. It has been suggested that the large-magnitude earthquakes could have been originated by thrust–wrench tectonic interference (Rosas *et al.* 2016). In addition to the seismogenic tsunamis in SW Iberia, three large tsunamigenic landslides have been reported in the SWIM – the Marquês de Pombal, North Gorringe and South Hirondelle (Fig. 2; Terrinha *et al.* 2003; Lo Iacono *et al.* 2012; Omira *et al.* 2016). Landslides, tsunami-impacted areas, tsunami sources and submarine volcanoes (Fig. 2) are sources of natural hazards with severe consequences to society. The maps provided by EMODNET Geology allow inspection of various parameters that allow a better view and therefore policy planning.

The same applies to occurrences of mineral resources. The map in Figure 2 results from overlapping various layers – mineral occurrences, fluid escape structures, volcanic centres, Quaternary faults, landslides and tsunami-impacted sites – on top of EMODNET Bathymetry. In addition to the visual information, lists of attributes add direct information on the geological features and re-direct readers to specific literature and other pertinent information allocated on external websites.

The MTR is also the location of several occurrences of Fe–Mn deposits with interesting contents of critical and noble metals (Co, Ni, Ce, Te and Pt; Muiños *et al.* 2013). The EMODNET data portal also shows Fe–Mn crusts occurrences on several seamounts over the Azores plateau, on the SWIM and on isolated seamounts on the WIM (https://www.emodnet-geology.eu/map-viewer/?bmagic=y&baslay=baseMapGEUS&layers=emodnet_mineral_occurrences). This information is useful when selecting places to search for more deposits along the whole length of the MTR and along the entire Iberia margin. The mineralogical and chemical composition of these crusts indicate that they are of prevailing hydrogenetic origin, formed by the direct precipitation of colloidal hydrated metal oxides over hard-rock substrates, typical of continental margins. The resource potential of the Fe–Mn crusts along the NE Atlantic is evaluated as comparable to those found in the central Pacific, therefore having potential as future resource of some critical raw elements (Muiños *et al.* 2013; Muiños 2015). Ferromanganese nodule fields and hardgrounds have also been discovered in the Cadiz Contourite Channel in the Gulf of Cadiz (850–1000 m). This channel is part of a large contourite depositional system generated by the Mediterranean Outflow Water. The genesis and growth of ferromanganese deposits, strongly enriched in Fe v. Mn (average 39 v. 6%) in this contourite depositional system result from the combination of hydrogenetic and diagenetic processes (González *et al.* 2012).

Phosphorites have been dredged (Fig. 2, https://www.emodnet-geology.eu/map-viewer/?p=marine_minerals) on slope breaks where the Mediterranean Outflow Water shapes the margin by carving moats and depositing contourite drifts (Teixeira *et al.* 2019). The phosphorites found here have a carbonated matrix and an authigenic/diagenetic composition dominantly carbonate-fluor-apatite to francolite, and their formation seems to be associated with coastal upwelling and sea-bottom water with oxygen minima (Gaspar 1984).

The EMODNET dataset shows numerous seabed fluid escape structures (https://www.emodnet-geology.eu/map-viewer/?p=geological_events_and_probabilities: EMODNET fluid emissions layer), such as mud volcanoes (MV), some of them bearing methane hydrates (Gardner 2001; Pinheiro *et al.* 2003; Somoza *et al.* 2003; Van Rensbergen *et al.* 2005; León *et al.* 2012; Hensen *et al.* 2015; Toyos *et al.* 2016), hydrocarbon-derived authigenic carbonates (MDAC) (Díaz-del-Río *et al.* 2003; Magalhães *et al.* 2012; Palomino *et al.* 2016) and crater-like depressions, pockmarks and collapse features (Fox and Opdyke 1973; Baraza and Ercilla 1996; León *et al.* 2010; Duarte *et al.* 2016; Palomino *et al.* 2016). Pockmarks on the Estremadura Spur continental shelf were reported by Duarte *et al.* (2016) (Fig. 2c). High-resolution seismic reflection shows fluids trapped in the upper sedimentary sequences and Quaternary tectonic deformation in the area. However, remotely operated vehicle dives and sampling did not show evidence of present-day fluid flow or MDAC formation. Hydrates and hydrocarbon gases sampled from MV sediments include both biogenic and thermogenic components (Mazurenko *et al.* 2002; Stadnitskaia *et al.* 2006; Nuzzo *et al.* 2009, 2019). In the upper-middle continental slope, MV fluids are strongly influenced by clay-mineral dehydration and leaching of Upper Triassic evaporites (Haffert *et al.* 2013) and Miocene diapirism of the Cadiz Allocthonous Unit (Medialdea *et al.* 2004, 2009; Fernández-Puga *et al.* 2007). Seaward, the MV fluid signature changes and fluid interaction with the underlying oceanic crust takes place in MVs of the lower continental slope such as in Porto MV (Hensen *et al.* 2015). The MVs located at water depths from −2500 to −4500 m are closely linked to active strike-slip faults (Duarte *et al.* 2013), which provide deeply rooted fluid pathways (Hensen *et al.* 2007) for deep-seated fluids sourced in oceanic crust older than 140 Ma (Hensen *et al.* 2015). Up to the present there are 79 recognized MVs in the SW Iberia Margin, clustered into four main fields (Fig. 2): (1) the Moroccan field offshore the Moroccan Atlantic margin; (2) the Guadalquivir Diapiric Ridge field located on the southern Atlantic Iberian margin; (3) the TASYO field in the central Gulf of Cadiz west off Gibraltar Strait; and (4) the Deep

Portuguese Margin field in the lower continental slope of the Gulf of Cadiz. These features and their distribution are relevant as they hold unique habitats, they are an interesting tool for the exploration of deep hydrocarbons, they contain fluids rich in methane that can promote a strong greenhouse effect and they can induce changes in the physical–chemical properties of the sediments, causing instabilities that can promote landslides (McKenzie 1972; Miranda *et al.* 2015; Toyos *et al.* 2016; Rincón-Tomás *et al.* 2019).

Widespread magmatism affected the Iberia Margin after the rifting phases and before the onset of the main tectonic compressive and uplift events. This episode that lasted at least 30 myr (from *c.* 100 to 69 Ma; Féraud *et al.* 1986; Schärer *et al.* 2000; Geldmacher *et al.* 2006; Merle *et al.* 2006; Miranda *et al.* 2009; Grange *et al.* 2010) may have had important implications for the existence and location of geological resources. For instance, it may have controlled hydrocarbon maturation and trapping, as well as occurrences of rare earth element placers related to erosion of uplifted alkaline rocks. Recent magnetic surveys off the Portuguese coast have brought important insights into the extension, location and intrusion mechanisms of alkaline magmatism. A magnetic anomaly map resulting from high-resolution surveying is shown in Figure 2b, where previously unknown magnetic anomalies NW of the Sines magmatic complex are now revealed. Magnetic anomalies maps are an example of data prone to be included in the EMODNET data portal in the near future.

The artificial nourishment of beaches with high-quality sand is an important societal issue as it involves economic activities of the first importance, such as tourism (Pinto *et al.* 2018). The beaches along the rocky coast of SW Portugal sculpted in Paleozoic slates are classified as coastline segments of 'imperceptible change' with regard to erosion or accretion (between 38° and 39° N in Fig. 2). However, these beaches need to be artificially nourished. The large area mapped as 'marine aggregates' (Fig. 2) is just a reference. Detailed studies have been recently carried out using ultra-high-resolution seismic reflection, multibeam bathymetry, backscatter and seafloor sampling (Noiva *et al.* 2017). These aim at recognizing sedimentary traps for metallic placers, sand (marine aggregates), like the fluvial incisions shown in Figure 4a, and various levels of sea-level stand stills that control the deposition of littoral sands (Fig. 4b).

Geology and geomorphology of the North Iberian and the West Galicia margins

The North Iberia and Galicia margins represent two morpho-structural regions that have been the object

(a)

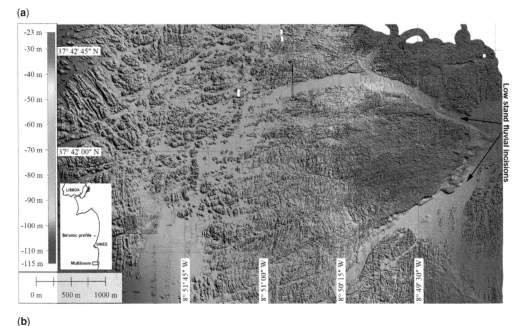

(b)

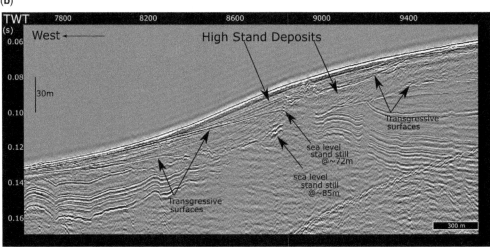

Fig. 4. Mapping for mineral resources in the continental shelf. (**a**) Morphology of a segment of the Alentejo shelf off SW Portugal; note the fluvial incision of cold Quaternary periods and the contrast of different outcropping terrains (multibeam map location in the inset). (**b**) Ultra-high-resolution seismic reflection profile showing sedimentation traps conditioned by palaeocoastal escarpments (seismic profile location in the inset). The joint use of multibeam bathymetry, backscatter and ultra-high-resolution reflection seismics allows for locating adequate sites for ground truthing the mobile deposits for mineral content characterization.

of numerous studies since the 1970s, including three DSDP and ODP Legs (Figs 5 & 6). The north–south West Galicia margin is considered an example of non-volcanic hyperextended rifted margin, whereas the NIM records the imprint of the convergence and partial subduction of the Bay of Biscay seafloor under the NIM.

The NIM

The NIM (also known as Cantabrian margin) together with the Armorican margin border the Bay of Biscay, a V-shaped oceanic basin open to the North Atlantic Ocean (Fig. 5). This bay was formed by Triassic to Jurassic continental rifting and

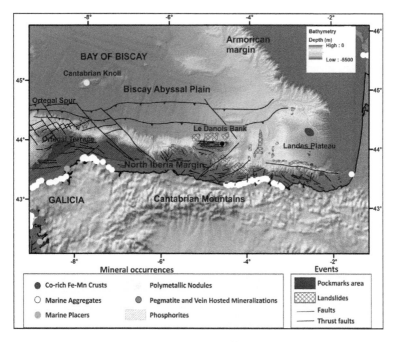

Fig. 5. Bathymetric map of the North Iberia Atlantic Margin (http://www.emodnet-bathymetry.eu) and onshore altimetry. Information on landslides, structures related to fluid emissions (pockmarks fields) and mineral deposits (Fe–Mn crusts and nodules, polymetallic nodules, phosphorites, placers, aggregates, pegmatite and vein-hosted mineralizations) are also shown. GC, Giant Craters area.

seafloor spreading from Aptian (118 myr) to Campanian times (80 myr, anomaly C33) (Boillot and Malod 1988; Sibuet and Collette 1991), giving rise to the formation of hyperextended basins and oceanic crust (Gallastegui *et al.* 2002; Tugend *et al.* 2014). Since the Late Cretaceous, evolution has been determined by NW–SE convergence during Late Cretaceous to Miocene times between the Iberian and European plates (Álvarez-Marrón *et al.* 1997; Rosenbaum *et al.* 2002). This motion resulted in continental collision in the Pyrenees, the Cantabrian mountains and the partial subduction of the Bay of Biscay beneath Iberia (Boillot and Malod 1988; Álvarez-Marrón *et al.* 1997; Gallastegui *et al.* 2002; Tugend *et al.* 2014; Cadenas *et al.* 2018).

The narrow continental shelf of this margin widens from east to west, reaching 50 km wide (Figs 1 & 5). Erosive processes dominate in the shelf, allowing the cropping out of a band of Paleozoic to Eocene rocks parallel to the coastline (Boillot *et al.* 1979; Medialdea and Terrinha 2016). The seaward is covered by Eocene to Quaternary deposits dominated by hemi-pelagic to turbiditic sediments.

The NIM shelf break and the continental slope are incised by a network of gullies and feeding turbidite systems in the Bay of Biscay (Zaragosi *et al.* 2001; Ercilla *et al.* 2008, 2011; Iglesias 2009;

Iglesias *et al.* 2010; Fig. 5). The slope, especially the abrupt lower slope, and canyon walls evidence important gravitational processes. Locally, along-slope processes related to bottom currents have an outstanding role. In the central part of the margin, at the upper slope, the interaction of the Mediterranean Outflow Waters with the Le Danois Bank promotes the development of a contourite system on its southern flank (Fig. 5), characterized by several drifts, sediment waves and moats (Van Rooij *et al.* 2010).

The continental slope connects with the Biscay Abyssal Plain at about 4800–5000 m water depth, which is divided into two basins by the east–west Charcot Seamounts (Fig. 6), which consist of east–west ridges rising 2400 m above the abyssal plain. The Charcot Seamounts correspond to the mid-oceanic ridge related to the opening of the Bay of Biscay, a subsidiary rift arm of the North Atlantic ridge. The Coruña seamount shows a complex relief consisting of NW–SE and east–west ridges systems, that recorded the two different directions of simultaneous spreading of the North Atlantic and the Bay of Biscay that occurred during the Early to Late Cretaceous transition (Somoza *et al.* 2019). The boundary between the two ridge systems marks the fossil track of the triple junction between the Eurasian, Iberian

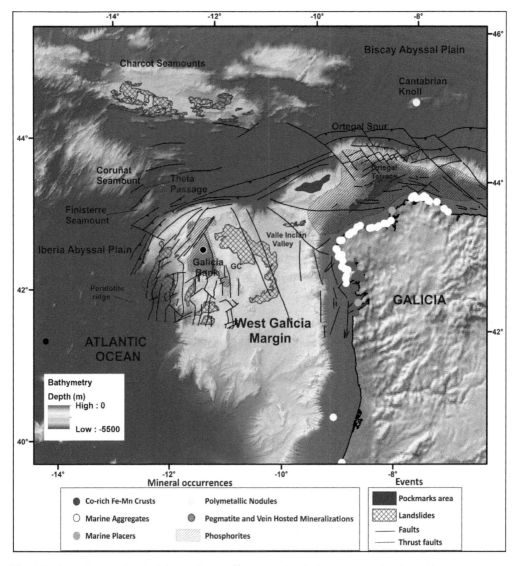

Fig. 6. Bathymetric map of the Galicia Margin (http://www.emodnet-bathymetry.eu) and onshore altimetry. Information on landslides, structures related to fluid emissions (pockmarks fields) and mineral deposits (Fe–Mn crusts and nodules, polymetallic nodules, phosphorites, placers, aggregates, pegmatite and vein-hosted mineralizations) are also shown.

and North American plates from chron M0 (118 Ma, Early Aptian) to chron A33 (80 Ma, Campanian), following magnetic anomalies reported by Sibuet *et al.* (2004).

The imprint of Cenozoic convergence is also reflected in the reactivation and inversion at the seamounts at the continental margin and the abyssal plain. Oceanic basement highs as the Charcot Seamounts are affected by reverse faults (Fig. 6). Similarly, structural highs such as the Le Danois Bank and the Finisterre Bank in the continental slope are

underlain by northwards-directed thrusts (Boillot *et al.* 1979; Malod *et al.* 1980, 1982; Medialdea *et al.* 2009).

The West Galicia Margin

The West Galicia Margin (WGM) is a broad continental margin, up to 350 km wide, that is mainly related to its poly-phase rifting history; rifting events progressively shifted westwards until continental breakup in Late Aptian times (Malod and Mauffret

1990; Srivastava *et al.* 2000; Péron-Pinvidic *et al.* 2007; Sibuet *et al.* 2007; Tucholke and Sibuet 2007). The architecture of the WGM is characterized by listric detachment faults and tilted blocks that determined a thin continental crust and a 500 km-long serpentinized ridge of peridotites between the continental and oceanic domain (e.g. Malod and Mauffret 1990; Srivastava *et al.* 2000; Péron-Pinvidic *et al.* 2007; Sibuet *et al.* 2007; Tucholke and Sibuet 2007). This ridge has been interpreted as part of the sub-continental exhumed mantle. Structures inherited from the Early Cretaceous rifting include north–south and NW–SE normal faults and NE–SW transfer faults that determined the morphostructure of the continental margin at a regional scale (Vázquez *et al.* 2008). The later evolution of the margin is linked to the Cenozoic convergence between the plates of Eurasia and Iberia. The Cenozoic convergence gave rise to the tectonic inversion of pre-existent extensional faults, such as gentle folding and associated thrusting, as well as the uplift of the Galicia Bank (Vázquez *et al.* 2008).

The WGM includes an inner sector constituted by the continental shelf and upper slope separated from the outer sector by the NNW–SSE Valle Inclán Valley, also named the Galicia Interior Basin (Fig. 6). The outer sector includes the Galicia Bank and the north–south-trending structural valleys and ridges and the peridotite ridge. The upper continental slope is dissected by east–west submarine canyons that converge downslope into a major channel that runs along the Valle Inclán Valley from 2900 to 4600 m water depth. This valley is bounded by a NW–SE structural scarp (Fig. 6). To the west, the Galicia Bank, the main structural high of the NW Iberian margin, stands out at water depths of 700–1800 m. Its flanks are bounded by faults that give rise to abrupt slopes rising from 4130 and 1600 to 730 m water depth at its western and eastern flanks respectively (Somoza *et al.* 2019). On top of the Galicia Bank there are north–south elongated contourite deposits, cold water coral mounds and living patches of Scleractinia corals (Somoza *et al.* 2014), as well as phosphorite and ferromanganese hardgrounds (González *et al.* 2016). The Bank is bound on its northern and southern sides by several tablemounts. The northwestern flank of the Galicia Bank corresponds to the steep, arc-shaped scarp-oriented ENE–WSW to NNE–SSW, clearly marked on the bathymetric map (Fig. 6). Seawards the Finisterre Seamount shows an abrupt scarp to the north that bounds the Theta Passage, an elongated and deep depression that connects the Biscay and the Iberia Abyssal Plains. The western scarp descends to a set of north–south ridges and asymmetric valleys at 1600–4700 m water depths. The seafloor is incised by channels and landslides scars (Fig. 6).

The North Iberian and the West Galicia margins: natural hazards and mineral occurrences

Although the NIM and the West Galicia margin have not been traditionally considered as a potentially risky region for geological events as tsunamis, fluid emissions, landslides and earthquakes, there are several areas where some of these events are quite relevant. The joint analysis of the EMODNET-Geology thematic maps (https://www.emodnet-geology.eu/map-viewer/?p=geological_events_and_probabilities: geomorphology, landslides, volcanic centres, fluid emissions, faults layers) and their correlation support understanding of the origin and genesis of various geological events such as submarine landslides and fluid emissions. In the NIM, landslides are widely distributed at the abrupt lower slope and at the canyon walls that incised the continental slope, most of them probably related to tectonic activity (Cadenas *et al.* 2018). Large landslides are also found at the east and west flanks of the Galicia Bank sourced from the bank and at the reactivated edges of most of the seamounts such as the Charcot Seamounts and the Finisterre seamount.

The EMODNET data portal evidences the relationship between fault systems, earthquakes and submarine landslides in the region of Galicia (Fig. 1), where three main earthquakes swarms are identified: (1) a cluster of seismicity with magnitudes around 5 concentrates around the Coruña seamount, NW of the Galicia Bank, related to the present day NW–SE-oriented compression (Somoza *et al.* 2019); (2) an earthquake swarm related to the Galicia Bank region is associated with a NW–SE-trending fault system; and (3) a third earthquake swarm is located on the northern Galicia margin associated with NW–SE strike-slip faults.

Seabed fluid emissions mapped in the EMODNET Geology Fluid Emissions Map (https://www.emodnet-geology.eu/map-viewer/?p=geological_events_and_probabilities) are evidenced by pockmarks and giant crater fields (Fig. 6). The main field of pockmarks is located in the uppermost sector of the Ortegal Terrace in the upper slope. It extends 100 km along-slope and 70 km downslope below the shelf break to 400 m depth. Single pockmarks appear as typical circular to sub-circular 300–150 m-wide and 10 m-deep depressions. According to Jané *et al.* (2010) the 500 m-thick organic-rich layer late Aptian–Senonian sequence ('Black-Shales') is the source for the expulsion of fluids at the seafloor. These pockmarks are related to NE–SW to ENE–WSW or NW–SE to WNW–ESE faults (Verreydt 2011). The first direction fits the extensional faults that formed the structural terraces in this sector of the margin. The second direction is related to the strike-slip Ortegal Fault (Figs 1 & 6). Another field of giant craters has been named the 'Burato' field at the eastern flank of the Galicia Bank. The most outstanding feature is the 'Gran Burato Hole' (Vázquez *et al.* 2009), which is a crater up to 2–5 km in diameter and 300 m deep with slopes up to 12°. At least three more giant craters have also been identified in the area following a NW–SE fault trend mapped in the EMODNET data portal (Fig. 6). It has been suggested that that the origin of these craters might be related to hydrocarbon leakages sourced from the 'Black-Shales' unit, as has been proposed in the Ortegal Spur affected by diapiric movements and triggered by NW–SE compressional deformation (e.g. Vázquez *et al.* 2009). The same source has been invoked for the pockmark fields in the Landes Plateau (Fig. 5; Iglesias *et al.* 2010).

In the WGM and North Iberian Atlantic margins, Fe–Mn crusts and nodules and phosphorites have been found in the Ortegal Spur, Galicia Bank and nearby seamounts and in the Le Danois Bank. Most studies have been focused on the Galicia Bank (Fig. 7), where phosphorites appear at 750–1450 m water depths accompanied by Fe–Mn mineralizations (González *et al.* 2016). Mineral deposits include phosphorite slabs up to 10 cm thick and nodules, ferromanganese crusts and stratabound layers consisting mainly of Mn and Fe which impregnated and replaced the phosphorite, Co-rich Mn nodules and Fe-rich nodules up to 5 cm in diameter (Fig. 7). These Fe–Mn crusts cover the nodules, phosphate pavements and basement formed from igneous, metamorphic and sedimentary rocks. Several phosphatization events since late Oligocene to early Miocene and diagenetic and hydrogenetic growth are proposed for the genesis of these deposits. The different types of mineralizations have been related to episodes of active tectonism and palaeoceanographic events affecting the North Atlantic Ocean during Cenozoic times and metal-rich hydrothermal fluid-transport processes along reactivated faults (González *et al.* 2016). Furthermore, the EMODNET-Geology portal also shows mineral occurrences in the abyssal plains. Cobalt-rich Fe–Mn crusts are reported from seamounts west of the Galicia Bank (Fig. 7) and Fe–Mn nodules are located on the Cantabrian Knoll, a small seamount in the Biscay Abyssal Plain (Fig. 6). Analyses of the conditions already determined for the generation of these mineralizations together with the information provided by the EMODNET-Geology datasets are a valuable help and useful tool for defining potential exploration targets and discovering these mineral deposits.

Phosphorite deposits are one of the main mineral occurrences in the WGM (Fig. 6) that have been reported as large crusts on several structural terraces (e.g. Ortegal terrace, Fig. 5) that form the continental slope of the northern Galicia Margin (Lamboy and Lucas 1979).

Pegmatite mineralizations, aggregates and placer deposits are typically found close to the coastal zones. The southern Galicia coast, named 'Rias Bajas', shows numerous occurrences of pegmatite and vein-hosted mineralizations related to the Variscan granites (Fig. 6). Placer deposits are restricted to shallow waters of the continental shelf in close relationship with Variscan rocks and onshore drainage patterns. Evidence of tungsten placers is related to deposition as fan deltas within the 'Rias' estuarine systems of Galicia. These deposits cover an area of 336 km^2 in south Galicia and include minerals such as ilmenite, magnetite, garnet, rutile, monazite, turmaline and zircon (https://www.emodnet-geology. eu/map-viewer/?bmagic=y&baslay=baseMapGEUS &layers=emodnet_mineral_occurrences). Aggregate deposits of gravels and sands are distributed in small basins scattered on the Paleozoic and Mesozoic outcrops (Figs 5 & 6). In the northern West Galicia coast, aggregates are mainly related to large beach-barrier systems developed in the open sea and their extraction aims at beach nourishment.

Study of the coastal behaviour has progressively become a matter of great interest owing to concern for the environment, increased population in coastal areas, infrastructure and tourism. The Cantabrian Mountains are very close to the coastline of the NIM, which is quite stable according to the coastal behaviour map (Figs 1, 5 & 6). The coastline is very irregular and it is dominated by abrupt cliffs and capes together with bays, rias and estuaries. In north Iberia, retreat rates of 0.03–2.78 m a^{-1} have been quantified in granitic cliffs. Retreat can be produced by rotational slides, rock falls at intensively fractured cliffs and cliff collapses linked to structurally controlled sea caves (Gómez-Pujol *et al.* 2014).

Discussion and conclusions

Despite the excellent datasets and seminal science produced by early works, some of which are reported here, data processing demanded a great team effort (e.g. see figure 3 and acknowledgements in Purdy 1975). The information that could be extracted was very limited when compared with the maps that can be made interactively by clicking in and out of the EMODNET Geology portal (compare maps in Figs 2 & 3). Now, just by selecting the right layers of information, at the click of the computer mouse, maps of bathymetry, geology, faults, seismicity, fluid emissions, landslides, volcanoes, etc. (with lists of attributes) can be visualized and all of this information can be interpreted, cross-correlated and downloaded.

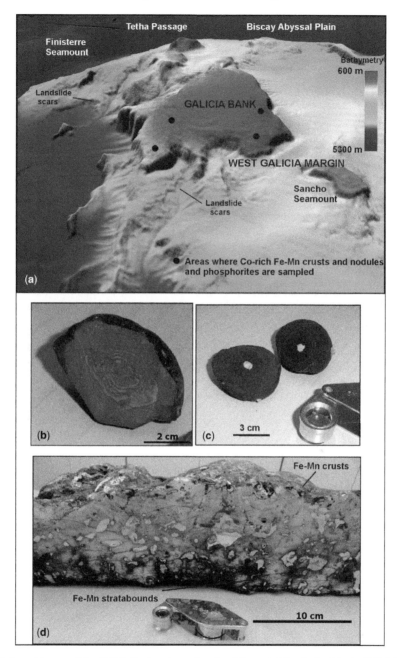

Fig. 7. (**a**) Bathymetric map of the Galicia Bank and samples of nodules and slabs obtained in the Galicia Bank; (**b**) Fe-rich nodule; (**c**) phosphorite nodule; and (**d**) phosphorite slab.

The EMODNET information allows a comprehensive inspection of the vast area off the Atlantic Margin of Iberia. Societal concerns such as geological hazards and mineral resources of marine origin can now be analysed in the interactive EMODNET Geology portal by selecting different layers of information. The map services provided by EMODNET Geology allow the reader to select georeferenced information that is usually not easily available. EMODNET Geology maps and their associated tables of attributes and metadata always carry a reference that allows the reader to inspect the original data source. They have added value as they allow cross-checking of a wide variety of layers of georeferenced information, some of most of which is not found together in the literature.

A good example, among others, is the location of Fe–Mn crusts and nodules and phosphorites with respect to the MTR seamounts and the continental slope morphology (https://www.emodnet-geology. eu/map-viewer/?bmagic=y&baslay=baseMapGEUS &layers=emodnet_mineral_occurrences). The freely available EMODNET Bathymetry allows inspection of the relationship of the mineral occurrences with respect to the submarine relief and thus with the oceanic currents with which these deposits are associated. This provides a valuable insight for future exploration targets, which is a valuable asset for both the scientific and economy communities.

Last but not least, the interconnection of information and data of the various EMODNET portals (Geology, Bathymetry, Chemistry, Physics, Seabed habitats, Biology, and Human activities of the European Seas) allows an integrated view of the European seas. This information is a valuable asset for supporting all fields of activity in the ocean with scientific knowledge in the various marine Earth systems, such as the water column, seafloor and sub-seafloor, interaction between living and non-living resources and engagement in the protection of the marine environment simultaneously with the development of the marine economy. Recent examples are exploration projects on aggregates, namely on finding areas for sand extraction for artificial nourishment of beaches. These actions involve the economy (extraction, tourism, fishing, etc.), societal needs and habitat protection.

Acknowledgements Thanks are due to Halliburton for providing the seismic interpretation software through the Landmark Universities software grant programme, and Geosoft for a research licence to Oasis Montaj to process magnetic data. M.N. acknowledges the Fundação para a Ciência e Tecnologia for post-doc fellowship SFRH/BPD/96829/2013. E.M. Marino acknowledges the Ministry of Science, Innovation and Universities of Spain for its FPU scholarship (FPU014/06774). To C4G-Co-Laboratory for Geosciences.

Funding This research was funded by EMODNet-Geology project (ref. EASME/EMFF/2016/1.3.1.2-Lot 1/ S12.750862) by the Portuguese Foundation for Science and Technology (FCT) through projects UIDB/50019/ 2020 – IDL and UIDB/04683/2020 – ICT (Institute of Earth Sciences) and by the projects GeoERA-MINDeSEA (No. 731166, GeoE.171.001) and EXPLOSEA (CTM2016-75947-R). MINEPLAT project (ALT20-03-0145-FEDER-000013) provided data and J.N. with a PhD grant.

Author contributions PT: conceptualization (lead), funding acquisition (lead), investigation (lead), project administration (equal), supervision (equal), validation (equal), writing – original draft (lead), writing – review & editing (lead); **TM**: conceptualization (equal), data curation (lead), investigation (lead), project administration (lead), validation (lead), visualization (lead), writing – original draft (equal), writing – review & editing (supporting); **LS**: data curation (lead), investigation (equal), project administration (equal), validation (lead), visualization (lead), writing – review & editing (equal); **VM**: investigation (equal), writing – original draft (equal); **FJG**: investigation (lead), writing – original draft (lead); **JN**: funding acquisition (equal), project administration (lead), writing – original draft (equal); **AL**: data curation (equal), investigation (equal), project administration (supporting), validation (equal); **MR**: data curation (equal), writing – review & editing (supporting); **EM**: data curation (lead), funding acquisition (equal), methodology (lead), project administration (supporting), validation (lead), visualization (lead), writing – review & editing (supporting); **MN**: data curation (equal), investigation (equal), writing – original draft (supporting); **CR**: investigation (equal), methodology (lead), writing – original draft (supporting).

Data availability statement The datasets generated during and/or analysed during the current study are available in the EMODNET repository, https://www.emod net-geology.eu/; https://www.emodnet-bathymetry.eu/.

References

Álvarez-Marrón, J., Rubio, E. and Torné, M. 1997. Subduction-related structures in the North Iberian Margin. *Journal of Geophysical Research: Solid Earth*, **102**, 22497–22511, https://doi.org/10.1029/97jb01425

Asch, K. 2007. *The 1:5 Million International Geological Map of Europe and Adjacent Areas*. BGR (Hannover).

Auzende, J.M., Ceuleneer, G. *et al.* 1984. Intraoceanic tectonism on the Gorringe Bank: observations by submersible. *Geological Society, London, Special Publications*, **13**, 113–120, https://doi.org/10.1144/gsl.sp.1984. 013.01.10

Baraza, J. and Ercilla, G. 1996. Gas-charged sediments and large pockmark-like features on the Gulf of Cadiz slope (SW Spain). *Marine and Petroleum Geology*, **13**, 253–261, https://doi.org/10.1016/0264-8172(95) 00058-5

Batista, L., Hübscher, C., Terrinha, P., Matias, L., Afilhado, A. and Lüdmann, T. 2017. Crustal structure of the Eurasia–Africa plate boundary across the Gloria Fault, North Atlantic Ocean. *Geophysical Journal International*, **209**, 713–729, https://doi.org/10.1093/gji/ggx050

Boillot, G. and Capdevila, R. 1977. The pyrenees: subduction and collision? *Earth and Planetary Science Letters*, **35**, 151–160, https://doi.org/10.1016/0012-821X(77) 90038-3

Boillot, G. and Malod, J. 1988. The north and north-west Spanish continental margin: a review. *Revista de la Sociedad Geológica de España*, **1**, 295–316.

Boillot, G., Dupeuble, P.A. and Malod, J. 1979. Subduction and tectonics on the continental margin off northern

Spain. *Marine Geology*, **32**, 53–70, https://doi.org/10.1016/0025-3227(79)90146-4

Buforn, E., Udías, A. and Colombás, M.A. 1988. Seismicity, source mechanisms and tectonics of the Azores-Gibraltar plate boundary. *Tectonophysics*, **152**, 89–118, https://doi.org/10.1016/0040-1951(88)90031-5

Cadenas, P., Fernández-Viejo, G., Pulgar, J.A., Tugend, J., Manatschal, G. and Minshull, T.A. 2018. Constraints imposed by rift inheritance on the compressional reactivation of a hyperextended margin: mapping rift domains in the North Iberian Margin and in the Cantabrian Mountains. *Tectonics*, **37**, 758–785, https://doi.org/10.1002/2016tc004454

Custódio, S., Dias, N.A. *et al.* 2015. Earthquakes in western Iberia: improving the understanding of lithospheric deformation in a slowly deforming region. *Geophysical Journal International*, **203**, 127–145, https://doi.org/10.1093/gji/ggv285

Díaz-del-Río, V., Somoza, L. *et al.* 2003. Vast fields of hydrocarbon-derived carbonate chimneys related to the accretionary wedge/olistostrome of the Gulf of Cádiz. *Marine Geology*, **195**, 177–200, https://doi.org/10.1016/S0025-3227(02)00687-4

Dourado, F., Cezario, A.P., Omira, R. and Baptista, M.A. 2018. Mathematical modeling analysis of 1755 tsunami propagation in Pernambuco and Paraiba states coast-Brazil, for present days. *AGU Fall Meeting Abstracts*, 10–14 December, Washington, USA.

Duarte, J.C., Rosas, F.M., Terrinha, P., Gutscher, M.-A., Malavieille, J., Silva, S. and Matias, L. 2011. Thrust–wrench interference tectonics in the Gulf of Cadiz (Africa–Iberia plate boundary in the North-East Atlantic): insights from analog models. *Marine Geology*, **289**, 135–149, https://doi.org/10.1016/j.margeo.2011.09.014

Duarte, J.C., Rosas, F.M., Terrinha, P., Schellart, W.P., Boutelier, D., Gutscher, M.-A. and Ribeiro, A. 2013. Are subduction zones invading the Atlantic? Evidence from the southwest Iberia margin. *Geology*, https://doi.org/10.1130/G34100.1

Duarte, D., Magalhães, V.H. *et al.* 2016. Identification and characterization of fluid escape structures (pockmarks) in the Estremadura Spur, West Iberian Margin. *Marine and Petroleum Geology*, https://doi.org/10.1016/j.marpetgeo.2017.02.026

Duarte, J., Riel, N. *et al.* 2019. Delamination of oceanic lithosphere in SW Iberia: a key for subduction initiation? *Geophysical Research Abstracts. Vol. 21, EGU2019-6001, 2019. EGU General Assembly 2019*, Vienna.

Ercilla, G., Casas, D. *et al.* 2008. Morphosedimentary features and recent depositional architectural model of the Cantabrian continental margin. *Marine Geology*, **247**, 61–83, https://doi.org/10.1016/j.margeo.2007.08.007

Ercilla, G., Casas, D. *et al.* 2011. Imaging the recent sediment dynamics of the Galicia Bank region (Atlantic, NW Iberian Peninsula). *Marine Geophysical Research*, **32**, 99, https://doi.org/10.1007/s11001-011-9129-x

Féraud, G., York, D., Mével, C., Cornen, G., Hall, C.M. and Auzende, J.-M. 1986. Additional ^{40}Ar–^{39}Ar dating of the basement and the alkaline volcanism of Gorringe

Bank (Atlantic Ocean). *Earth and Planetary Science Letters*, **79**, 255–269, https://doi.org/10.1016/0012-821X(86)90184-6

Fernández-Puga, M., Vázquez, J., Somoza, L., Díaz del Rio, V., Medialdea, T., Mata, M.P. and León, R. 2007. Gas-related morphologies and diapirism in the Gulf of Cádiz. *Geo-Marine Letters*, **27**, 213–221, https://doi.org/10.1007/s00367-007-0076-0

Fox, P.J. and Opdyke, N.D. 1973. Geology of the oceanic crust: magnetic properties of oceanic rocks. *Journal of Geophysical Research*, **78**, 5139–5154, https://doi.org/10.1029/JB078i023p05139

Fukao, Y. 1973. Thrust faulting at a lithospheric plate boundary the Portugal earthquake of 1969. *Earth and Planetary Science Letters*, **18**, 205–216, https://doi.org/10.1016/0012-821X(73)90058-7

Gallastegui, J., Pulgar, J.A. and Gallart, J. 2002. Initiation of an active margin at the North Iberian continent-ocean transition. *Tectonics*, **21**, 15-11–15-14, https://doi.org/10.1029/2001tc901046

Gardner, J.M. 2001. Mud volcanoes revealed and sampled on the Western Moroccan Continental Margin. *Geophysical Research Letters*, **28**, 339–342, https://doi.org/10.1029/2000gl012141

Gaspar, L.C. 1984. *Geochemical characterization of phosphorites from the Portuguese Margin*. Master's thesis, University of Aveiro.

Geldmacher, J., Hoernle, K., Klügel, A., Bogaard, P.V.D., Wombacher, F. and Berning, B. 2006. Origin and geochemical evolution of the Madeira–Tore Rise (eastern North Atlantic). *Journal of Geophysical Research: Solid Earth*, **111**, B09206, https://doi.org/10.1029/2005JB003931

Girardeau, J., Cornen, G. *et al.* 1998. Extensional tectonics in the Gorringe Bank rocks, Eastern Atlantic ocean: evidence of an oceanic ultra-slow mantellic accreting centre. *Terra Nova*, **10**, 330–336, https://doi.org/10.1046/j.1365-3121.1998.00209.x

Gómez-Pujol, L., Pérez-Alberti, A., Blanco-Chao, R., Costa, S., Neves, M. and Del Río, L. 2014. Chapter 6 the rock coast of continental Europe in the Atlantic. *Geological Society, London, Memoirs*, **40**, 77–88, https://doi.org/10.1144/m40.6

González, F.J., Somoza, L. *et al.* 2012. Ferromanganese nodules and micro-hardgrounds associated with the Cadiz Contourite Channel (NE Atlantic): palaeoenvironmental records of fluid venting and bottom currents. *Chemical Geology*, **310–311**, 56–78, https://doi.org/10.1016/j.chemgeo.2012.03.030

González, F.J., Somoza, L. *et al.* 2016. Phosphorites, Co-rich Mn nodules, and Fe–Mn crusts from Galicia Bank, NE Atlantic: reflections of Cenozoic tectonics and paleoceanography. *Geochemistry, Geophysics, Geosystems*, **17**, 346–374, https://doi.org/10.1002/2015gc005861

Grange, M., Scharer, U., Merle, R., Girardeau, J. and Cornen, G. 2010. Plume–lithosphere interaction during migration of cretaceous alkaline magmatism in SW Portugal: evidence from U–Pb Ages and Pb–Sr–Hf Isotopes. *Journal of Petrology*, **51**, 1143–1170, https://doi.org/10.1093/petrology/egq018

Grevemeyer, I., Matias, L. and Silva, S. 2016. Mantle earthquakes beneath the South Iberia continental margin and Gulf of Cadiz – constraints from an onshore–offshore seismological network. *Journal of Geodynamics*, **99**, 39–50, https://doi.org/10.1016/j.jog.2016.06.001

Gutscher, M.A., Malod, J., Rehault, J.P., Contrucci, I., Klingelhoefer, F., Mendes-Victor, L. and Spakman, W. 2002. Evidence for active subduction beneath Gibraltar. *Geology*, **30**, 1071–1074, https://doi.org/10.1130/0091-7613(2002)030<1071:Efasbg>2.0.Co;2

Haffert, L., Haeckel, M. *et al.* 2013. Fluid evolution and authigenic mineral paragenesis related to salt diapirism – the Mercator mud volcano in the Gulf of Cadiz. *Geochimica et Cosmochimica Acta*, **106**, 261–286, https://doi.org/10.1016/j.gca.2012.12.016

Hein, J.R., Mizell, K., Koschinsky, A. and Conrad, T. 2013. Deep-ocean mineral deposits as a source of critical metals for high- and green-technology applications: comparison with land-based deposits. *Ore Geology Reviews*, **51**, 1–14, https://doi.org/10.1016/j.oregeorev.2012.12.001

Hensen, C., Nuzzo, M., Hornibrook, E., Pinheiro, L.M., Bock, B., Magalhães, V.H. and Brückmann, W. 2007. Sources of mud volcano fluids in the Gulf of Cadiz – indications for hydrothermal imprint. *Geochimica et Cosmochimica Acta*, **71**, 1232–1248, https://doi.org/10.1016/j.gca.2006.11.022

Hensen, C., Scholz, F. *et al.* 2015. Strike-slip faults mediate the rise of crustal-derived fluids and mud volcanism in the deep sea. *Geology*, https://doi.org/10.1130/g36359.1

Iglesias, J. 2009. *Sedimentation on the Cantabrian Continental Margin from Late Oligocene to Quaternary*. PhD thesis, Universidad de Vigo.

Iglesias, J., Ercilla, G., García-Gil, S. and Judd, A.G. 2010. Pockforms: an evaluation of pockmark-like seabed features on the Landes Plateau, Bay of Biscay. *Geo-Marine Letters*, **30**, 207–219, https://doi.org/10.1007/s00367-009-0182-2

Ivanov, M.K., Kenyon, N. *et al.* 2000. Scientific Party, *TTR-9 Cruise Goals and principal results of the TTR-9 cruise*. IOC/UNESCO Work Report **168:3–4**.

Jané, G., Maestro, A. *et al.* 2010. Occurrence of pockmarks on the Ortegal Spur continental margin, Northwestern Iberian Peninsula. *Marine and Petroleum Geology*, **27**, 1551–1564, https://doi.org/10.1016/j.marpetgeo.2010.04.001

Jiménez-Munt, I., Fernàndez, M., Torné, M. and Bird, P. 2001. The transition from linear to diffuse plate boundary in the Azores–Gibraltar region: results from a thin-sheet model. *Earth and Planetary Science Letters*, **192**, 175–189, https://doi.org/10.1016/S0012-821X(01)00442-3

Johnston, A.C. 1996. Seismic moment assessment of earthquakes in stable continental regions – III. New Madrid 1811–1812, Charleston 1886 and Lisbon 1755. *Geophysical Journal International*, **126**, 314–344, https://doi.org/10.1111/j.1365-246X.1996.tb05294.x

Kaskela, A., Kotilainen, A.T. *et al.* and EMODNET Geology Partners. 2019. Picking up the pieces – harmonizing and collating seabed substrate data for European Maritime Areas. *Geosciences*, **9**, 84, https://doi.org/10.3390/geosciences9020084

Lamboy, M. and Lucas, J. 1979. Les phosphorites de la marge Nord de l'Espagne. Étude géologique et pétrographique. *Oceanologica Acta*, **2**, 325–337.

Lenôtre, N., Thierry, P., Batkowski, D. and Vermeersch, F. 2004. EUROSION project The Coastal Erosion Layer WP 2.6 BRGM/PC-52864-FR.

León, R., Somoza, L., Medialdea, T., Hernández-Molina, F.J., Vázquez, J.T., Díaz-del-Rio, V. and González, F.J. 2010. Pockmarks, collapses and blind valleys in the Gulf of Cádiz. *Geo-Marine Letters*, **30**, 231–247, https://doi.org/10.1007/s00367-009-0169-z

León, R., Somoza, L. *et al.* 2012. New discoveries of mud volcanoes on the Moroccan Atlantic continental margin (Gulf of Cádiz): morpho-structural characterization. *Geo-Marine Letters*, **32**, 473–488, https://doi.org/10.1007/s00367-012-0275-1

Lo Iacono, C., Gràcia, E. *et al.* 2012. Large, deepwater slope failures: implications for landslide-generated tsunamis. *Geology*, **40**, 931–934, https://doi.org/10.1130/g33446.1

Magalhães, V.H., Pinheiro, L.M. *et al.* 2012. Formation processes of methane-derived authigenic carbonates from the Gulf of Cadiz. *Sedimentary Geology*, **243–244**, 155–168, https://doi.org/10.1016/j.sedgeo.2011.10.013

Maldonado, A., Somoza, L. and Pallarés, L. 1999. The Betic orogen and the Iberian–African boundary in the Gulf of Cadiz: geological evolution (central North Atlantic). *Marine Geology*, **155**, 9–43, https://doi.org/10.1016/S0025-3227(98)00139-X

Malod, J., Boillot, G. *et al.* 1980. Plongées en submersible au sud du golfe de Gascogne: stratigraphie et structure de la pente du banc Le Danois. *Société Géologique de France Compte Rendues, XXII*, 3, 73–76.

Malod, J.A. and Mauffret, A. 1990. Iberian plate motions during the Mesozoic. *Tectonophysics*, **184**, 261–278, https://doi.org/10.1016/0040-1951(90)90443-C

Malod, J.-A., Boillot, G. *et al.* 1982. Subduction and tectonics on the continental margin off northern Spain: observations with the submersible Cyana. *Geological Society, London, Special Publications*, **10**, 309–315, https://doi.org/10.1144/gsl.sp.1982.010.01.20

Martínez-Loriente, S., Gràcia, E. *et al.* 2013. Active deformation in old oceanic lithosphere and significance for earthquake hazard: seismic imaging of the Coral Patch Ridge area and neighboring abyssal plains (SW Iberian Margin). *Geochemistry, Geophysics, Geosystems*, **14**, 2206–2231, https://doi.org/10.1002/ggge.20173

Mazurenko, L.L., Soloviev, V.A., Belenkaya, I., Ivanov, M.K. and Pinheiro, L.M. 2002. Mud volcano gas hydrates in the Gulf of Cadiz. *Terra Nova*, **14**, 321–329, https://doi.org/10.1046/j.1365-3121.2002.00428.x

McKenzie, D. 1972. Active tectonics of the Mediterranean region. *Geophysical Journal of the Royal Astronomical Society*, **30**, 109–185, https://doi.org/10.1111/j.1365-246X.1972.tb02351.x

McKenzie, D. 1978. Some remarks on the development of sedimentary basins. *Earth and Planetary Science Letters*, **40**, 25–32, https://doi.org/10.1016/0012-821X(78)90071-7

Medialdea, T. and Terrinha, P. 2016. *Mapa Geológico de España y Portugal a escala 1:1.000.000, Margen Continental*.

Medialdea, T., Vegas, R. *et al.* 2004. Structure and evolution of the 'Olistostrome' complex of the Gibraltar

Arc in the Gulf of Cádiz (eastern Central Atlantic): evidence from two long seismic cross-sections. *Marine Geology*, **209**, 173–198, https://doi.org/10.1016/j. margeo.2004.05.029

Medialdea, T., Somoza, L. *et al*. 2009. Tectonics and mud volcano development in the Gulf of Cádiz. *Marine Geology*, **261**, 48–63, https://doi.org/10.1016/j.mar geo.2008.10.007

Merle, R., Schärer, U., Girardeau, J. and Cornen, G. 2006. Cretaceous seamounts along the continent–ocean transition of the Iberian margin: U–Pb ages and Pb–Sr–Hf isotopes. *Geochimica et Cosmochimica Acta*, **70**, 4950–4976, https://doi.org/10.1016/j.gca.2006. 07.004

Merle, R., Jourdan, F., Marzoli, A., Renne, P.R., Grange, M. and Girardeau, J. 2009. Evidence of multi-phase Cretaceous to Quaternary alkaline magmatism on Tore–Madeira Rise and neighbouring seamounts from ^{40}Ar/^{39}Ar ages. *Journal of the Geological Society*, **166**, 879–894, https://doi.org/10.1144/0016-76492008-060

Merle, R., Jourdan, F. and Girardeau, J. 2018. Geochronology of the Tore–Madeira Rise seamounts and surrounding areas: a review. *Australian Journal of Earth Sciences*, **65**, 591–605, https://doi.org/10.1080/ 08120099.2018.1471005

Miranda, J.M., Matias, L. *et al*. 2015. Marine seismogenic-tsunamigenic prone areas: The Gulf of Cadiz. *In*: Favali, P. *et al*. (eds) *Seafloor Observatories*. Springer Praxis Books, pp. 105–125, https://doi.org/10.1007/978-3-642-11374-1_6

Miranda, R., Valadares, V. *et al*. 2009. Age constraints on the Late Cretaceous alkaline magmatism on the West Iberian Margin. *Cretaceous Research*, **30**, 575–586, https://doi.org/10.1016/j.cretres.2008.11.002

Muiños, S. 2015. *Ferromanganese crusts from the seamounts north of the Madeira Island: composition, origin and paleoceanographic conditions*. PhD thesis, Christian-Albrechts-Universität Kiel.

Muiños, S.B., Hein, J.R. *et al*. 2013. Deep-sea Fe–Mn Crusts from the northeast Atlantic Ocean: composition and resource considerations. *Marine Georesources & Geotechnology*, **31**, 40–70, https://doi.org/10.1080/ 1064119X.2012.661215

Nance, R.D., Gutiérrez-Alonso, G. *et al*. 2012. A brief history of the Rheic Ocean. *Geoscience Frontiers*, **3**, 125–135, https://doi.org/10.1016/j.gsf.2011.11.008

Neres, M., Bouchez, J.L. *et al*. 2014. Magnetic fabric in a Cretaceous sill (Foz da Fonte, Portugal): flow model and implications for regional magmatism. *Geophysical Journal International*, **199**, 78–101, https://doi.org/ 10.1093/gji/ggu250

Neres, M., Carafa, M.M.C., Fernandes, R.M.S., Matias, L., Duarte, J.C., Barba, S. and Terrinha, P. 2016. Lithospheric deformation in the Africa–Iberia plate boundary: improved neotectonic modeling testing a basal-driven Alboran plate. *Journal of Geophysical Research: Solid Earth*, **121**, 6566–6596, https://doi. org/10.1002/2016JB013012

Neves, M.C., Terrinha, R., Afilhado, A., Moulin, M., Matias, L. and Rosas, F. 2009. Response of a multi-domain continental margin to compression: study from seismic reflection-refraction and numerical modelling in the Tagus Abyssal Plain. *Tectonophysics*, **468**, 113–130, https://doi.org/10.1016/j.tecto.2008.05.008

Noiva, J., Ribeiro, C., Terrinha, P., Neres, M., Brito, P. and Survey Team, M. 2017. Exploring the alentejo continental shelf for minerals and Plio-Quaternary environmental changes: Preliminary results of the mineplat survey. *Comunicacoes Geologicas*, **104**, 61–67.

Nuzzo, M., Hornibrook, E.R.C. *et al*. 2009. Origin of light volatile hydrocarbon gases in mud volcano fluids, Gulf of Cadiz – evidence for multiple sources and transport mechanisms in active sedimentary wedges. *Chemical Geology*, **266**, 350–363, https://doi.org/10.1016/j. chemgeo.2009.06.023

Nuzzo, M., Tomonaga, Y. *et al*. 2019. Formation and migration of hydrocarbons in deeply buried sediments of the Gulf of Cadiz convergent plate boundary – insights from the hydrocarbon and helium isotope geochemistry of mud volcano fluids. *Marine Geology*, **410**, 56–69, https://doi.org/10.1016/j.margeo.2019.01.005

Omira, R., Ramalho, I., Terrinha, P., Baptista, M.A., Batista, L. and Zitellini, N. 2016. Deep-water seamounts, a potential source of tsunami generated by landslides? The Hirondelle Seamount, NE Atlantic. *Marine Geology*, **379**, 267–280, https://doi.org/10. 1016/j.margeo.2016.06.010

Palomino, D., López-González, N., Vázquez, J.-T., Fernández-Salas, L.-M., Rueda, J.-L., Sánchez-Leal, R. and Díaz-del-Río, V. 2016. Multidisciplinary study of mud volcanoes and diapirs and their relationship to seepages and bottom currents in the Gulf of Cádiz continental slope (northeastern sector). *Marine Geology*, **378**, 196–212, https://doi.org/10.1016/j.margeo. 2015.10.001

Peirce, C. and Barton, P.J. 1991. Crustal structure of the Madeira–Tore Rise, eastern North Atlantic – results of a DOBS wide-angle and normal incidence seismic experiment in the Josephine Seamount region. *Geophysical Journal International*, **106**, 357–378, https://doi.org/10.1111/j.1365-246X.1991.tb03898.x

Péron-Pinvidic, G., Manatschal, G., Minshull, T.A. and Sawyer, D.S. 2007. Tectonosedimentary evolution of the deep Iberia–Newfoundland margins: evidence for a complex breakup history. *Tectonics*, **26**, https://doi. org/10.1029/2006TC001970

Pinheiro, L.M., Whitmarsh, R.B. and Miles, P.R. 1992. The ocean–continent boundary of the western continental margin of Iberia – II. Crustal structure in the Tagus Abyssal Plain. *Geophysical Journal International*, **109**, 106–124, https://doi.org/10.1111/j.1365-246X. 1992.tb00082.x

Pinheiro, L.M., Ivanov, M.K. *et al*. 2003. Mud volcanism in the Gulf of Cadiz: results from the TTR-10 cruise. *Marine Geology*, **195**, 131–151, https://doi.org/10. 1016/S0025-3227(02)00685-0

Pinto, C.A., Silveira, T.M. and Teixeira, S.B. 2018. *Beach nourishment practice along the Portuguese coastline: framework and review (1950–2017)*. Technical Report. Portuguese Environment Agency, 57. https://doi.org/ 10.13140/RG.2.2.12655.89767

Ponte Lira, C., Nobre Silva, A., Taborda, R. and Freire de Andrade, C. 2016. Coastline evolution of Portuguese low-lying sandy coast in the last 50 years: an integrated approach. *Earth System Science Data*, **8**, 265–278, https://doi.org/10.5194/essd-8-265-2016

Purdy, G.M. 1975. The eastern end of the Azores–Gibraltar plate boundary. *Geophysical Journal of the Royal*

Astronomical Society, **43**, 973–1000, https://doi.org/10.1111/j.1365-246X.1975.tb06206.x

Rincón-Tomás, B., Duda, J.P. *et al.* 2019. Cold-water corals and hydrocarbon-rich seepage in Pompeia Province (Gulf of Cádiz) – living on the edge. *Biogeosciences*, **16**, 1607–1627, https://doi.org/10.5194/bg-16-1607-2019

Rosas, F., Duarte, J., Schellart, W., Tomás, R. and Terrinha, P. 2016. Seismic potential of Thrust-Wrench tectonic interference between major active faults offshore SW Iberia. *In*: Duarte, J.C. and Schellart, W.P. (eds) *Plate Boundaries and Natural Hazards, AGU Geophysical Monograph*, **219**, 193–218, https://doi.org/10.1002/9781119054146.ch9

Rosenbaum, G., Lister, G.S. and Duboz, C. 2002. Relative motions of Africa, Iberia and Europe during Alpine orogeny. *Tectonophysics*, **359**, 117–129, https://doi.org/10.1016/S0040-1951(02)00442-0

Ryan, W.B.F., Hsü, K.J. *et al.* 1973. Gorringe Bank; Site 120. *Initial reports of the Deep Sea Drilling Project; covering Leg 13 of the cruises of the Drilling Vessel Glomar Challenger*, Lisbon, Portugal to Lisbon, Portugal, August–October 1970, **13**, 19, https://doi.org/10.2973/dsdp.proc.13.102.1973

Sallarès, V., Martínez-Loriente, S. *et al.* 2013. Seismic evidence of exhumed mantle rock basement at the Gorringe Bank and the adjacent Horseshoe and Tagus abyssal plains (SW Iberia). *Earth and Planetary Science Letters*, **365**, 120–131, https://doi.org/10.1016/j.epsl.2013.01.021

Sartori, R., Torelli, L., Zitellini, N., Peis, D. and Lodolo, E. 1994. Eastern segment of the Azores–Gibraltar Line (Central-Eastern Atlantic) – an Oceanic plate boundary with diffuse compressional deformation. *Geology*, **22**, 555–558, https://doi.org/10.1130/0091-7613(1994)022<0555:Esotag>2.3.Co;2

Sawyer, D., Withmarsh, R., Klaus, A. and Party, S.S. 1994. Leg 149. *Proceedings of the Ocean Drilling Program, Initial Reports*. Ocean Drilling Program, College Station, TX, **719**.

Schärer, U., Girardeau, J., Cornen, G. and Boillot, G. 2000. 138–121 Ma asthenospheric magmatism prior to continental break-up in the North Atlantic and geodynamic implications. *Earth and Planetary Science Letters*, **181**, 555–572, https://doi.org/10.1016/S0012-821X(00)00220-X

Schettino, A. and Turco, E. 2009. Breakup of Pangaea and plate kinematics of the central Atlantic and Atlas regions. *Geophysical Journal International*, **178**, 1078–1097, https://doi.org/10.1111/j.1365-246x.2009.04186.x

Schettino, A. and Turco, E. 2011. Tectonic history of the western Tethys since the Late Triassic. *Geological Society America Bulletin*, **123**, 89–105, https://doi.org/10.1130/B30064.1

Sibuet, J.C. and Collette, B.J. 1991. Triple junctions of Bay of Biscay and North Atlantic: new constraints on the kinematic evolution. *Geology*, **19**, 522–525, https://doi.org/10.1130/0091-7613(1991)019<0522:TJOBOB>2.3.CO;2

Sibuet, J.-C., Monti, S., Loubrieu, B., Mazé, J.-P. and Srivastava, S. 2004. Bathymetric map of the NE Atlantic Ocean and Bay of Biscay: kinematic implications. *Bulletin de la Societe Geologique de France*, **175**, 429–442, https://doi.org/10.2113/175.5.429

Sibuet, J.-C., Srivastava, S. and Manatschal, G. 2007. Exhumed mantle-forming transitional crust in the Newfoundland–Iberia rift and associated magnetic anomalies. *Journal of Geophysical Research*, **112**, B06105, https://doi.org/10.1029/2005jb003856

Silva, S., Terrinha, P. *et al.* 2017. Micro-seismicity in the Gulf of Cadiz: is there a link between micro-seismicity, high magnitude earthquakes and active faults? *Tectonophysics*, **717**, 226–241, https://doi.org/10.1016/j.tecto.2017.07.026

Somoza, L., Díaz del Río, V. *et al.* 2003. Seabed morphology and hydrocarbon seepage in the Gulf of Cádiz mud volcano area: acoustic imagery, multibeam and ultra-high resolution seismic data. *Marine Geology*, **195**, 153–176, https://doi.org/10.1016/S0025-3227(02)00686-2

Somoza, L., Ercilla, G. *et al.* 2014. Detection and mapping of cold-water coral mounds and living Lophelia reefs in the Galicia Bank, Atlantic NW Iberia margin. *Marine Geology*, **349**, 73–90, https://doi.org/10.1016/j.margeo.2013.12.017

Somoza, L., Medialdea, T. *et al.* 2019. Morphostructure of the Galicia continental margin and adjacent deep ocean floor: from hyperextended rifted to convergent margin styles. *Marine Geology*, **407**, 299–315, https://doi.org/10.1016/j.margeo.2018.11.011

Srivastava, S., Roest, W., Kovacs, L., Oakey, G., Lévesque, S., Verhoef, J., and Macnab, R. 1990. Motion of Iberia since the Late Jurassic: results from detailed aeromagnetic measurements in the Newfoundland Basin. *Tectonophysics*, **184**, 229–260, https://doi.org/10.1016/0040-1951(90)90442-B

Srivastava, S.P., Sibuet, J.C., Cande, S., Roest, W.R. and Reid, I.D. 2000. Magnetic evidence for slow seafloor spreading during the formation of the Newfoundland and Iberian margins. *Earth and Planetary Science Letters*, **182**, 61–76, https://doi.org/10.1016/S0012-821X(00)00231-4

Stadnitskaia, A., Ivanov, M.K., Blinova, V., Kreulen, R. and van Weering, T.C.E. 2006. Molecular and carbon isotopic variability of hydrocarbon gases from mud volcanoes in the Gulf of Cadiz, NE Atlantic. *Marine and Petroleum Geology*, **23**, 281–296, https://doi.org/10.1016/j.marpetgeo.2005.11.001

Stich, D., Mancilla, F.D.L. and Morales, J. 2005. Crust-mantle coupling in the Gulf of Cadiz (SW-Iberia). *Geophysical Research Letters*, **32**, https://doi.org/10.1029/2005gl023098

Teixeira, M., Terrinha, P., Roque, C., Rosa, M., Ercilla, G. and Casas, D. 2019. Interaction of alongslope and downslope processes in the Alentejo Margin (SW Iberia) – implications on slope stability. *Marine Geology*, **410**, 88–108, https://doi.org/10.1016/j.margeo.2018.12.011

Terrinha, P., Pinheiro, L.M. *et al.* and the TTR10 Shipboard Scientific Party. 2003. Tsunamigenic-seismogenic structures, neotectonics, sedimentary processes and slope instability on the southwest Portuguese Margin. *Marine Geology*, **3266**, 1–19, https://doi.org/10.1016/S0025-3227(02)00682-5

Terrinha, P., Pueyo, E.L., Aranguren, A., Kullberg, J.C., Kullberg, M.C., Casas-Sainz, A. and Azevedo, M.D.R. 2018. Gravimetric and magnetic fabric study of the Sintra Igneous complex: laccolith-plug emplacement in the Western Iberian passive margin. *International Journal of Earth Sciences*, **107**, 1807–1833, https://doi.org/10.1007/s00531-017-1573-7

Terrinha, P., Kullberg, J.C. *et al.* 2019*a*. Rifting of the Southwest and West Iberia Continental Margins, ch 6. *In*: Quesada, C. and Oliveira, J.T. (eds) *The Geology of Iberia: A Geodynamic Approach*. Springer, Cham, **3**: The Alpine Cycle, 251–284.

Terrinha, P., Neres, M. *et al.* 2019*b*. Imaging the Azores–Gibraltar Fracture Zone and the Madeira–Tore Rise intersection with multichannel seismics. *The PROPEL cruise (PROPagation of the Eurasia–Africa pLate boundary East of the Gloria Fault). Geophysical Research Abstracts*. Vol. 21, EGU2019-11788, 2019. EGU General Assembly.

Terrinha, P., Ramos, A. *et al.* 2019*c*. Alpine orogeny: deformation and structure in the West and Southwest Iberia margins. *In*: Quesada, C. and Oliveira, J. (eds) *The Geology of Iberia: A Geodynamic Approach*. Springer, Cham, **3**: The Alpine Cycle, 251–284.

Toyos, M.H., Medialdea, T., León, R., Somoza, L., González, F.J. and Meléndez, N. 2016. Evidence of episodic long-lived eruptions in the Yuma, Ginsburg, Jesús Baraza and Tasyo mud volcanoes, Gulf of Cádiz. *Geo-Marine Letters*, **36**, 197–214, https://doi.org/10.1007/s00367-016-0440-z

Tucholke, B.E. and Sibuet, J.-C. 2007. Leg 210 synthesis: Tectonic, magmatic, and sedimentary evolution of the Newfoundland-Iberia rift. *Proceedings of the Ocean Drilling Program, Scientific Results*. Ocean Drilling Program, College Station, TX, 1–56.

Tugend, J., Manatschal, G., Kusznir, N.J., Masini, E., Mohn, G. and Thinon, I. 2014. Formation and deformation of hyperextended rift systems: Insights from rift domain mapping in the Bay of Biscay–Pyrenees. *Tectonics*, **33**, 1239–1276, https://doi.org/10.1002/2014tc003529

Van Rensbergen, P., Depreiter, D. *et al.* 2005. The El Arraiche mud volcano field at the Moroccan Atlantic slope, Gulf of Cadiz. *Marine Geology*, **219**, 1–17, https://doi.org/10.1016/j.margeo.2005.04.007

Van Rooij, D., Iglesias, J. *et al.* 2010. The Le Danois Contourite Depositional System: interactions between the Mediterranean Outflow Water and the upper Cantabrian slope (North Iberian margin). *Marine Geology*, **274**, 1–20, https://doi.org/10.1016/j.margeo.2010.03.001

Vázquez, J.T., Medialdea, T. *et al.* 2008. Cenozoic deformational structures on the Galicia Bank Region (NW Iberian continental margin). *Marine Geology*, **249**, 128–149, https://doi.org/10.1016/j.margeo.2007.09.014

Vázquez, J.T., Ercilla, G. *et al.* 2009. El colapso BURATO ERGAP: Un rasgo morfo-tectónico de primera magnitud en el Banco de Galicia, 205–208, https://doi.org/10.13140/2.1.5161.6963

Vergés, J., Kullberg, J. *et al.* 2019. An introduction to the alpine cycle in Iberia. *In*: Quesada, C. and Oliveira, J. (eds) *The Geology of Iberia: A Geodynamic Approach. Regional Geology Reviews*. Springer, Cham. https://doi.org/10.1007/978-3-030-11295-0_1

Verreydt, W. 2011. *Late Cenozoic sedimentary processes on the outer edge of the NW Iberian shelf, Cabo Ortegal*. MSc thesis, Ghent University.

Zaragosi, S., Le Suavé, R., Bourillet, J.-F., Auffret, G., Faugères, J.-C., Pujol, C. and Garlan, T. 2001. The deepsea Armorican depositional system (Bay of Biscay), a multiple source, ramp model. *Geo-Marine Letters*, **20**, 219–232, https://doi.org/10.1007/s003670100061

Zitellini, N., Gracia, E. *et al.* 2009. The quest for the Africa–Eurasia plate boundary west of the Strait of Gibraltar. *Earth and Planetary Science Letters*, **280**, 13–50, https://doi.org/10.1016/j.epsl.2008.12.005

The Pliocene deposits of the Black Sea Shelf east of the Danube River Delta

Petro F. Gozhik[1] and Valery E. Rokitsky[2]*

[1]Institute of Geological Sciences NAS of Ukraine, 55-b O. Honchara str., Kyiv, Ukraine, 0160

[2]Prichornomorske State Regional Geological Enterprise (PrichornomorskeSRGE), 1, Inglezi St, Odessa, 65070, Ukraine

 VER, 0000-0003-1034-2269
*Correspondence: roki57@ukr.net

Abstract: This paper provides analysis of the published materials on the occurrences of the Dacian and Cimmerian molluscs in the Danube River valley as well as the results of Pliocene sediments study based on core material of the boreholes drilled at the Black Sea Shelf east of the Danube River Delta.

In the early Pontian time, the Dacian Basin was a large sub-basin of Paratethys which, due to an abrupt drop in sea level, separated into the Euxinian, Dacian and Caspian basins. At the end of the Bosphorus time, the discharge of the Dacian Basin waters into the Euxinian Basin formed a wide valley from the Galati-Reni region to the east through the Galati gateway. During the Cimmerian transgression, a vast bay existed on the site of the modern Danube Delta, from which mutual migrations of the Dacian and Cimmerian molluscs took place along the runoff valley. The cessation of runoff occurred during the regressive phase of the Late Cimmerian. The rhythmically bedded thick strata originated during the existence of the runoff valley. These strata were identified as the Pridanubian Formation (Suite). The cryptogenic form of *Tulotoma Tulotoma* (=*Viviparus*) *ovidii nasonis* (Bogachev) is characteristic of the lower and middle parts of the suite. The presence of the Dacian and Cimmerian molluscs in this suite became the basis for the correlation of sediments of the Dacian and Cimmerian regional stages. The Duabian molluscs were registered in the Cimmerian deposits of the Transcaucasus (the Duabian layers), Priazovye and the Kerch–Taman region. The migration of these molluscs took place during the regressive phases due to the circular current in the Euxinian Basin similar to the one existing in the Black Sea today.

The Pliocene formation contains marine and continental deposits of the Lower and Upper Pliocene, which are represented by the Pridanubian Formation (Lower and Upper), Cimmerian deposits (non-subdivided Lower and Middle Cimmerian), Lower Kujalnician deposits, Upper Poration deposits, complex of red-coloured palaeosols (the Upper Miocene–Lower Pliocene non-subdivided).

The formation of the Pliocene sediments on the Black Sea Shelf, east of the Danube Delta, was controlled by the inter-basin connectivity of the Eastern Paratethys.

During the Cenozoic Alpine orogeny, Tethys separated into several basins, which all together were part of the Paratethys. In the early Pontian time, the Dacian Basin was a large sub-basin of the Paratethys, but due to the fall of its sea level by 50–100 m (Jipa and Olariu 2013), the formation of independent basins (Caspian, Euxinian, Dacian and Pannonian) took place, in which endemic brackish and freshwater fauna originated (Orszag-Sperber 2006). The bio- and magnetostratigraphic studies of the Mediterranean Neogene, in the 1990s–2000s interval, have shown that brackish water and freshwater fauna occurred at the top of evaporites and locally at the base of the Zanklian transgression sediments. This became the basis for assigning these deposits to the Lower Pliocene or the pre-Pliocene, as well as before the 'Lago Mare event' (Ruggieri 1967; Orszag-Sperber 2006). In the Euxinian Basin this stage corresponds to the erosional contact between the Pontian and Cimmerian deposits not only on the shelf, but also in the Indolo–Kuban Trough (Semenenko 1987), as well as in the Black Sea, where a thick Pliocene strata of coarse-grained deposits among the pelagic and semi-pelagic sediments, discovered by the Deep Sea Drilling Project (DSDP) borehole 380 at a depth between 883.5 and 864.5 m (Hsü and Federico 1979). The age is confirmed by the palaeontological data (*Diatomeae*) and the establishment of the lower boundary of the Gilbert Epoch at 1 m above the mentioned thick strata of coarse-grained sediments.

In the late Pontian (at the end of the Bosphorian time), the volume of the Dacian Basin decreased due to a lowering of the sea level, which was recorded by erosional surface. Andreescu (2009, p. 323, fig. 3) showed at one of the geological sections that the Gettian deposits overlie the Novorossiysk beds with deep erosion, while also cutting off

From: Asch, K., Kitazato, H. and Vallius, H. (eds) 2022. *From Continental Shelf to Slope: Mapping the Oceanic Realm.* Geological Society, London, Special Publications, **505**, 117–130.
First published online November 20, 2020, https://doi.org/10.1144/SP505-2019-102

the Upper Bosphorian rocks. The sands (at the base with pebbles) of the Berbesti Formation fill the incised valleys. Upwards stratigraphically, one more erosional surface is overlain by the sands of the Candesti Formation (Ghenea 1970). These sands, referred to the Parskovian (Andreescu and Papaianopol 1975), cut off the tops of the Berbesti Formation. There is also an erosional surface between the Parskovian and early Romanian, which we compare with the erosion, which took place before the accumulation of the Upper Poratian sediments (Chepalyga 1967; Bukatchuk et al. 1983; Gozhik 2006).

The intensification of erosion at the end of the Bosphorian time is also observed in the northwestern part of the Dacian Basin. So, in the Turnu–Severin area, the Messinian erosion surface is overlain by marine clayey sediments of the Bosphorian age, in which nanofossils of the zone No. 12 were recorded (Clauzon et al. 2005, p. 450, fig. 10). Conglomerates and sands overlie marine clays, and themselves in turn are overlain by horizontally layered Dacian deposits. According to the bio- and chrono-stratigraphic data they belong to chron SZp.4p with an age of 5.28 Ma (Popescu 2001).

The filling of the Pannonian and Dacian basins (based on seismic data) suggest the formation of a large river (palaeo-Danube) during Dacian time (Olariu et al. 2018) which might have been triggered by a sea level fall. This palaeogeography is confirmed by the presence of erosion of the Pontian deposits and the formation of the thick strata of rhythmically constructed marine and continental (mainly alluvial) deposits, traversed by the boreholes near the town of Reni. At the base of the strata, Bogachev (1961) identified the molluscs characteristic of the Dacian deposits. He described marine and freshwater molluscs, which contain not only typical Dacian forms, but also species characteristic of the Cimmerian deposits of the Euxinian Basin, such as: *Limnocardium fervidum orsa* Ebers., *Congeria supramoquica* Gabun., *C. caucasica* Senin., *C. aff. croatica* Brus., *Melanopsis abchasica* Senin., *Prosodacna munieri* Sabba, *Dreissena weberi* Senin., *Viviparus bifarcinatus* Bielz., *V. ovidii nasonis* Bog., *V. turgidus* Bielz., *V. cyrmatophorus* Brus., *V. nodaso-costatus* Halav., *V. aff. woodwardi* Sabba, *Neritina scripta* Sabba, *Tylopoma pilari* Neum., *T. bruinai* Sabba, *T. plicata* Sabba, *Unio* sp.

The Dacian mollusc fauna was also discovered outside the Dacian Basin in the Danube River Delta south of Izmail town. Here, *Prosodacna orientalis* Sabba, *Pr. heberti* Cobalc., *Dreissena rimestriensis* Font. were discovered at a depth of 66–77 m in sandy-clay sediments. These facts were the basis to show the distribution of the Dacian deposits in the Danube River Delta in the atlas for the Neogene lithofacies (Atlas litofacial 1969).

The deposits with the Dacian molluscs and *Tulotoma ovidii nasonis* (Bog.) were recognized later during the geological surveys in the southwestern part of Moldova (Bukatchuk and Gozhik 1986) and the southern part of the left bank of the Lake Yalpug in the Danube Valley (Gozhik and Chyrka 1973; Gozhik 2006). Based on the study of ostracods Sinegub (1969) assigned the rhythmically bedded strata with the Dacian molluscs to the Bosphorian Regional Stage. Roshka and Hubka (1982) referred these strata together with the Lower Poratian sediments to the Danube Formation, which was included in the stratigraphic schemes for the southern Moldova and southwestern Ukraine. Bukatchuk (1985) subdivided the Danube Formation into the lower, middle and upper subsuites. The Lower Subsuite lies above an erosion surface on the Pontic deposits, and the Upper Poratian sediments (Bukatchuk and Gozhik 1986) with *Pristinunio procumbens* (Fuchs.) lie on the Upper Subsuite also above an erosional surface. The following species of marine and freshwater molluscs are characteristic of the Lower Pridanubian Subsuite: *Parapachidacna ex gr. cobalcescui* (Font.), *Pachyprionopleure haueri* (Cobalc.), *Stylodacna heberti* (Cobalc.), *Prosodacna semisulcata* (Rouss.), *Tulotoma (=Viviparus) ovidii nasonis* (Bog.), *Viviparus (Protulotoma) sadleri* (Partsch.), *V. (P.) cyrtomaphorus* (Brusina), *Lithoglyphus herpaeformis* Cobalc, *Tulopoma (=Bulinus) plicata* (Sabba), *Melanopsis (Mingreliciana) pterochila* Brus., *M. (M.) decolbata* Stol., *Theodoxus quadrofasciatus* Bielz., *Unio rournus* Tourn.

The most common fossils in the Middle Pridanubian Subsuite are: *Prosodacna stenopleura* (Sabba), *Horiodacna rumana* (Sabba), *Limnocardium fervidum orsa* Ebersin, *T. ovidii nasonis* Bog., *T. argesiensis* (Sabba), *Viviparus (Protulotoma) bifarcinata* Bielc., *M. (M.) decollata* Stol., *M. (M.) pterochila* Brus., *M. (Lyrcea) onusta* Stef., *Melanoides abchasicus* Senin.

The marine molluscs were not found in the Upper Pridanubian Subsuite, but the freshwater ones were recorded. These are *V. (P.) bifarcinatus* Bielz., *V. (V.) craiovensis* Tourn. and unionoids of the Lower Poratian with *Bittneriella (=Potomida) sandbergeri* (Neum.) etc.

Material and methods

Geological survey work on the Black Sea Shelf east of the Danube Delta was carried out by the PrichornomorskeSRGE and accompanied by drilling. A total of 35 boreholes were drilled to depths of up to 142 metres (borehole 358), in which the Quaternary, Pliocene and Upper Miocene deposits were discovered (Fig. 1. Legend Fig. 11). The Quaternary formations are represented by alternating

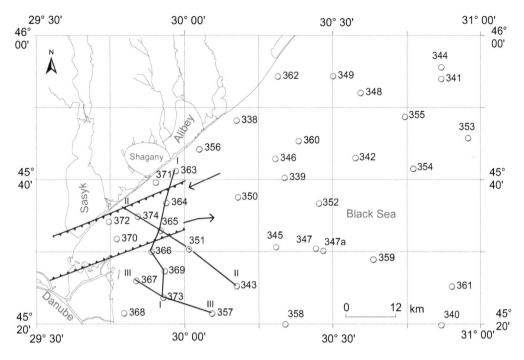

Fig. 1. (**a**) Location of the boreholes and erosional valley. (**b**) The contemporary landscape of the study area.

sequences with strata of non-marine and marine deposits. At the base of such sequences alluvial sands, sometimes with gravel and clays with shells of freshwater molluscs, occurred. They are overlain by silts, clays and detrital sands with shells of marine molluscs (Fig. 2, borehole 345). Outside the valleys (in watersheds), the subaerial loams, lake clays and marine deposits alternate in the section. The vertical alternation of marine to non-marine sequences is a reflection of climatically conditioned regressive and transgressive sedimentation in the Pleistocene history of the Black Sea. The focus of this study is on Pliocene sediments characteristics and the resulting stratigraphy.

Pliocene stratigraphy

The newly acquired data from drilling in combination with the adjacent on-shore data allow us to distinguish the following subdivisions in the complex of Pliocene formations: Pridanubian Formation (N_2^1pd), Lower and Upper Subformation, Cimmerian deposits (N_2^1km) (non-subdivided Lower and

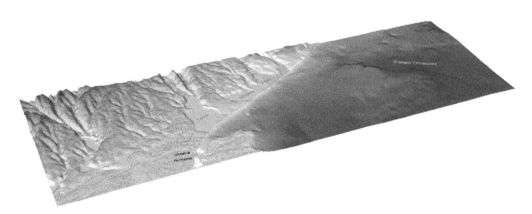

Fig. 1.1. The contemporary landscape of the study area.

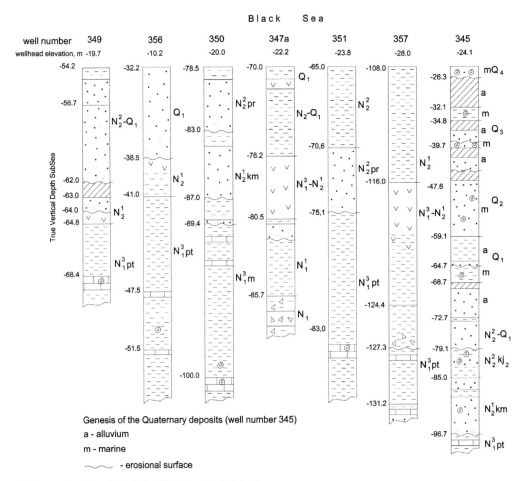

Fig. 2. Borehole logs (349, 356, 350, 347a, 351, 357, 345).

Middle Cimmerian) in the Lower Pliocene; the Lower Kujalnician deposits ($N_2^2kj_1$), Upper Poration deposits (N_2^2pr) in the Upper Pliocene; complex of red-coloured palaeosols (the Upper Miocene–Lower Pliocene non-subdivided) (Table 1).

Pridanubian Formation

The Pridanubian Formation (boreholes 364, 365, 370, 372 and 374) occurs only within the erosional valley (Fig. 1). The borehole 370 (Fig. 3) was located in the sea at a depth of 14.5 m. In the interval −59.5 − −78.5 m occur two units of Lower Pleistocene alluvium, separated by a layer of greenish-grey clay. We assigned the lower strata, with fragments of viviparids, to the Lower Pleistocene and correlated it with the Upper Kujalnician (Gozhik 2019).

The greenish-grey clays occur in the interval of − 78.5 – −81.5 m, they are separated by sands

with *Rytia* (=*Pelendunio*) *bielzi* (Czek) *Pristinunio ex gr. procumbens* (Bielz.), *Tylopoma* (=*Bulimus*) *brusinae* (Sabba), *T. pilari* (Neum.), *Melanopsis* (*Mingreliciana*) *pterochila* Brus., *Theodoxus rumanus* Sabba, *Lithoglyphus* sp. The presence of *R. bielzi*, a characteristic form of the Upper Poratian deposits (Pavlov 1925; Chepalyga 1967) of the Lower Romanian (Pelendavian), attributed to the NSM$_{11c}$ Subzone (Andreescu 2009; Andreesku *et al.* 2010), with an upper boundary of 3.2 million years, as well as its occurrence in the Lower Kujalnician deposits of the Sea of Azov region allows us to refer the abovementioned strata to the Upper Poratian and compare them with the Lower Kujalnician deposits (N_2^2 kj$_1$).

Below the mentioned strata (−81.5 – −83.5 m) there occur greenish-grey medium-grained sands with gravel, fragments and detritus of molluscs: *Parapachydacna ex gr. cobalcescui* (Font.), *Limnocardium* sp., *Prosodacna* sp., *Stylodacna* sp.,

Table 1. *Subdivision chart of the Pliocene deposits*

ISC System	ISC Stage	Dacian basin Stage	Dacian basin Substage	Region of investigation	Species	South-western Black Sea region — Terrace	Climatolite	Azov-Black Sea basin — Stage	substage	beds
Pleistocene	Calabrian	Argedavaian	Milcovanian	Alluvium sediments with *V.subconcinus*		IX Boshernitsa	kr / br	Gurian (Q₁¹)		Upper / Middle / Lower
	Gelasian			Alluvium sediments with *U.kujalnicensis* (Q₁¹)		X Ferladany	bv			Dzhankoian beds / Tamanian beds with *B.tamanensis*
Pliocene	Piacentian	Romanian	Valachian	Lower Kujalnician (Kj₁) with *L.limanicum, V.v.kubanica*		XI Vady Lujvode	sv / bd	Kujalnician Kj (N₂²)		
				Upper Poratian (N₂²pr): *V.turritus* / *R.bielzi*		XII Runkashov	kz / jr			Galizgian beds / Bachterian beds
			Pelendavian	Lower Poratian *B.sundbergeri* / Sadiments with *V.bifarcinatus*		XIII Kuchurgan terrace with *Pl.flabelatiformis*	aj / st	Cimmerian Km (N₂²)	km₃ / Middle km₂	Duabian beds (*Ph.planum, P.crassatelata, P.cobalcescui, P.multistriata, Dr.theodori, Ps.zlatarskii, Ps.donacoides, Pr.prionopleura, L.velutina, H.rumana, P.crassidens, C.escheri*)
	Zanclian	Dacian	Siensian	Pridanubian formation (N₂¹pd): Lower (N₂¹pd₁) / Upper (N₂¹pd₂)	*P.ex.gr.cobalcescui, T.ovidii nasonis; M.decolata; Cimmerian (km₁ and km₂) Dr.theciori, L.squamulosum, P.crassatelata, P.multistriata, Ps.donacoides, Ps.zlatarskii*				Lower km₁	
Miocene	Messinian	Pontian	Parscovian / Getian / Bosphorian	Red coloured formation (N₁³-N₂¹)				Pontian	Bosphorian	Beds with *Limnicardium*

N₂¹ - Lower Pliocene, N₂² - Upper Pliocene.

Tulotoma (=Viviparus) ovidii nasonis (Bog.), *T. bogachevi* Dats., *Melanopsis (Lyrcea) onusta* Sabba, *M. (M.) pterochila* Brus., *M. (M.) ex gr. decolata* Stol. and *Tylopoma plicata* (Sabba). The sands are underlain by greenish-grey silts with interlayers of clays and pelitomorphic limestone with fragments of *Tulotoma* sp. (−83.5 – −86.0 m), below which (−86.0 – −88.5 m) are inequigranular sands with

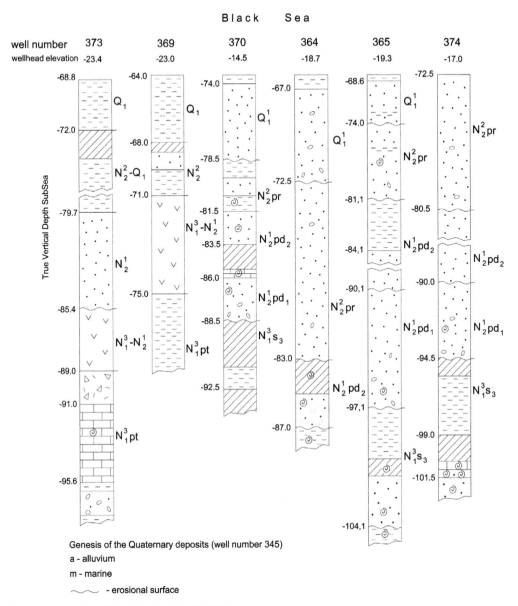

Fig. 3. Borehole logs (373, 369, 370, 364, 365, 374).

gravel and small pebbles. The molluscs that they contain are equivalent to those mentioned above in the greenish-grey medium-granular sands. Cryptogenic form *Tulotoma ovidii nasonis* (Fig. 4) occurs only in the Lower and Middle Pridanubian Subformations, the age of which is determined due to the Dacian molluscs. The presence of the typical representative of the early Dacian *Parapachydacna ex. gr. cobalcescui* undoubtedly indicates the Dacian age of the sediments in the range of −81.5 – −88.5 m and naturally belongs to the

Pridanubian Formation (N_2^1pd). The Pridanubian Formation erosively overlies the Upper Sarmatian deposits ($N_1^3S_3$) with numerous remnants of *Mactra bulgarica* Toula.

In the borehole 372 at a water depth of 10.7 m (Fig. 5) under the Lower Pleistocene alluvium (Q_1), medium-grained sands with gravel (−76.8 – −81.7 m) and fragments of shells *Teodoxus* sp., *Viviparus ex gr. turritus* Bog., *Melanopsis (M.) pterochilla* Brus., *Tylopoma (B.) brusinae* Sabba occur. Given the presence of *V. ex gr. turritus* and the

Fig. 4. *Tulotoma ovidii nasonis*. Photo by V.E. Rokitsky.

position of alluvium in the Pliocene section of the study area, the marked alluvium strata is assigned to the Upper Pliocene and similar to the Upper Poratian layers.

The underlying greenish-grey clays (−81.7 – −93.0 m) with *M. pterochila* Brus. and *Teodoxus* sp. are assigned to the Pridanubian Formation. The underlying clays with limestone interlayers include shells of *Mactra bulgarica* Toula. and *Mactra* sp. and are assigned to the Upper Sarmatian.

At the borehole 374, located to the east of the borehole 372 at a water depth of 17.0 m, were discovered the Lower Pleistocene deposits (− 66.0 – − 72.5 m) of medium- and fine-grained sands with gravel and small pebbles. The presence of *Viviparus ex gr. subconcinus* Sinz., gives the basis to correlate the strata of sands with the Upper Kujalnician (Q_1kj_2) and IXth terrace of the Danube and Dniester rivers (Bukatchuk *et al.* 1983; Gozhik 2006). In the range of −72.5 – −80.5 m there occur greenish-grey sands with gravel and fragments of limestone. They contain *Viviparus ex gr. turritus* Bog., *Melanopsis* (*L.*) *onusta* Sabba, *M.* (*M.*) *pterochilla* Brus., *Unio* sp. and are considered Upper Poratian (Gozhik 2006) and Lower Kujalnician ($N_2^2kj_1$) in age. Below, down to the depth −94.5 m, inequigranular poorly sorted sands with gravel occur (a boulder of limestone was found at a depth of −84.5 m.) These sands contain *Limnocardium* sp., *Prosodacna* sp., *Parapachydacna* sp., *Tulotoma ovidii nasonis* Bog., *Melanopsis* (*L.*) *onusta* Sabba, *M.* (*M.*) *pterochilla* Brus., and *Tylopoma pilari* (Neum.). That gives the grounds to determine the Dacian age of these sediments and assign them to the Pridanubian Formation (N_2^1pd).

The Pridanubian Formation was also discovered by borehole 365. In the interval of −81.1 – −97.1 m, the greenish-grey clays (−81.1 – −84.1 m) and two units of medium-grained sands with gravel and detritus at the base (−84.1 – − 90.1 m and −90.1 – −97.1 m) contain *Tulotoma ovidii nasonis* Bog., *Prosodacna* sp., *Parapachydacna* sp., *Lithoglyphus acutus* Cobalc., as well as fragments of thick-walled unionids. We correlate the lower sandy strata with the Lower Pridanubian Subformation, and the upper one with the Middle Pridanubian Subformation. It is overlain (−74.0 – −81.1 m) by fine-medium-grained sands with the greenish-grey clay interlayers, conventionally attributed to the Upper Poratian deposits. The Lower Pridanubian Subformation is underlain by greenish-grey clays, siltstones and sands with numerous shells and fragments of *Mactra bulgarica* Toula and *Mactra ex gr. caspia* Eichw.

The borehole 364, positioned at a water depth of 18.7 m, revealed in the interval of −72.5 – −83.0 m medium-grained sands with gravels and small pebbles, as well as fragments of unionids and viviparids, which, by analogy with the borehole 365, were attributed to the Upper Poratian. Below, beneath silts of about 2 m thick, the medium-grained sands with *Prosodacna* sp., *Congeria* sp., *Melanopsis* (*L.*) *onusta* Sabba, *Tulotoma* sp. occur and were assigned to the Middle Pridanubian Subformation (−85.0 – −87.0 m). The same mollusc fragments were also found in the underlying grey silty clays.

The geometry of the Pridanubian deposits

The Pridanubian Formation occurs in an erosional valley extending east of the Danube Delta (Fig. 1). The base of the Pridanubian Formation is located between −83.0 – −88.0 m and − 93.0 – −97.0 m, that means the bottom of the erosional depression is uneven, with maximum depth −97.0 m. Its width varies between 10 and 13 km, with a depth of 40 m (Fig. 6, 7). Taking into account the available data of drilling and occurrence of regional subdivisions in the lower reaches of the Danube River, its delta and this study area on the shelf, it can be concluded that the erosional valley is the depression formed by the connection between the Dacian Basin and the Black Sea. The erosion of the upper part of the Pridanubian Formation took place after the accumulation of the Pridanubian Formation, during the erosional phase between the Upper Dacian (Parskovian) and Lower Romanian (Pelendavian).

The northern boundary of the valley was defined by the data from boreholes 363, 356 and 371. In borehole 363 at a depth of −40.5 m Pontian limestone was discovered, and in borehole 356 Pontian clays with limestone interlayers were recorded at the depth of −41.5 m.

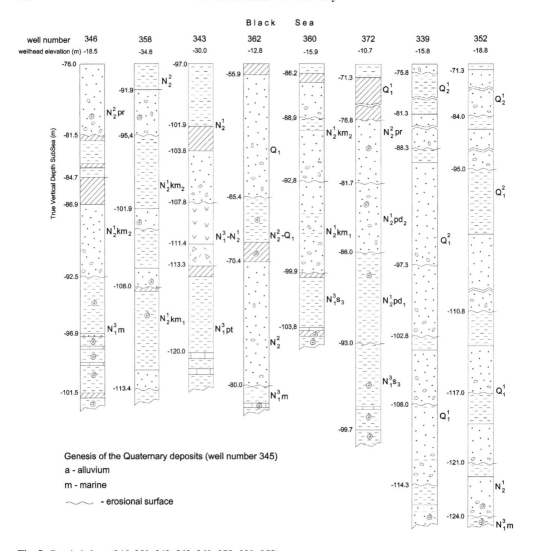

Fig. 5. Borehole logs (346, 358, 343, 362, 360, 372, 339, 352).

The southern boundary of the valley is defined by borehole 366. Here, the Pontian deposits occur at a depth of −59.0 m under the red-coloured eluvium. To the north of the erosional valley, Pontian sediments occur in areas that have been preserved from erosion at depths of −38 − −50 m. To the south of it, the surface of the Pontian sediments varies from −60 to −90 and −113 m (boreholes 366, 373, 343). It should be noted that these deposits do not have complete thickness in all of the boreholes.

During the Pliocene, the erosional valley was shaped as a deep and wide bay of the sea, the mouth of which was located east of the meridian of the Alibey Lake (Fig. 1). The available material does not permit us to define a more accurate distribution of the Pridanubian valley due to the small number of boreholes and low per cent core recovery and, consequently, discontinuous palaeontological sampling. Unfortunately, this also affected the detailed subdivision of the Cimmerian and Dacian deposits.

East of the supposed mouth of the valley at the depths of −88.00 − −99.00 m, (boreholes 341, 344, 345, 358) occur the sediments with the Cimmerian molluscs. It is likely that the mouth of the valley migrated along the east–west direction during the regression–transgression processes linked with the evolution of the Cimmerian basin. With a decrease in the level of the Cimmerian Sea and the discharge of the waters of the Dacian Basin, the Dacian

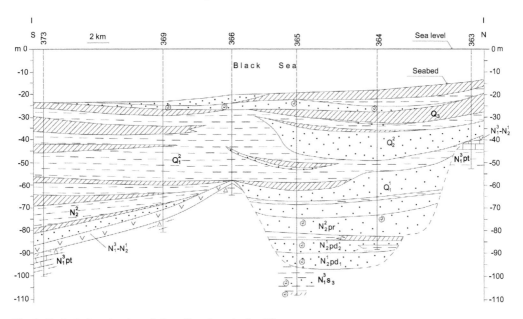

Fig. 6. Geological section through the valley along the line I-I.

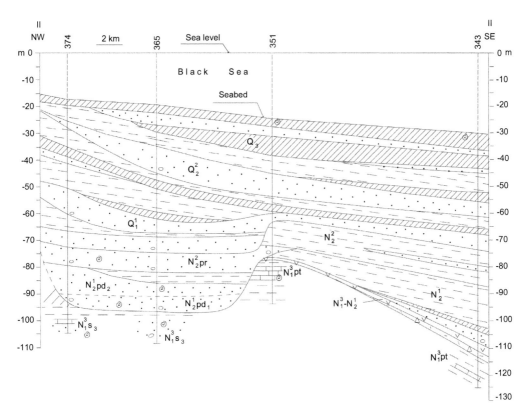

Fig. 7. Geological section through the valley along the line II-II.

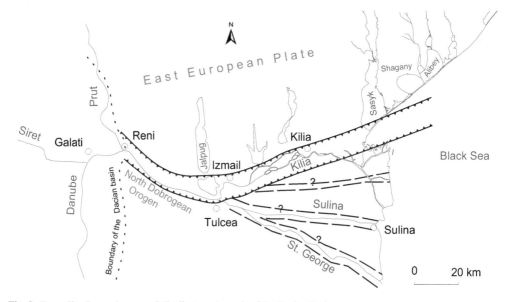

Fig. 8. Run-off valley and supposed distributary channels of the Dacian Basin.

molluscs migrated to the Euxinian Basin, while an increase in the Cimmerian sea-level resulted in the Cimmerian molluscs migrating from the Euxinian to the Dacian Basin along the valley.

The boundary of the distribution of the Dacian deposits to the east of the city of Reni, as well as the position of the northern slopes of the Dobrudja Mountains indicates that there was a single wide waterway connection of the Dacian waters down towards the area of the present day city of Tulcea. Eastward of Tulcea there were several channels in the area of Kiliya, Sulina and probably another one in the south (Fig. 8), similar to what was described by (Gillet 2004; Popescu *et al.* 2004)'s data on the presence of distributary channels east of the Danube Delta.

Cimmerian and Lower Kujalnician deposits

Of particular interest are the marine Pliocene deposits outside the limits of the erosional valley. Their bedding conditions and composition change laterally and with depth and was determined by the palaeorelief of sedimentation and subsequent hydrodynamic and hydrological processes. In many places the Pliocene deposit thickness is insignificant due to erosion caused by the fall of the sea level. The Cimmerian and Lower Kujalnician ages of these deposits were distinguished by analysis of mollusc fauna.

The Cimmerian deposits of the study area were recorded at the depths of the Pridanubian Formation occurrence and are considered estuarine deposits of the eastern end of the erosional valley. In boreholes 345 and 360 the base of the Cimmerian is located at depths of −96.7 to −99.9 m. To the south and SE, it reaches −113.4 m (borehole 358) and to the NE it increases to −69.0 m (boreholes 341 and 344). The Cimmerian deposits are represented by clays and sands with interlayers of clays. Depending on the completeness of the section, its thickness varies from 7 to 15 m. The following molluscs were identified in the Cimmerian sediments: *Limnocardium squamulosum* (Desh.), *Limnocardium* sp., *Prosodacna* sp., *Pontalmyra multistriata* (Rouss.), *P. crassatelata* (Desh.), *P. crassetelloides* (Andr.), *P. pseudomultistriata* (David.), *Pseudocatillus danacoides* (Andr.), *Ps. ex gr. lebedinzevi* (Andr.), *Ps. zlatarskii* (Andr.), *Ps. ex gr. subpolemonis* Ebers., *Dreissena theodori* Andr., *Dr tenuissima* Sinz., *Dr polymorpha* Pall., *Didacnomya* sp., *Hydrobia* cf. *melanoides* Sinz., *Viviparus achatinoides* Pall., *Melanopsis* sp., *Congeria* sp., *Micromelania* sp., *Unio* sp.

According to the given list of molluscs, the subdivision of the Cimmerian Regional Stage into substages is problematic and therefore the deposits were shown in the sections as undifferentiated Lower–Middle Cimmerian. However, in some sections (boreholes 358 and 345) at the base of the Cimmerian there are species also known from the Pontian: *Pontalmyra incerta, Valvata biformis* and *Eupatorina littoraris*. The fossil assemblage was the basis for assigning these deposits conditionally to the Lower Cimmerian. The basis to distinguish the Upper Cimmerian deposits is absent. Apparently, at the time of their formation, there was a sea level fall, and continental sediments accumulated on the marine rocks (boreholes 346 and 345).

The Upper Pliocene is represented by the Lower Kujalnician deposits, overlying the Cimmerian. These are clays and sands of 5–6 m thick. In borehole 345 they contain *Pontalmura* sp., *Prosodacna* sp., *Lithoglyphus neumayri michaeli* Cobalc., *Valvata aff. vinciana kubanika* Krest., *V. piscinaloides* Mich., *Dreissena* sp. and *Micromelania* sp. The Upper Kujalnician sediments are mainly represented by alluvial, lagoonal sediments with *Viviparus subconcinus* Sinz., which are correlated with alluvium of the IXth terrace of the Dniester and Danube rivers (Bukatchuk *et al.* 1983; Gozhik 2006) and are classified as the Lower Pleistocene.

Complex of red-brown palaeosols

The red-coloured eluvial formation, described by drilling as red-brown (Scythian) clays (Fig. 9), represents a complex of red-brown palaeosols, the formation of which took place since the Upper

Fig. 9. Red-brown (Scythian) clay (core sample photo, borehole 343, 108 m below sea-level). Photo by V.E. Rokitsky.

Miocene and continued through the Early Pliocene. By analogy with the sections of the red-coloured formations distributed on-shore, the Lubimov,

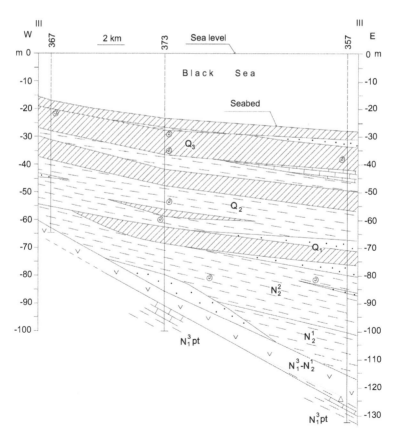

Fig. 10. Geological section through the valley along the line III-III.

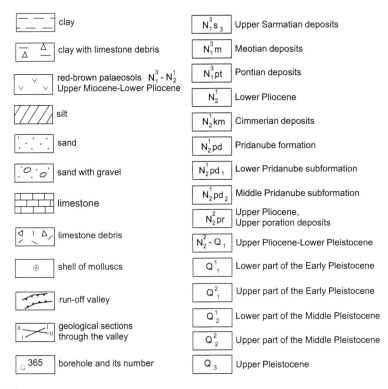

Fig. 11. Legend.

Sevastopol palaeosols (Veklich 1982; Gozhik 2019) can occur in the drilled deposits. However, considering the thickness of the eluvial formations and the conditions of their bedding, we cannot assert the continuity of their accumulation. It seems likely that only in the sections where they occur on the weathering crust of the Pontian limestones (Fig. 10) or with a gradual transition from greenish-grey clay, one can speak of a relatively complete depositional section. This is supported by the data from boreholes 343, 373, 357, 347a and 369. The other boreholes show the thickness of eluvial formations from 1 to 2 m, and they are erosively overlain by the alluvial deposits of the Early Pleistocene. The subdivision of these rocks by non-palaeopedologists is difficult, so in our sections we used $N_1^3 - N_2^1$ indexing. The surface of the red-coloured eluvial formations decreases southwards from $-40 - -50$ m north of the erosional valley to -116 m (borehole 357) to the south of it.

The fall in sea-level at the end of Cimmerian time caused the formation of deep valleys and accumulation of alluvium, comparable to the XI terrace of the Dniester and Danube rivers (Bukatchuk *et al.* 1983; Gozhik 2006, 2019). The alluvium contains the Late Poratian and Early Kujalnician molluscs (*Viviparus turritus* and *Rytia bielzi*).

The fall in sea level was the most significant reason for sedimentation during the Early Gurian time (1.8–1.6 Ma). According to various sources, the sea level fall varied from 150 to 200 m below the current sea level (Semenenko 1984). Indirectly, this is evidenced by the occurrence of the Gurian molluscs on the continental slope of the Black Sea.

In the study area, this regression caused deep erosion of the underlying Cimmerian and even Pontian (not Pantikapian) sediments. An illustrative section in this respect is borehole 339, where the basal horizon of alluvium occurs at depths of more than -110.0 m, and the thickness of alluvium is 20–23 m. In the borehole 352 the basal horizon of alluvium occurs at depths of -124 m. The age of this alluvium is identified by the presence of *Viviparus subconcinus* Sinz., which is a characteristic form of the Late Kujalnician.

To recreate the history of the dynamics of the Eastern Paratethys basins, the section of the Duab layers in the Transcaucasian is very important. At the base of this section, above the *Moquicardium* layers, there is a gap in marine sedimentation, fixed by the erosion facies and comparable with the erosion phase at the end of the Bosphorian time in the Dacian Basin. In the lower part of the section (Lower Cimmerian) there occur the species characteristic of the Gettian deposits (*Parapachy-dacna cobalescui*) and viviparids of the Middle Palludian deposits of Slavonia (*Viviparus (Protulotoma) sadleri, V. (P.) dezmaniana, V. (P.) cyrtomaphora. (P.)*) (Anistratenko and Gozhik 1995). These species were also found in the Lower Pridanubian Subformation. *Pontalmyra crrassatellata, P. multistriata* and *Limnocardium fervidum orsa*, as well as the Dacian *Caladacna steindachneri* and *Carticoncha* sp. were registered in the Middle Cimmerian deposits of the Duab section. The upper part of the Duab beds contains the molluscs characteristic of the Pantikapaean Regional Substage of the Cimmerian, according to Eberzin (1940).

Conclusion

Analysis of the data obtained from the core of the boreholes on the western Black Sea Shelf east of the Danube Delta, as well as the data on the structure of the Pliocene formations of adjacent on-shore deposits, allow the reconstruction of sedimentation on the Black Sea Shelf during the Pliocene.

First of all, we note that the sedimentation of marine Pliocene deposits was formed under the influence of the transgressive–regressive stages of the Euxinian and Dacian basins. During the regressive stages, when non-marine environments were established on large areas of the shelf, eluvial red-coloured formations (palaeosols) and alluvial deposits were formed.

The valley, which was incised due to the water flow from the Dacian Basin spilling into the Black Sea, began to form during the regression of the basins of the Eastern Paratethys at the end of the Bosphorian time of the Pontian and continued throughout the entire Early Pliocene. Lowering the level of the Euxinian Basin intensified erosion activity (forming a 40 m deep and 10–13 km wide valley), while sedimentation prevailed during transgression (infilled the valley).

The available material indicates that the runoff of the Dacian Basin into the Euxinian Basin is compared with the time of the formation of the coarse-grained sediments among the pelagic deposits (Hsü and Federico 1979) and the formation of the erosion facies between the Pontian and Cimmerian sediments (Semenenko 1987). Formation of the runoff valley and accumulation of the Pridanubian Formation was carried out in the northeastern part of the Dacian Basin in the Galati-Reni area (Galati gateway) where, due to the persistent subsidence during the Pliocene, the palaeo-rivers that predated the Danube, Prut, Siret and Buzan converged.

The presence of the Dacian and Cimmerian molluscs in the Pridanubian Formation indicates intensive fauna migration between the basins: during the

regression, the occupation of the Euxinian Basin by the Dacian molluscs, and during the transgression, moving of the Cimmerian molluscs into the Dacian Basin. At the time of the maximum of the Cimmerian transgression, the entire present Danube Delta was an extensive bay of the sea. This facilitated the mutual migration of the Cimmerian and Dacian molluscs (Eberzin 1940; Bogachev 1961; Andrusov 1963; Papaianopol 1975; Semenenko 1987).

The Cimmerian deposits outside the valley are represented by undifferentiated Lower–Middle Cimmerian sediments, of variable thickness due to the later erosions. There is a steady decrease (lowering) in their surface southwards, caused by neotectonic subsidence.

The Lower Kujalnician deposits overlay the Cimmerian ones conformably (without a noticeable gap) and are replaced by the alluvial sediments of the XIth terrace along the strike.

The eluvial red-coloured formations are a complex of superimposed palaeosols of red-brown colour.

Acknowledgements We would like to express our gratitude to the marine geologists of Prichornomor-skeSRGE M. G. Sibirchenko (senior geologist), A. B. Guslia, N. V. Dzemidko, who participated in marine research, as well as the drilling technicians and the crew of the research vessel *Diorite*.

Author contributions PFG: conceptualization (equal), investigation (equal); **VYR**: conceptualization (equal), investigation (equal).

Funding Funding for this research came from the state budget of Ukraine.

Data availability The datasets analysed during the current study are stored in the Prichornomorske SRGE Funds pgrgp@ukr.net in electronic form and in hard copy and are available upon reasonable request. Online access is not supported. You can also contact the authors.

References

Andreescu, I. 2009. When did the Danubes become the Danube? *Muzeul Oltenici Craiova, Oltenia. Studii şi comunicări Ştintele Naturii.*, **XXV**, 319–328.

Andreescu, I. and Papaianopol, I. 1975. Dacian. Stratotypes of the Neogene Mediterranean Stages. *Bratislava*, **2**, 57–70.

Andreesku, I., Codrea, V., Lubenescu, V., Petculescu, A. and Stiuca, E. 2010. New developments in the Upper Pliocene–Pleistocene stratigraphic units of the Dacian Basin (Eastern Paratethys), Romania. Quaternary stratigraphy and paleontology of the Southern Russia: Connections between Europe, Africa and Asia. *2010*

Annual Meeting INQUA – SEQS, 21–26 June, Rostov-on-Don, Russia, 21–23.

Andrusov, N.I. 1963. *Upper Pliocene of the Black Sea basin. The Selected Papers.* Academy of Science of USSR, Moscow, **2**, 583–607 [in Russian].

Anistratenko, V.V. and Gozhik, P.F. 1995. Molluskans of the Families *Neritidae, Viviparidae, Lithoglyphidae, Pyrgulidae* from Kimmerian Deposits of the Abchasia. *Vestnik zoologii*, **1**, 3–13 [in Russian].

Atlas litofacial 1969. *VI – Neogene. – 1:2 000 000.* Institutul Geologic, Bucuresti, 13 volumes.

Bogachev, V.V. 1961. *Materials for the History of the Freshwater Fauna of Eurasia.* NAS of Ukraine, Kyiv [in Russian].

Bukatchuk, P.D. 1985. New data on the fauna of mollusks of the Middle–Upper Pliocene deposits of the South-Western Moldavian. *In: Fauna and Flora of the Late Cenozoic of Moldavia.* Kishinev, Shtiinsa, 116–120 [in Russian].

Bukatchuk, P.D. and Gozhik, P.F. 1986. Middle–Upper Pliocene deposits of the south-west of Moldavia. *Geological Journal*, **46**, 21–26 [in Russian].

Bukatchuk, P.D., Gozhik, P.F. and Bilinkis, G.M. 1983. On the correlations of the alluvial deposits of the Dniester, Prut and Lower Danube. *In: Geology of the Quaternary Deposits of the Alluvial Deposits of the Moldavia.* Kishinev, Shtiinsa, 35–70 [in Russian].

Chepalyga, A.L. 1967. *Anthropogenous Freshwater Mollusks of the Russian Plain South and their Stratigraphic Significance.* Nauka, Moskva [in Russian].

Clauzon, G., Suc, J.-P., Popescu, Sp.-M., Marunteanu, M., Rubino, J.-L., Marinescu, F. and Melinte, M.C. 2005. Influence of Mediterranean Sea-level changes on the Dacic Basin (Eastern Paratethys) during the late Neogene: the Mediterranean Lago Mare facies deciphered. *Basin Reseach*, **17**, 437–462, https://doi.org/10.1111/j.1365-2117.2005.00269.x

Eberzin, A.G. 1940. Middle and Upper Pliocene of the Black Sea basin. *Stratigraphy of the USSR*, **12**. Neogene, M., L., 477–566.

Ghenea, C. 1970. Stratigraphy of the Upper Pliocene–Lower Pleistocene Interval in the Dacic Basin (Romania). *Palaeogeography, Palaeoclimatology, Palaeoecology*, **8**, 165–174, https://doi.org/10.1016/0031-0182(70)90090-X

Gillet, H. 2004. *La stratigraphie tertiaire et la surface derision messinienne sur les merges occidentals de la Mar Noise: stratigraphie sismique haute resolution.* Thesis, L'Universite de la Bretagne Occidental.

Gozhik, P.F. 2006. *Fresh-Water mollusks from the late Cenozoic in the south of Eastern Europe. Part 1. Superfamilia Unionidea.* Kyiv [in Russian].

Gozhik, P.F. 2019. On the lower boundary of the Quaternary System in the Azov–Black Sea basin. *Journal of Geology, Geography and Geoecology*, **28**, 292–300, https://doi.org/10.15421/111929

Gozhik, P.F. and Chyrka, V.G. 1973. *New data on the Prut and the Danube downstreams Pliocene sediments and*

the questions on their correlations. *About lower boundary of Quaternary sediments.* Naukova dumka, K, 66–72 [in Russian].

Hsü, K.J. and Federico, G. 1979. Messinian event in the Black Sea. *Palaeogeography, Palaeoclimatology, Palaeoecology*, **29**, 75–93, https://doi.org/10.1016/0031-0182(79)90075-0

Jipa, D. and Olariu, C. 2013. Sediment routing in a semi-enclosed epicontinental sea: Dacian, Paratethys domain, Late Neogene, Romania. *Global and Planetary Change*, **103**, 193–206, https://doi.org/10.1016/j.gloplacha.2012.06.009

Olariu, C., Krezsek, C. and Jipa, D. 2018. The Danube River inception: evidence for a 4 Ma continental-scale river born from segmented ParaTethys Basins. *Terra Nova*, **30**, 63–71, https://doi.org/10.1111/ter.12308

Orszag-Sperber, F. 2006. Changing perspectives in the concept of 'Lago-Mare' in Mediterranean Late Miocene evolution. *Sedimentary Geology*, **188–189**, 259–277, https://doi.org/10.1016/j.sedgeo.2006.03.008

Papaianopol, I. 1975. Dreissenide ale Stratelor du Duaba in decinul superior din Muntenia. Dari de seama ale sedintelir. 3. *Paleontologie*, **61**, 111–123 [in Romanien].

Pavlov, A.P. 1925. Neogene and Post-Tertiary deposits of Southern and Eastern Europe//Memoirs of the Geological Department of the Society of Natural History, Anthropology and Ethnography Lovers, 217p.

Popescu, S.-M. 2001. Repetitive changes in Early Pliocene vegetation revealed by high-resolution pollen analysis: revised cyclotratigraphy of southwestern Romania. *Review of Palaeobotany and Palynology*, **120**, 181–202, https://doi.org/10.1016/S0034-6667(01)00142-7

Popescu, I., Lericolais, G., Panin, N., Normand, A., Dinu, C. and Le Drezen, E. 2004. The Danube submarine canyon (Black Sea): morphology and sedimentary processes. *Marine Geology*, **206**, 249–265, https://doi.org/10.1016/j.margeo.2004.03.003

Roshka, V.H. and Hubka, A.N. 1982. An outline of the stratigraphy of Neogene deposits between the Dniester and the Prut rivers. *In: Biostratigraphy of the Anthropogen and Neogene of the Southwest of the USSR.* Kishinev, Shtiinca, 78–106 [in Russian].

Ruggieri, G. 1967. The Miocene and later evolution of the Mediterranean Sea. *In:* Adams, C.G. and Ager, D.V. (eds) *Aspects of Tethyan Biogeography.* Systematics Association, London, **7**, 283–290.

Semenenko, V.N. 1984. The Pliocene series. *Geology of the UkrSSR shelf.* Naukova Dumka, Kiev, 141–153.

Semenenko, V.N. 1987. *Stratigraphic Correlation on the Upper Miocene and the Pliocene of the Eastern Paratethys and Tethys.* Naukova Dumka, Kiev [in Russian].

Sinegub, V.V. 1969. The Pliocene. *Stratigraphy of the USSR*, **XVII**, MoldSSR, 171–188 [in Russian].

Veklich, M.F. 1982. *The Paleostages and Stratotypes of Soil Formation of the Upper Cenozoic.* Naukova Dumka, Kiev [in Russian].

Integrated geophysical and sedimentological datasets for assessment of offshore borrow areas: the CHIMERA project (western Portuguese Coast)

Mário Mil-Homens[1]*, Pedro Brito[2,4,6], Vitor Magalhães[2,6], Marcos Rosa[2], Marta Neres[2,6], Marta Silva[4], Emília Salgueiro[2,3], Teresa Drago[2,6], Ana Isabel Rodrigues[1], Miriam Tuaty Guerra[1], Maria José Gaudêncio[1], Eveline Almeida[4], Mariana Silva[4], Mafalda Freitas[2], Celso Aleixo Pinto[5], Cidália Bandarra[1] and Pedro Terrinha[2,6]

[1]Divisão de Oceanografia e Ambiente Marinho, Instituto Português do Mar e da Atmosfera, I.P., Rua Alfredo Magalhães Ramalho, 6, 1495-006 Lisboa, Portugal

[2]Divisão de Geologia e Georrecursos Marinhos, Instituto Português do Mar e da Atmosfera, I.P., Rua C ao Aeroporto, 1749-077 Lisboa, Portugal

[3]CCMAR – Centre of Marine Sciences, Universidade do Algarve, Faro, 8005-139, Portugal

[4]Xavisub – Mergulhadores Profissionais, LDA. Rua José Silva Mariano, 3830-688, Gafanha da Nazaré - Ílhavo, Aveiro, Portugal

[5]Núcleo de Monitorização Costeira e Risco, Departamento do Litoral e Proteção Costeira, Agência Portuguesa do Ambiente, Rua da Murgueira, 9/9A Zambujal, Ap. 7585, 2610-124 Amadora, Portugal

[6]Instituto Dom Luiz (IDL), Faculdade de Ciências da Universidade de Lisboa, Campo Grande Edifício C1, Piso 1, 1749-016 Lisboa, Portugal

MM-H, 0000-0002-1570-2641; VM, 0000-0002-3205-771X; MN, 0000-0003-3939-4636; TD, 0000-0003-3496-3270; MTG, 0000-0003-0178-310X; MJG, 0000-0002-6303-9995; EA, 0000-0001-8769-4583; MF, 0000-0002-9953-4146; PT, 0000-0002-6824-6002

*Correspondence: mario.milhomens@ipma.pt

Abstract: Coastal erosion impact on low-lying sandy shorelines represents a worldwide problem, which is particularly felt in various segments of the Portuguese coast where this geomorphological type represents 42% of its total length. Beach nourishment is a viable engineering alternative for shore protection and the assessment of offshore sources of beach-fill material is an essential aspect when implementing this mitigation strategy. The CHIMERA project carried out a multidisciplinary inspection on four segments of the west Portuguese coast to assess their potential as offshore borrow areas for beach nourishment. Altogether, these segments covered an area of $c.$ 35 km^2, at water depths between 20 and 42 m. They were surveyed using multibeam, sub-bottom profiler, ultra-high resolution multichannel seismics and a set of 126 surface samples and 72 vibrocores (with 3 m long each). To comply with the Portuguese legislation, sand types were assessed by granulometric and chemical analyses for evaluating the quality of sediments in terms of contamination. High-resolution magnetic surveys were conducted to find potential archaeological artefacts. The adopted methodology proved to be adequate to quantify and describe the spatial distribution of useful sediment volumes, supporting the ongoing Integrated Coastal Sediment Strategy for mainland Portugal.

Supplementary material: Classification of sediment's quality according to the Portuguese legislation (Ordinance 1450/2007) in the FFLV area is available at https://doi.org/10.6084/m9.figshare.c.5007266

Portugal has a shoreline 987 km long, geomorphologically dominated by low-lying sandy beaches (42%), rocky cliffs (48%), soft cliffs (2%) and low-lying rocky coast (8%). Sandy beaches have always attracted humans and human activities, having an important economic, social and recreational value. However, in some areas, beaches are becoming extremely vulnerable and at risk of disappearance.

Besides that, beaches are important natural barriers that provide improved protection to upland structures and infrastructure from the effect of storms. The Portuguese coast is undergoing environmental pressures caused by the increase of the population density (major Portuguese cities are located along the coast), by the expansion of the construction of heavy infrastructures (e.g. harbours) and of coastal hard

From: Asch, K., Kitazato, H. and Vallius, H. (eds) 2022. *From Continental Shelf to Slope: Mapping the Oceanic Realm.* Geological Society, London, Special Publications, **505**, 131–153.
First published online July 20, 2020, https://doi.org/10.1144/SP505-2019-100

engineering structures (e.g. groins and revetments behind the jetties), and by the growing of touristic maritime activities. Some sectors of the Portuguese western coast, particularly sandy beaches and soft rock cliffs are affected by the increase in the rates of coastal erosion. The present-day negative sedimentary budget of the Portuguese coast result in a generalized retreat of the coastline (Andrade *et al.* 2015) along approximately 20% of its total length. According to Ferreira and Matias (2013), three main causes contribute to the current coastal erosion scenario in Portugal: increasing storminess, lack in sediment supply and interventions at river basins, coastal engineering structures. These effects are being enhanced climate change, in particular, global warming. One of the main consequences of climate change relevant to the coast are sea level rise and expected changes in wave climate (i.e. increased storminess and changes in prevailing wave direction), that will generate impacts, such as increased coastal erosion and frequency and magnitude of overtopping/coastal flooding. Consequently, the protection of coastlines is a topic of growing concern for the international community (ICES 2016). Beach nourishment is an environmentally acceptable and viable engineering alternative for shore protection and restoration to mitigate coastline retreat and to preserve the socioeconomic occupation of the coast. It is used in emergency situations as a local and short-term solution (i.e. mitigation of short-term erosion induced by storms), or as a regional and long-term management strategy, that is mitigation of installed erosion tendency and vulnerability to sea-level rise (Hamm *et al.* 2002; US AID 2009). In addition to providing protection to valuable areas of the territory from an environmental and strategic point of view, artificial beach nourishment also conserves the natural state of the beach, while enhancing its recreational use. In certain situations, beach nourishment has the sole objective of improving the comfort of its users, either by increasing the area available to beach activities (Vera-Cruz 1972) or by changing the grain size of its sediments (Anthony *et al.* 2011). Beach nourishment can be assured by terrestrial sources (e.g. dredged estuarine and lagoon sediments) or by extracting marine aggregates, which have grown in importance due to the increase of land-use constraints and depletion of terrestrial aggregate resources (Van Lancker *et al.* 2017). It involves the addition of sand from an offshore borrow area to expand an eroding coastal segment, respecting the natural hydrodynamic regime. The feasibility of this type of intervention is controlled by local geomorphological (e.g. variations in the orientation of the coastline, sediment grain size), morphodynamical (e.g. long-term erosion, wave-energy magnitude, degree of beach exposure and location, length, cross-shore and longshore sediment transport rates) and anthropogenic

constraints (e.g. coastal engineering structures, beach nourishment at adjacent beaches) (Davis *et al.* 2000). However, it is important to highlight that the benefits of using marine sand and gravel for beach nourishment must be balanced with the potentially significant environmental impacts (ICES 2016).

As a result, the degree of success of a beach nourishment project varies widely and is site-specific, lasting from only a few months to several decades (Pinto *et al.* 2020). Additionally, the sand supply must fulfil the mandatory physical, chemical and environmental quality requirements of the Portuguese regulations.

To meet the requirements, detailed exploration and characterization of the source material must be done, supported by geological models based on a diverse geophysical and sampling dataset (bathymetry, reflection seismic, magnetics, sedimentology and geochemistry). This leads to an understanding of the internal variability of the aggregate deposits, not only at the seafloor, but also their variability in depth. The later requires the characterization of the borrow area in terms of their 3D geometry, physical and chemical composition, and quantification of the volume of sand and gravel that can be dredged and effectively mobilized.

The aim of the present study is to identify and to characterize (e.g. quality and quantity) the sedimentary resources on the continental shelf with a potential to be nourishing areas for future beach nourishments in adjacent coastal sectors that are eroding. This was reached through the use and integration of diverse methodological contributions to assess the sand deposits available offshore. Four case study areas along the western Portuguese inner shelf, namely Espinho-Torreira (ET), Barra-Mira (BM), Figueira da Foz-Leirosa (FF) and Costa da Caparica (CC) were selected, and their potential as sand resources (natural finite resource that must be used responsibly and appropriately to be sustainable (ICES 2016)) were evaluated. These four areas correspond to areas previously identified as potentially submitted to high-magnitude beach nourishment interventions (Andrade *et al.* 2015). These potential nourishment areas were subject to increasingly detailed surveys and studies to characterize (in terms of dimension and sedimentological characteristics) the deposit as a sand resource. This study was carried out in the context of the CHIMERA project, developed by the Portuguese Environment Agency (APA, IP) and co-funded by the POSEUR (Operational Programme for Sustainability and Efficient Use of Resources, established through an Execution Decision from the European Commission on 16 December 2014) programme (POSEUR-09-2016-48-FC-000030). The data obtained were included in an interactive spatial data infrastructure (webGIS), which provides georeferenced multi-layer cartography at a scale of the study area, using data visualization tools to be utilized as a support tool in the implementation of these proposed beach nourishment programmes.

Geological setting

The four study areas located in the north and central Portuguese continental shelf, where the sedimentary transport and deposition are the result of a complex interplay between continental and oceanic factors (Fig. 1). The sources of sediments to these study areas are the Precambrian and Paleozoic igneous and metamorphic rocks of the Variscan Belt and the Meso-Cenozoic sedimentary rocks from the Lusitanian Basin, the Mondego and the Lower Tagus Basins (Mougenot 1988). The hydrographic basins that drain and supply sediments to this part of the shelf have a temperate climate, with the rivers' discharge peaks in winter. North of about 41° N, the Douro and other rivers have large discharges and steep gradients, and because of that, much of the sediment reaches the continental shelf directly. The Portuguese coast is divided into eight cells according to its geomorphological characteristics and sedimentary dynamics (Andrade et al. 2002, 2015). The ET and BM are in distinct sectors of sub-cell 1b and, FF and CC borrow areas are in sub-cell 1c and cell 4 (Andrade et al. 2015). In the west coast, wave regime predominantly from west–NW, induce net littoral southward drift, with a potential longshore sediment transport from 1 to 2×10^6 m^3 a^{-1} (e.g. Oliveira et al. 1982; Vicente and Clímaco 2012). This potential transport is variable along the west coast, according to the coastline orientation, decreasing southward due to lower wave energy (Ferreira and Matias 2013). This is the main sediment transport regime adjacent to the ET, BM and FF borrow areas. South of 39° N, the fluvial transport is less vigorous, and sediments have longer residence time, especially in estuarine traps at the river mouths (such as the Tagus and Sado rivers). Adjacent to the CC borrow area, sediment dynamics is more complex due to the influence of the Tagus river ebb-delta (Taborda and Andrade 2014) with a net longshore drift towards the north in order of 10^5 m^3 a^{-1} (Taborda et al. 2014). The tides along the west coast are semi-diurnal and mesotidal, having the average neap tidal range of 1.0 m and average spring tidal range of 2.8 m (Ferreira and Matias 2013).

The western Portuguese continental shelf is relatively narrow and deep (Fig. 1) with an average width of about 45 km and an average shelf-break depth of 160 m (Mougenot 1988). Along the different study areas, the shelf ranges between 45 and 58 km in ET, BM and FF areas and is narrower (25 km) in the CC area. Water depths (relative to Hydrographic zero) in ET, BM, FF and CC range from 24–30, 25–34, 30–42 and 20–30 m, respectively. The shelf has, in general, a gently dipping surface. The west Portuguese shelf break is indented by several submarine canyons, notably the Nazaré canyon (located south of the FF area, Fig. 1), which nearly extends to the shore and acts as a barrier for the southward alongshore sedimentary transport, as most of the sediments are caught within the canyon head and are transported to the offshore.

The sedimentary deposits of the West Portuguese continental shelf are characterized by the dominance of sand-sized detrital grains up to 80 m MWD and biogenic carbonate grains that predominate below (Dias and Nittrouer 1984; Dias and Neal 1990). The nearshore deposits of the inner shelf, down to 30 m water depth, are dominated by well-sorted fluvial sands, that are predominantly transported southward by the dominant waves and currents. The mid-shelf, between 30 and 80 m mean water depth (MWD), is dominated by coarse sand and gravel deposits resulting from littoral processes during the Holocene transgression (Dias et al. 2002). The outer shelf deposits (>80 m MWD), are carbonate-rich sands with shell fragments dominating landward and finer oozes dominating seaward. The shelf break deposits (>150 m MWD) are dominated by very fine, well-sorted foraminifera sand (Dias and Nittrouer 1984). As the average sea-level was stabilized at about 3.5 ka BP (Dias et al. 2000) the evolution of the shelf sedimentation processes and coastline location has been essentially conditioned by the sedimentary balance (Andrade et al. 2015). Periods of accretion/progradation or erosion/retreat are associated with excess or deficit of sediments, leading to the migration of the coastline towards the sea or towards inland, respectively.

Material and methods

Seafloor characterization

Multibeam bathymetry and backscatter data were acquired with a Teledyne RESON T50-P multi-beam system on board of IPMA research vessels 'Noruega' and 'Diplodus'. The positioning of the vessel was controlled by an integrated system (Applanix POSMV Ocean Master) that combines the satellite positioning data received via 2 GNSS antennas and data from one inertial unit (IMU) mounted close to the multibeam transducers. During the surveys, real-time RTK corrections of the SERVIR network of CIGeoE (centimetric precisions) were used, complemented with Fugro MarineStar DGPS corrections. Navigation control was done using the Teledyne PDS2000 software. The correction for the Portuguese hydrographic vertical datum, located 2 m below the mean sea-level, was parameterized in the PDS2000 program. All depth measurements in the survey areas refer to this vertical datum.

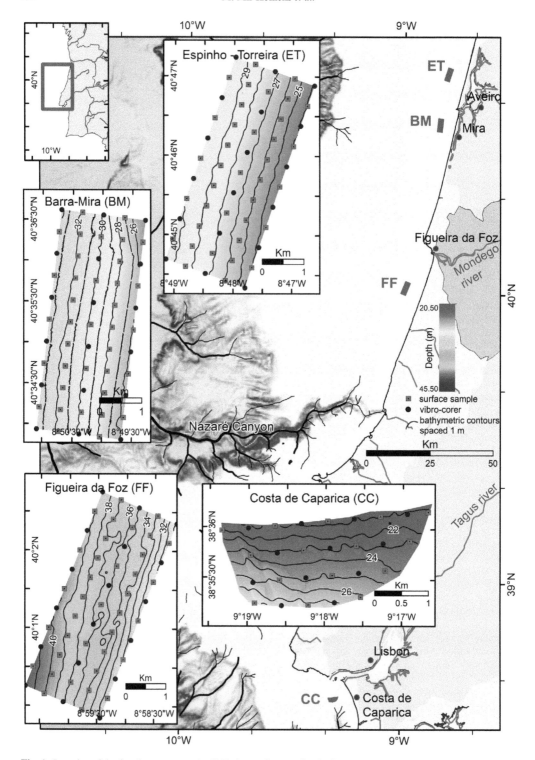

Fig. 1. Location of the four borrow areas. (**a–d**) Bathymetric maps for the four areas.

A Teledyne RESON T50-P multi-beam system was used for the acquisition of bathymetry and backscatter data. Sound velocity profiles were acquired during the surveys, at least twice per each survey session, using a manually deployed Odom Digibar-S SVP. The multibeam system was calibrated following standard patch test procedures.

To ensure the high resolution requested for the bathymetric survey (average resolution of 16 point/m^2), a maximum angular swath width between 110 and 100° was set, adjusted according to depth (shallower depths, greater the aperture). This parameterization maximized the acoustic signal emission rate (between 12 and 13 pulse/s) to obtain an average resolution of 25 cm along track (survey speeds between 4 and 5 kn). The resolution obtained along the swath varied between 15 and 20 cm. The main reason for the high-resolution request was to obtain a detailed seabed characterization.

The overlap between successive lines covered by the swath of the multibeam was 10% or higher. The bathymetric surveys achieved IHO Special Order specifications. The Reson T50-P multibeam system features the 'normalized snippets' functionality, which automatically normalizes the backscatter data, generating a magnitude signal compensated for the characteristics of the sonar. According to the experience already obtained in the operation of this multibeam system in former cruises made by IPMA, the frequency of 350 kHz was adopted for the acoustic signal emitted (CW) for all the surveys, allowing the acquisition of high resolution bathymetric data and backscatter data with strong signal dynamics.

Bathymetry and backscatter data processing. The navigation/position data were post-processed in POSPAC software, using the publicly available Portuguese network of GNSS base stations in order to further improve the position accuracy.

The bathymetric data was processed with Teledyne PDS2000 software. The collected SVP data was applied to each survey line, artefacts and outliers where identified and removed and the final gridded bathymetry maps (25 cm resolution) were produced. The backscatter data was processed with Fledermaus FMGeocoder Toolbox software. The processing flow included the following main steps:

- backscatter data and processed bathymetric data exported from Reson PDS files to GSF format;
- definition of sonar type and characteristics (e.g. signal frequency, beam count, gain levels);
- automatic backscatter coverage geometry calculations (using the final, corrected, bathymetry);
- automatic filtering;
- statistics calculations;
- 2 m resolution mosaic production.

Subsurface imaging

Subsurface imaging was obtained with a parametric echosounder (PE) and an ultra-high resolution multi-channel seismics (UHRS). A total of about 360 km of 2D seismic lines were acquired with each system, to survey the four study areas with quadrangular 250 m spaced survey grids. The PE used was an Innomar SES-2000 Standard unit, deployed on a vessel side pole with the PE transducer and the inertial motion unit (IMU) mounted side by side at about 1.3 m below the water line. Heave compensation and roll correction based on the IMU information, were applied on-line to the seismic data. The UHRS system used was a Geo Marine Survey Systems spread, with a sparker GEO-SOURCE LW200 tips, a 1 kJ GEO-SPARK pulsed power supply and a 24 channel GEO-SENSE LW streamer with 3.125 m single element group spacing.

Seismic data processing. The implemented procedures for the seismic data processing considered two phases. The first phase consisted in the quality control (QC), carried out on-board for each line, aiming to discriminate if the acquired data qualified for further processing or if a line rerun was needed. The second processing phase carried out at the office, aimed to produce seismic sections suitable for the interpretation of the upper sedimentary layers' architecture and to identify the top of the bedrock basement.

On-board QC. The QC of the seismic data focused on assuring not only the quality of the seismic signal but also the navigation. The quality of the navigation was controlled by exporting the positioning information stored in the seismic files and overlaying it on the planed seismic lines maps. For the UHRS the vessel navigation information was integrated with the seismic spread positioning information retrieved from the source and receiver DGPS buoys. This integration resulted in filtered and smoothed navigation data, discriminating the positioning for each shot of the seismic source, the several receiver channels and the streamer feathering angle. After the navigation processing, shell scripts were used to generate plots for each line to access variations in the feathering angle, shooting interval, vessel course and tidal corrections. To control the seismic signal quality of the PE data, the ISE2 software (from Innomar) was used to apply a processing flow that comprises a noise filter, static corrections and a smoothing algorithm. To control the seismic signal quality of the UHRS data, the Deco Geophysical RadExPro software was used. The processing followed general industry standard QC procedures (e.g. Dondurur 2018) considering a flow with the following eight steps:

(1) importing of SEG-Y files and positioning data;
(2) geometry attribution and common depth point (CDP) binning;

(3) assessment of missed shots;

(4) validation of computed offsets by overlaying it with the direct arrival;

(5) assessment of the streamer depth and its nominal correction;

(6) interactive velocity analysis;

(7) band pass filter, amplitude correction, normal moveout (NMO) correction and stack;

(8) analysis of the brute stack frequency spectrum, static tidal correction and exportation of files with relevant data for further processing.

Data processing for interpretation. The seismic data processing for interpretation of both systems (PE and UHRS) was made with the software RadExPro. The processing flow applied to the UHRS data, followed the general flow proposed by Duarte *et al.* (2017). Vertical and horizontal resolutions of the datasets prepared for interpretation were assessed both graphically by minimum reflection individualization and analytically by estimation of the central frequency and the sound velocity. Estimated vertical and horizontal resolutions, sub-bottom signal penetration and other relevant characteristics of the PE non-migrated and UHRS migrated datasets prepared for interpretation are shown in Table 1.

Interpretation methodology. The aim of the seismic interpretation was twofold: (1) to establish the sedimentary architecture and to evaluate the volume of the sub-bottom uppermost unconsolidated sediment layers, and (2) to cross correlate seismic lines with vibrocores' description and evaluate volumes suitable for beach nourishment. The combined interpretation of the PE and UHRS datasets allowed the use of PE data to image and map the upper sediment layers (up to 3–5 m deep) and correlate them with the sampling data, while relying on the UHRS to image the deeper sedimentary structures until the bedrock substrate. Seismic stratigraphic interpretation was done accordingly with the general classical principles presented in Payton (1977). The number of

Table 1. *Main characteristics of the interpreted PE (non-migrated) and UHRS (migrated) seismic datasets prepared for interpretation*

Characteristic	Estimated value	
	PE	UHRS
Shooting rate (Hz)	8–12	2
Sampling rate (kHz)	83	10
CDP bin (m)		1.56
Central frequency (kHz)	10	0.8–1.2
Sub-bottom penetration (m)	3–7	35–120
Vertical resolution (m)	0.04–0.25	0.1–0.5
Horizontal resolution (m)	1.7–2.9	1.6–3.2

seismic units (SU) was the minimum to ensure internal and stratigraphic coherence across each study area. Estimation of the depth and volume of the interpreted SU, were made from computation between surfaces interpolated from the picked seismic horizons, converted from two-way-time (ms) to space (m) using a mean sound spreading velocity of 1507 and 1700 ms^{-1} for the water and sediment columns, respectively.

Sediment analysis

In Portugal, sediments of borrow areas must comply with the national regulation system (Ordinance 1450/2007, 12 November) for the assessment of dredged materials that involves the determination of a set of physical and compositional parameters (grain size, organic carbon (C_{org}), total solids content and density), including contaminants (trace metals (As, Cd, Cr, Cu, Hg, Pb, Ni and Zn) and persistent organic pollutants (POPs) (hexachlorobenzene (HCB), polychlorinated biphenyls (PCBs), polycyclic aromatic hydrocarbons (PAHs))). Extracted material is classified into five categories depending on the degree of contamination: class 1, clean dredged material; class 2, slightly contaminated; class 3, moderately contaminated; class 4, contaminated; and class 5, very contaminated. These classes represent increasing concentrations on the trace metals and POPs (see Table S1 in the Supplementary material).

Sediment sampling. Surface sediment samples and cores were collected using a Van Veen grab with a sampling area of 0.1 m^2 and a vibrocoring system, respectively. The vibrocorers were recollected with the support of professional divers at the seafloor. All samples were collected in a regular mesh, spaced 0.5 km for surface samples and spaced 1 km for vibrocores, without overlap between them (Fig. 1). The sample density of surface and vibrocores is 22 and 7.2 sample/km^2, respectively.

On board, the surface samples were photographed, described, subsampled for POPs, grain size, C_{org}, density and CaCO$_3$, trace metals and total solids content. Sub-samples were stored in refrigerated conditions. Vertical cores were sectioned in 1.5 m sections, sealed and labelled. They were transported horizontally to the laboratory and maintained in refrigerated conditions until opened, described, photographed and subsampled every 25 cm for grain size and CaCO$_3$ content analyses. The archive half section of each core was stored and preserved in refrigerated conditions.

Grain-size analysis. Subsamples for grain-size analysis were dried at 100°C. Approximately 100 g of dried sample were sieved using an analytical sieve shaker (Retsch AS 200, for 2 min and at 0.9 mm of amplitude), with a set of 13 stainless-steel sieves (from 63 μm (4 phi) to 4 mm (−2 phi) with an interval of 0.5 phi). The data were processed using GRADISTAT software v 8.0 (Blott and Pye 2001). The program performs the calculation of the statistical parameters of the granulometric distribution of the sediment samples (mean grain size, standard deviation, skewness, kurtosis) using the graphical (Folk and Ward 1957) and moments (Friedman 1961) methods. Moreover, it provides the physical description of the textural group and its graphical representation in a triangular diagram, as well as the percentage distribution of the granulometric fractions of the samples according to a scale adapted from Udden (1914) and Wentworth (1922) and respective cumulative curves, allowing for the classification of the sediment types.

Density and total solids contents. The density of the sediments was determined by the ratio between the weight and the volume of the wet sediment samples (Flemming and Delafontaine 2000), corresponding to the average of the determinations in five replicates per sediment sample (results were expressed in g cm^{-3}). The content of total solids (in %), was determined by the ratio between the weight of the samples after oven drying at 100°C until reaching constant weight and their wet weight.

CaCO$_3$, C$_{org}$ and trace metal determinations. The subsamples for the determinations of the C$_{org}$, CaCO$_3$ and trace metals were frozen and lyophilized. They were then sieved through a 2 mm square mesh sieve; the lower fraction was milled using agate pots in a Fritsch Pulverisette 7 Classic Line planetary mill. The CaCO$_3$ content was determined for each sample (*c.* 2.5 g) using the volumetric method (Eijkelkamp calcimeter) by measuring the CO$_2$ volume released by the reaction of the sample with 7 mL of HCl.

The content of C$_{org}$ was determined using the equipment Leco Truspec micro-analyser CHNS. The content of C$_{org}$ is calculated by the difference between total carbon content of the ground sample and the carbon content of the calcinated sample. In summary, for each ground and calcinated sample the average values of three and two readings were considered, respectively. The content of inorganic carbon was obtained by combusting the organic matter of each sample in a muffle furnace for 3 h at 400°C.

For the analysis of As, Cd, Pb, Cu, Cr, Ni and Zn, the sediments were partially decomposed by microwave (CEM Mars-XPRESS) with a mixture of HNO$_3$ and H$_2$O$_2$ in several steps at 95°C according to EPA method 3050B (US EPA 1996). With this methodology we intended to access the most bioavailable fraction of trace metals in the sediments. Trace metal contents were determined by induced

plasma coupled mass spectrometry (ICP-MS; Thermo Elemental – X series). The elemental contents were determined by calibration curve interpolation. The limits of quantification (LQ) for this set of trace metals were obtained by regression of the points of the calibration curve. The determination of mercury (Hg) was carried out directly by thermal decomposition coupled to atomic absorption spectrometry (ET-AAS; Leco AMA 254 Mercury analyser) (Costley *et al.* 2000). The LQ for Hg was obtained considering the ratio between the sum of the average value of the blank reads and ten times the standard deviation of the readings of the blanks and the average mass of the sediment used. QC was ensured by blank analysis, certified reference material (CRM) MESS-3 (National Research Council of Canada) and duplicate samples that underwent the same type of preparation of the remaining samples. Blank samples were used to infer contamination during the analytical procedure, whereas the CRM and the duplicates allowed to evaluate the accuracy and precision of the analytical methodology used.

POPs analyses. The sub-sample for POP determinations were dried at 40°C. POPs were extracted using the ASE equipment (Dionex, ASE 200 Accelerated Solvent Extraction) with dichloromethane/hexane (1:1, v/v) at 100°C and 1500 psi; two cycles of 5 min. The following polycyclic aromatic hydrocarbon (PAHs) compounds were quantified: benzo(a)-pyrene (BaP), benzo(ghi)pyrene, indeno(1,2,3-cd) pyrene (InP), benzo[b]fluoranthene (BbF), benzo[k] fluoranthene (BkF), benzo(ghi)-perylene (BgP), anthracene (A), benzo[a]anthracene (BaA), phenanthrene (Phe), fluoranthene (Flu), pyrene (Pyr), chrysene (Chr) and naphthalene (N). For PAHs quantification, extracts were first treated with metallic copper for *c.* 12 h, then purified by column chromatography on silica/alumina (1:1), eluting with hexane and 9:1 and 4:1 hexane/dichloromethane mixtures. Then they were concentrated and injected into a gas chromatograph (Thermo, Trace GC ultra) coupled to a mass spectrometer operating in SIM mode (Selected Ion Monitoring mode) using a DB-5 capillary column (30 m × 0.25 mm, 0.25 mm). Compounds were identified by comparison of the retention times and the mass-to-charge ratio (m/z) of the compounds with those of a NIST-PAH standard solution (SRM 2260a) containing the same analytes. The quantification was performed using calibration lines with at least nine concentrations of that standard solution. For the quantification of PCBs (congeners CB26, CB52, CB118, CB118, CB138, CB153 and CB180) and HCB, the crude extracts were purified by column chromatography using Florisil as the solid phase, eluting with 15 ml of hexane (fraction I – PCBs) and 45 ml of hexane/dichloromethane

(70:30) (fraction II – HCB). The two fractions were treated with H_2SO_4 for about 12 h, then the purified extracts were concentrated under nitrogen flow and injected into a gas phase chromatograph (Agilent Technologies, 6890N network GC system) equipped with an electron capture detector (micro-ECD) Agilent Technologies, with a DB-5 capillary column (60 m × 0.25 mm, 0.25 µm) and an auto sampler. The quantification of the various compounds was carried out using calibration lines and the external standard method.

Magnetics

Magnetic surveys were conducted aiming at detecting eventual archaeological artefacts that could interfere with the future extraction operations. A total field scalar magnetometer G882 (Geometrics, Caesium vapour) was used to acquire data along lines and tie-lines spaced at 50 and 250 m, respectively, at 10 Hz acquisition frequency. The magnetometer was towed with 60 m layback, to avoid noise generated by the vessel (17 m long). Processing of magnetic data included: (1) noise removal; (2) IGRF (International Geomagnetic Reference Field) correction for subtraction of the principal magnetic field; (3) correction of the diurnal anomaly using base station data; (4) iterative data levelling using tie-line intersections; (5) calculation of residual magnetic anomaly by filtering anomalies with wavelengths larger than 500 m; (6) minimum curvature gridding. For the identification of notable anomalies potentially related to archaeological artefacts processing also included: (7) calculation and analysis of the analytic signal (Nabighian 1972) for anomaly enhancement and relocation by shifting to the top of respective causative bodies; and (8) estimation of source depths through Euler deconvolution (Thompson 1982) (structural index SI = 3.0) to identify sources with estimated depths compatible to a location close to the seafloor. The processed data allowed not only identifying possible archaeological artefacts but also investigating for a geological correlation with the other geophysical datasets (Neres *et al.* 2019).

Sand resources assessment methodology

Although the volumes of the various sedimentary units were defined, the employed strategy to calculate the volumes of useful material (coarse sand + medium sand) was based on the individualization of the first 3 m of the sedimentary package in six depth layers of intervals of 50 cm thickness, in which the sediments were characterized and their usability as nourishment material quantified. The sediments that can be employed for nourishment must correspond to classes 1 or 2 (clean and slightly contaminated dredged material), thus eliminating

potential impacts due to contamination by POPs and trace metals released from the sand to the water column during dredging operations (Pinto *et al.* 2020). In addition to this, the APA, IP defined a set of extra restrictive specifications to be considered in the estimation of the resource volume: carbonate content <30%, gravel content <15% and the fine fraction (silt + clay) <10%. The volume calculation for each layer was implemented in several calculation steps as described in Figure 2, using Microsoft Excel™ software and the Spatial Analysis and 3D-Analyst extensions of the ESRI ArcGIS™ software. The sediments that respect these specifications are classified as useful sediment. Within the CHIMERA project, the spatial distribution and content of the sum of coarse and medium sand that fulfil all the restrictive specifications was calculated and mapped, thus defining the useful material (Fig. 2). The volumes of useful material were calculated between 0 and 50 cm below seafloor, from 50 to 100 cm, 100 to 150, 150 to 200, 200 to 250 and 250 to 300 cm. The interpolation method used was Inverse Distance Weighted (IDW), with a raster resolution of 10 × 10 m. This method of deterministic (mathematical) spatial interpolation is based on the assumptions that:

- Unsampled point values can be predicted as the weighted average of known values within the vicinity.
- Sampled points closest to the unsampled point are more similar than those farther apart, with the weights inversely proportional to the distances among the forecast locations and the sampled locations.

Results

This article does not intend to show all the results obtained by the CHIMERA project but only a selection of illustrative examples of the distinct types of data obtained and their interpretations, showing the importance of the development of the integrated multidisciplinary approach followed here. This ensures a detailed physical and chemical characterization of the borrow sediments and of their useful volume.

Seafloor morphology and nature

The three northern areas (about 10 km² each) are rectangular in shape, oriented parallel to the bathymetric contours, with water depths ranging between 24 and 40 m (Fig. 1). The seafloor presents, in general, as undulations and soft slopes with maximum amplitudes of approximately 25 cm, oriented parallel to the general bathymetry contours. In these areas, only scarce topographic highs were detected, here

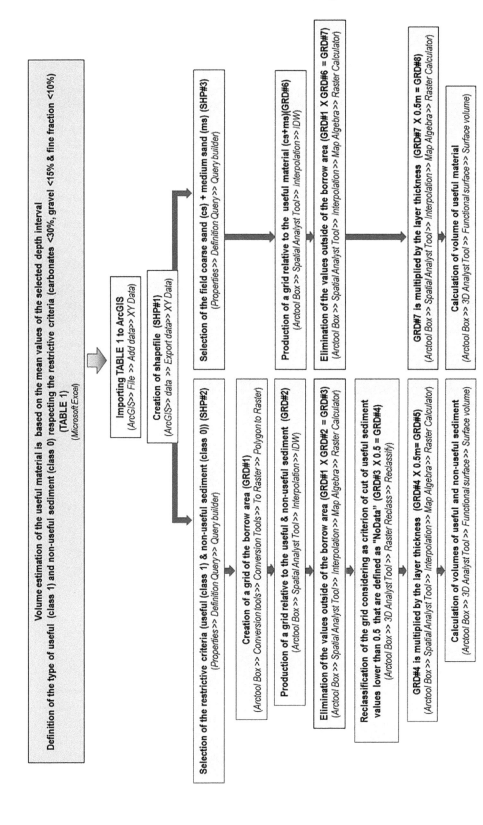

Fig. 2. Volume calculation of useful sediment and useful material (coarse sand + medium sand) workflow conducted in ArcGIS.

defined as 'sorted bedforms' (heterogeneous shelf seabed features indicators for hydrodynamic conditions marked by subtle bathymetric reliefs) characterized by decimeter order heights and/or by abrupt variations in grain size. The BM borrow area shows localized reliefs up to 60 cm high in a small zone close to the north limit. The CC borrow area has a near semicircle shape, of approximately 4.9 km², located off the coast of Lisbon between depths of 20 and 28 m (Fig. 1) with a general slope of 0.4%. This area presents a set of rippled scour depressions, generally limited by well-defined slopes with maximum amplitudes of the order of 70 cm.

Backscatter intensity and sedimentary characteristics

The backscatter intensity and bathymetric maps together with sedimentological data allowed seafloor sediment distribution patterns to be defined. The backscatter mosaics obtained for the four borrow areas reveal spatial coherent patterns including areas with lower and higher acoustic backscatter intensities. From the analysis of the bathymetric, backscatter and surface samples data, interpreted sediment textural distribution maps were produced for all areas. An example of the 350 kHz multibeam

backscatter mosaic obtained in the CC borrow area is shown in Figure 3. The histogram for the backscatter values presents a distribution with a main mode centred around −34 dB and a secondary mode at −24 dB. The main mode (−34 dB) corresponds to the very fine to medium sand, that covers most of the borrow area. Medium to coarse sands were observed in areas with high reflectivity, corresponding to the secondary mode (−24 dB), found in small depressions (areas of maximum slope gradients corresponding to 70 cm depressions of the seafloor) with half-moon (barchan-like) and rounded shapes (Fig. 3).

Surface sediments are mainly composed of sand, with fine fraction (<63 μm) contents lower than 4% in all the borrow areas. The two northernmost borrow areas, ET and BM, are characterized by low spatial variability, with more than 50% of the surface samples showing a mean grain size corresponding to coarse sand (Fig. 4a). The FF and CC borrow areas present the highest particle size variability. The mean grain size varies between fine and very coarse sand (Fig. 4b). In terms of sorting, the ET and BM surface samples show the highest variability (Fig. 4b). The median sorting values for all borrow areas lie in Folk's moderately sorted category (Fig. 4b). The surface samples have a median gravel content lower than 10% in the four borrow areas

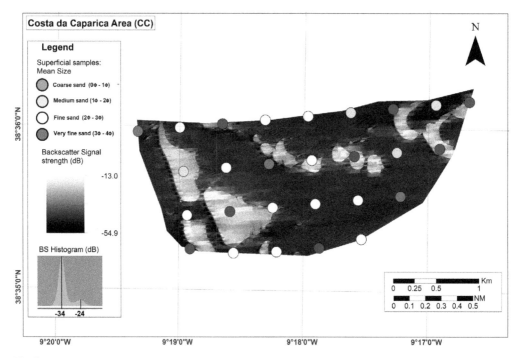

Fig. 3. The 350 kHz multibeam backscatter mosaic of the CC borrow area. Overlapped on the backscatter mosaic are the sediment types identified at each sample location. Inset: backscatter histogram with the two modes.

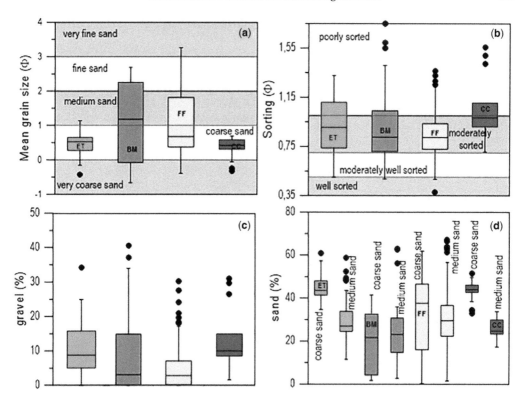

Fig. 4. Box-and-whisker plots representing the Folk and Ward Graphical methods (1957): (**a**) mean grain size; (**b**) sorting; (**c**) gravel; (**d**) coarse and medium sand contents in the four borrow areas. The plots show the minimum, maximum, median and lower and upper quartiles. The boxes represent the interquartile range that contains 50% of the values. The whiskers are the lines that extend from the boxes to the highest and lowest values. The lines across the boxes indicate the median. Outliers (defined as 1.5 × inter-quartile range) are represented by black filled symbols.

(Fig. 4c). The composition of the gravel fraction is dominated by shell debris in the CC borrow area and by quartz grains in the other three areas. The ET and BM borrow areas have the highest median contents of coarse sand, while the FF and CC show high contents of medium sand (Fig. 4d).

Regarding carbonates content, the CC borrow area presents both high variability and the highest median $CaCO_3$ contents (Fig. 5a). The area also presents the greater variability in C_{org} contents and density values (Fig. 5b). Nevertheless, it is important to highlight that the median contents of C_{org} are lower than 0.2%. The total solids content varies between 77 and 94% in all borrow areas, which is consistent with the mostly sandy composition (the median values of the sand fraction in ET, BM, FF and CC borrow areas are 90, 94, 99 and 98%, respectively (Fig. 5c).

A scatterplot of the mean grain size and sorting values from the ET and BM areas evidences that coarse sediments show high sorting values (Fig. 6a & b). This relationship was not observed for the FF and CC borrow areas (Fig. 6c & d).

Characterization of the degree of contamination

The evaluation of the degree of contamination is based on the comparison of the trace metal and POP contents with the ranges of concentrations of the five categories defined by the Portuguese Legislation (Ordinance 1450/2007, 12 November) (Table S1). All the selected surface sediment samples were classified as Class 1.

Subsurface morphosedimentary facies

Seismic interpretation of the deeper layers indicate that this detrital sediment package is present on all four areas up to depths that range between 4.7 and 57.3 m below seafloor (Table 2). This sedimentary package, not affected by tectonic deformation, lies unconformably on a clearly imaged deformed bedrock substrate. The sedimentary architecture of this sediment package was organized in four to six SUs bounded by discontinuities, usually of erosional nature, which were independently defined for each

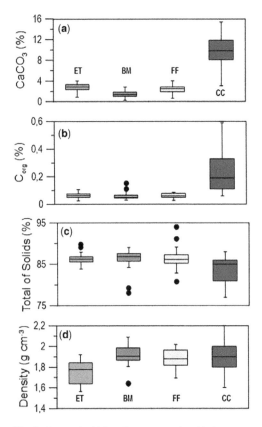

Fig. 5. Box-and-whisker plots representing (**a**) the CaCO₃, (**b**) the C$_{org}$ contents, (**c**) the total solids content and (**d**) the density in the surface sediments of the four borrow areas.

project area. The SUs exhibit complex internal geometries and inter-unit arrangements that vary significantly not only across the project areas but also locally inside the same area. However, the spacing defined for the survey grid allowed to image and delineate the 3D geometry of all these sedimentary bodies.

The outcropping SU that tops the sediment packages, being the easiest to mobilize and dredge, has a very similar geometry in all borrow areas. This unit corresponds to a thin sediment layer, 0.2–1.7 m thick, which extends throughout the four studied areas. The exception was found in small sections of FF area, where another SU outcrops. The unconformable base of this upper SU most frequently coincides with the base of sea bottom morphological features (e.g. sorted bedforms or rippled scour depressions). Beneath this superficial unit, lies a sequence of usually thicker sediment packages wherein two types of morphological features stand out, infilled channels and sediment bodies. They are interpreted as former coastal barriers, as already

proposed (e.g. Rodrigues *et al.* 1991; Dias *et al.* 2000, 2002).

Infilled channels' features, occasionally showing several successive cycles, are present in the four working areas. The channels' dimensions range from the seismic resolution limit to structures comprehending an entire SU (e.g. unit U6 of area BM) with a length of more than 3500 m, 750 m wide and 30 m of depth. Sometimes these features occur related to the landward side of sediment bodies interpreted as coastal barriers (e.g. units U3 and U4 of area BM). The most prominent sediment bodies associated with former coastal barriers are the ones from the FF and BM areas (units U3, in both areas), reaching thicknesses of more than 8 m, 3750 m of length and 1750 m of width. These features are characterized by high amplitude sigmoidal to oblique downlap, prograding internal reflections that have frequent internal discontinuities and are limited at their top by an erosive unconformity (Figs 7 & 8). They can be remnants of coastal barriers, associated with palaeocoastlines (e.g. delta, swash bar or other shoreface structures,) or detached coastlines structures (e.g. barrier islands system).

Absolute age control of the SU was not undertaken. However, sedimentary architecture and facies similarities with other coastal sectors and neighbouring inland geology, suggest that the bedrock substrate could be of Meso-Cenozoic age (probably Cretaceous to Miocene) and that the generation of the overlaying detrital sequence can be associated with the sea-level variation of the last Quaternary eustatic cycles.

Grain size down-core variability

The vibrocorer samples up to 3 m below seafloor confirmed the quartz rich sand/gravel nature of the upper sediment layers imaged in the PE data, characterized by minor percentages of fine-grained sediments (Fig. 9). The dominant dimensional classes vary among borrow areas (Fig. 9). The two most northern areas (ET and BM) are characterized by coarser sediments compared to the FF and CC borrow areas (Fig. 9). While ET sediments are essentially coarse to very coarse sands, medium to coarse sediments predominate in BM. The CC borrow area is dominated by fine to medium sands. Considering the example of the FF borrow area, the dominant dimensional classes in the sediment vibrocores are fine and medium sands (Fig. 9). The mean grain size and sorting coefficient (Graphical method, Folk and Ward 1957) do not generally exhibit great variability with depth. Some cores (e.g. FF-V_16 and FF-V_17) have a larger amount of coarse sand at the deeper levels. In terms of sorting the sediment samples, they vary mainly between moderately and poorly sorted (Fig. 10).

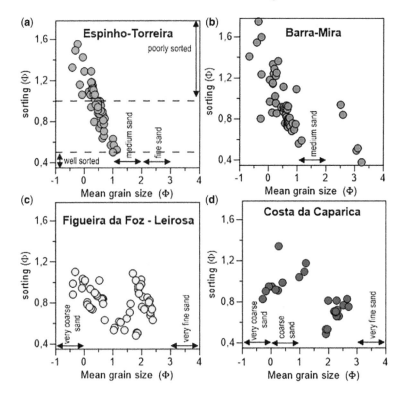

Fig. 6. Mean grain size v. sorting values (Graphical method, Folk and Ward 1957) of the surface samples collected in the four borrow areas: (**a**) ET; (**b**) BM; (**c**) FF; (**d**) CC.

Notable magnetic anomalies / potential archaeological structures

The identification of potential archaeological artefacts either at the seafloor or partially buried was made by analysing each acquired line and the processed grids, residual anomaly and analytic signal, for detecting significant dipolar isolated anomalies. For the identified anomalies, source depth was estimated by Euler deconvolution to check their

Table 2. *Number of upper basements individualized SUs, estimated sediment thickness for the upper basement, volumes for the upper basement and upper SU. The volumes are obtained from depth differencing of the SU horizons*

Area	Number of upper basements SU	Sediment thickness (m)	Sediment volume (10^6 m^3)	
		Upper basement	Upper basement	Upper SU
CC	6	23.4–57.3	230	5.6
BM	6	4.7–15.2	105	9.8
ET	4	7.5–30.9	151	9.3
FF	4	15.8–22.3	183	7.8

consistency with depths close to the seafloor. The criteria for defining target anomalies were: local dipole in the residual anomaly verified along line (*c.* 0.5 m resolution) or recorded in more than one line (line separation is 50 m; Figs 11 & 12); maximum analytic signal; and estimated source depth compatible to a location close to the seafloor.

Three target anomalies were identified, one for each borrow area, except FF. The identification of the BM target anomaly is shown in Figure 11. It shows a crop of the BM residual anomaly and the respective analytic signal grid. In the east part of the area, the black circle highlights a dipolar anomaly of 4.6 nT (peak to peak) also coincident with an analytic signal maximum. This target anomaly has an estimated source depth of 27 m below sea-level, compatible with an object buried 1–2 m below seafloor. Given the anomaly wavelength and the survey resolution, the causative body is expected to be less than 100 m long.

Subsurface structure: insight from magnetic data

The total magnetic anomaly showed varying characteristics among the four areas, with amplitudes

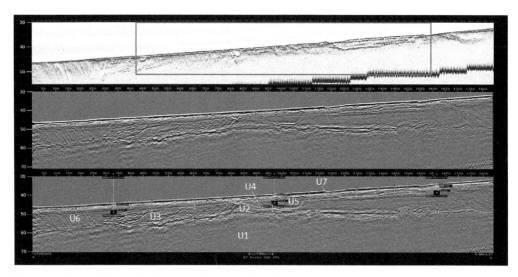

Fig. 7. Line BM_L08, PE (top), UHRS (middle) and UHRS with seismic interpretation and vibrocore sampling data locations superimposed (lower). Orange box outlines the area enlarged in Figure 8.

varying from 12 nT in FF to 200 nT in BM. In general, the total anomaly expresses long wavelength anomalies that are due to the deep geology, not observed in the structure imaged by the high-resolution seismic with shallow penetration (Neres *et al.* 2019). The residual magnetic anomaly is characterized by very low amplitudes: *c.* 2 nT for FF, ET and BM, and *c.* 10 nT for CC. The

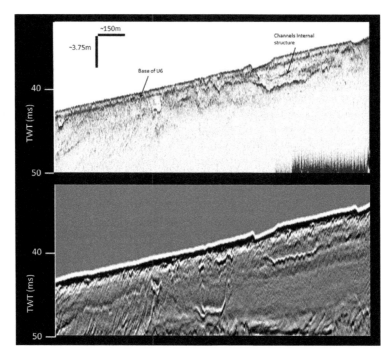

Fig. 8. Detail of PE (top) and UHR (bottom) profiles BM_L01, showing structures hardly imaged in the UHR data. Namely, the base of the upper most unit (U7) and the internal structure of unit U6 channels.

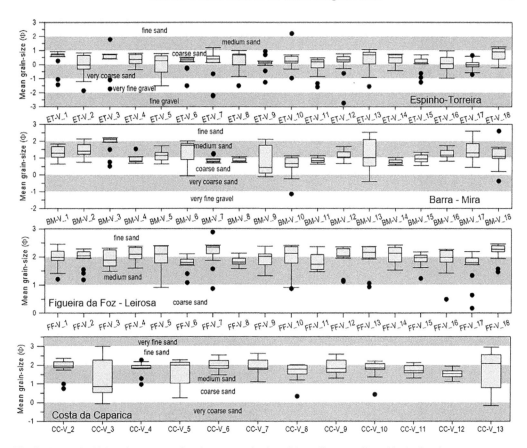

Fig. 9. Box-and-whisker plots representing the mean grain size of the sediments collected in the four borrow areas.

geological significance of the residual anomaly varies among the four study areas. According to the joint analysis with multichannel seismics, it reflects either the intra-basement structure, or some supra-basement sedimentary features such as buried channels and coastal barriers.

The magnetic results and combined analysis for the area ET is illustrated in Figure 11. The gridded total and residual magnetic anomalies, the gridded depth to the top of seismostratigraphic units U1 and U2 and the seismic interpretation of the MCS line X06. Outstanding features in the SUs are the channels in the north of the ET area that cut more than 15 m deep into the basement (marked as # in Fig. 11) and palaeoreliefs rising up to 10 m above the basement at the SW (marked as *). Joint analysis of the residual magnetic anomaly and sedimentary units show an outstanding magnetic signature coincident with the channels and the U2 reliefs. A quantification of magnetic properties is not possible with the available data, but it can be deduced that unit U2 has a higher magnetic susceptibility than U3.

This may indicate a more terrigenous sedimentary contribution for U2, and thus a later progression of the coastline to the east. The magnetic signature of the northern palaeochannels may be due to a contrast between the filling sediments and a higher susceptibility basement.

Discussion

Adopted methodology to assess sand resources

Reliable assessment of sand and gravel resources for beach nourishment and the planning of the dredging operations requires the identification and characterization of the sedimentary package and its 3D geometry, as well as the identification and depth distribution of the basement below the sedimentary cover. Only with all this information it is possible to estimate volumes of material that can be dredged for nourishment. This characterization is also important to select and to define the necessary

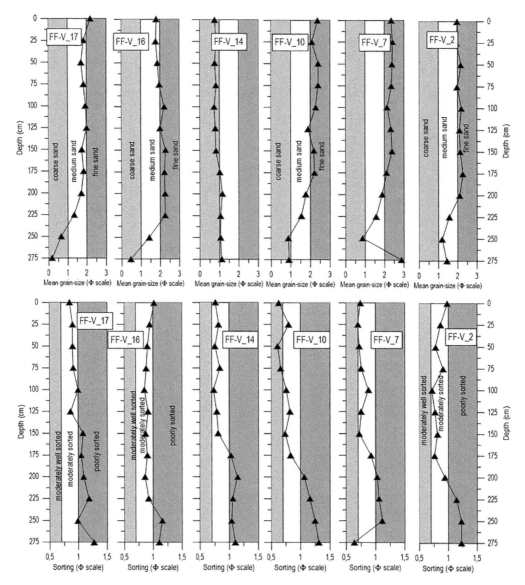

Fig. 10. Mean grain size and sorting (Graphical method) down-core variability in selected sediment cores collected in the FF borrow area.

and more adequate dredging techniques and to identify the potential operation difficulties.

The Portuguese inner continental shelf has a very complex geology, with the record of the interplay between erosion stages during the last glacial maximum and sedimentary infill episodes during transgressions, very frequently sculpted in various glacial–interglacial cycles (Rodrigues *et al.* 1991; Dias *et al.* 2000). The Portuguese inner shelf sedimentary deposits typically include continental meander-like sandy deposits, beach and beach

barriers that have migrated according to the transgressive cycles (Rodrigues *et al.* 1991; Dias *et al.* 2000), which are also identified in these four case studies (e.g. Figs 7 & 8). Consequently, the required image and characterization of the sedimentary deposits need to be based on the seafloor characterization by multibeam bathymetry, on the acoustic backscatter and seafloor ground-truthing for sedimentary characterization, and on a dense grid of high-resolution seismic profiles to image the subseafloor geometry of the sedimentary bodies.

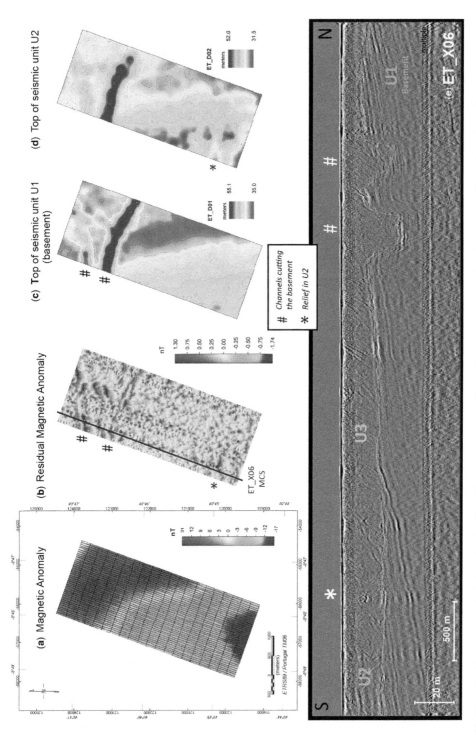

Fig. 11. Magnetic data analysis and relation with identified SUs: example for area ET. (**a**) Total magnetic anomaly grid and survey lines. The high amplitude, long wavelength anomalies are due to regional geology. (**b**) Residual magnetic anomaly grid. The residual magnetic anomaly is marked by low amplitude though outstanding magnetic lineations in the north of the area, and shorter lineations at SW. (**c**) Gridded depth to the top of seismostratigraphic unit U1 (top of basement). (**d**) Gridded depth to the top of seismostratigraphic unit U2. (**e**) Interpretation of the seismic profile X06 (location shown in b). * locates reliefs in unit U2 and # locates channels cutting through the basement and filled by U2 and U3 material.

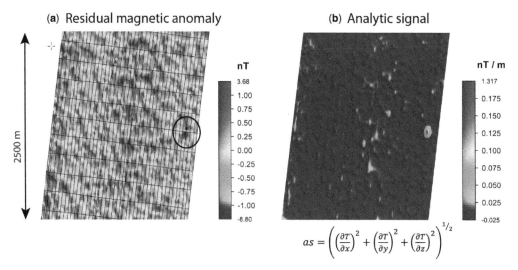

$$as = \left(\left(\frac{\partial T}{\partial x}\right)^2 + \left(\frac{\partial T}{\partial y}\right)^2 + \left(\frac{\partial T}{\partial z}\right)^2 \right)^{1/2}$$

Fig. 12. Identification of a magnetic target possibly attributed to an archaeological artefact (example for the area BM). (**a**) Residual magnetic anomaly and survey lines; (**b**) analytic signal. The target anomaly, to the east (black circle), is identified as a local dipole in the residual anomaly and a maximum of analytic signal.

In this work, the adopted methodology to fulfil the objectives included the following steps: (1) mapping the surface bathymetry and backscatter, and correlating the backscatter with the surface samples; (2) acquisition of high resolution seismics to define the Quaternary sedimentary package and SUs within; (3) definition of the sedimentary systems and main units of the Quaternary sedimentary package; (4) quantification of volumes of useful sediment and material content in the top 3 m of the Quaternary package that fulfil all the requirements for use. Thus, the seafloor bathymetry was acquired with full coverage, with more than 10% superposition of parallel swath coverages, ensuring a bathymetric resolution of 25 cm. The dense surface sampling grid allowed the characterization of the nature of the seafloor, in terms of granulometry, $CaCO_3$ contents and chemical characterization. The seafloor ground-truth allowed calibrating the swath backscatter classification and confirmed that the density of the sampling stations was both adequate and what was required to map the seafloor natural variability. Multichannel seismic profiles were acquired with 250 m spacing between parallel lines, ensuring that the basement was imaged and that all the expected geomorphologies in an inner-shelf setting were identified and mapped. The vibrocorer sampling of the top 3 m was done with a grid space of 500 and 1000 m. This 3 m depth was sufficient since this is the expected maximum depth of dredging. The sampling allowed for characterizing the different sedimentary bodies present in the four study areas. Since this project's

workflow defined that the vibrocorer sampling was to be done in a regular grid, the number of vertical sampling stations frequently over sampled the same sedimentary body. This allowed confirmation that the sedimentary structures exhibit very little lateral variability in terms of the required properties – granulometry, composition and contamination. The definition of the sedimentary systems and main units of the Quaternary sedimentary package were done by comparing and correlating the seismic stratigraphy with the results of the lithological characterization of the vibrocores, as described in the section below. The definition and quantification of volumes of 'useful sediment' and of 'useful material' was done for the top 3 m of the Quaternary package in the individual 50 cm thick layers, excluding the volumes of material that did not fulfil the $CaCO_3$,

Table 3. *Volumes in m^3 of non-useful and useful sediment (carbonates $<30\%$, gravel $<15\%$ and fine fraction $<10\%$), and useful material, for two sediment layers of the CC borrow area*

Depth Interval	Volume of non-useful sediment	Volume of useful sediment	Volume of useful material (coarse sand + medium sand)
0–50 cm	0.15×10^6	2.30×10^6	0.97×10^6
150–200 cm		2.13×10^6	1.09×10^6

grain-size and contaminants requirements for use as nourishment material.

Comparison of lithologic and seismic stratigraphy

Seismic interpretation of the multichannel high-resolution dataset was done to establish a seismic stratigraphic model of each site, defining units which are characterized by their geometries, external relationships and the internal configuration of reflectors. The depositional context of these SUs was inferred, and the lithological variations of the units was assessed and later confirmed by the vibro-core samples characterization. This methodology allowed to map the base of the Quaternary package in the four borrow areas as defined by the deeper reflectors within the sandy deposits, which reached maximum depths below seafloor of 55.1 (ET), 45.2 (BM), 59.7 (FF) and 84.2 m (CC). However, it is only the complementary correlation of the lithological stratigraphy with the seismic stratigraphy that leads to the inference that the sedimentary facies and the palaeoenvironment correspond to the top (3 m) SUs. In most cases the facies could be extrapolated to deeper parts of the units. In some regions of the four study areas, the fine-scale sedimentary variability was not clearly imaged by the seismic data. But in general, a good correlation between major (metre scale) sedimentary packages and the top seismic reflectors was achieved. Based

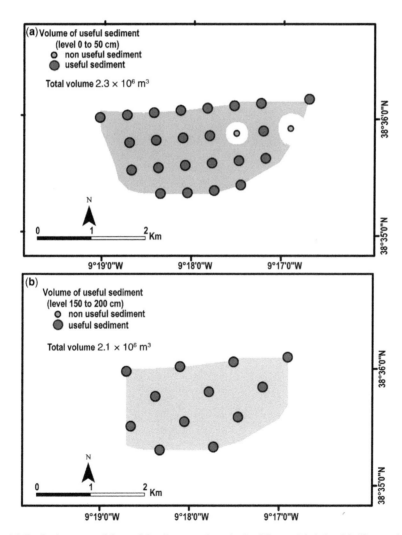

Fig. 13. Spatial distribution maps of the useful sediments volume in the CC area. (**a**) At level 0–50 cm and (**b**) 150–200 cm.

on the composition of the units within the top 3 m and on the geometry and characteristics of the seismic reflectors below the top 3 m, we could infer the sedimentary geometry and the sedimentary settings of these units.

Estimation of sand resources

The case study of the CC borrow area is presented here as an example of the adopted strategy to estimate the sand resources. The methodology implemented in this project is described for two depth intervals: 0–50 and 150–200 cm, and the volumes of useful sediment and useful material are shown in Table 3. In the top layer, two surface samples

(blue dots in Fig. 13) with gravel contents higher than 15% do not fulfil the required criteria and were therefore not considered for the calculation of useful sediment volume. In these two examples, it is possible to observe that only 42% (top interval) and 51% (deep interval) of the useful sediment is considered as useful material. This indicates a high abundance of other dimensional classes (e.g. fine sand) in this borrow area. The spatial distribution pattern of useful material varies in depth, with the lowest mean contents being found in the western half of the area for the most recent sediments, while in the sediment layer, 150–200 cm, they were mainly located in the southeastern limit (Fig. 14).

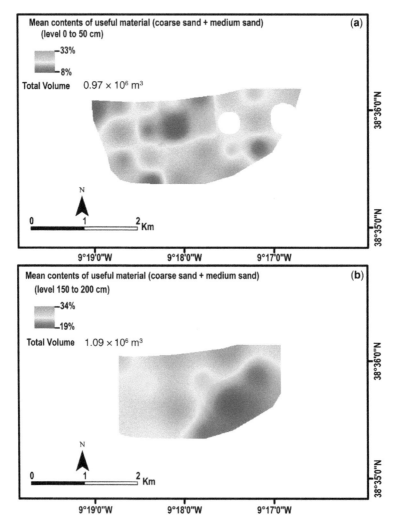

Fig. 14. Spatial distribution maps of the mean contents of useful material (medium sand + coarse sand) and respective volume units in the CC area. (**a**) At level 0–50 cm and (**b**) 150–200 cm.

The results obtained for the four borrow areas confirm the almost generalized existence of useful sediments (fines – silt and clay <10%; gravel <15%; carbonates <30%; no contaminants) dominating the sands dimensional class. For an exploration depth of 3 m, a volume of the order of 14×10^6 m^3 is estimated at ET and CC, and 28×10^6 and 29×10^6 m^3 at the BM and FF spots, respectively. The seismic-stratigraphic model inferred from geophysical data indicates bedrock between 6 and 57 m below the seabed in the four areas.

Sediment management strategy

The sediment management strategy allows for establishing the goals to manage and assure a sustainable exploitation of the sand resources. The results obtained show sedimentary resources (in a thickness of 3 m) compatible with a sediment management strategy based on sediment budget balance until the year of 2035 (ET) and until 2050 (BM, FF and CC), if the nourishment volumes are equal to the magnitude of the residual littoral drift in the reference situation at each site. Furthermore, there is potential to prolong the longevity of the borrow areas, since the sedimentary thickness interpreted on the reflection seismic data suggests the possibility of further increase of the exploration depth and therefore the total volume of useful material.

Final considerations

Offshore sand is a limited resource, whose renewal depends on various natural and anthropogenic environmental variables, the interaction of which can be quite unpredictable. Therefore, offshore sand resources must be properly managed to ensure sustainability of the ongoing shore protection strategy in Portugal – as in other European countries, such as the Netherlands, Denmark, the UK and Spain – based on beach nourishment. The latter aims to minimize or reverse large background erosion in coastal cells with sediment deficit.

The results obtained in this work show suitable sand (in quality and quantity) in the four selected areas, which enable a series of artificial beach nourishment interventions to be undertaken in the medium–long term. The decision process regarding the prioritization of places and type of nourishment, volumes involved and refill periodicity will depend on the results obtained by the business case and environmental impact assessment, but also by financial, political, administrative and social constraints.

Quantification and spatial distribution of both useful sediment volume and sand quality is of great importance for selecting/detailing the area to be dredged. This information is fundamental for management of the sedimentary resources for beach nourishment in areas of intense coastal erosion. Additionally, the detailed textural information on the nourishment material will allow for hydrodynamic modelling and thus prediction of the sustainability of the nourished beaches. Also, the baseline bathymetric and backscatter mapping of the borrow area allows for subsequent evaluation of the morphological evolution of the extraction pits upon implementation of monitoring programmes. One of the aims of the monitoring programmes is to integrate diverse (e.g. hydrodynamic, morphological, sedimentological and biological) information from the borrow and fill areas, during the different phases of the operation.

Both the methodology and the data acquired within the CHIMERA project ensure the understanding of the Quaternary sand deposits as well as its sedimentological characteristics, geometry and degree of contamination. This information is essential to assess the suitability of the deposits for beach nourishment. They also allow for identifying possible archaeological artefacts that might interfere with the future dredging operations. The quantification and mapping of the 3D sub-surface distribution of the useful sediment material was designed to guarantee selection of the appropriate dredging operations in these inner shelves complex systems with large spatial variability. The detailed characterization of the borrow site sediments allowed identification and selection of the most suitable useful material.

Acknowledgements The authors are grateful to António Pereira, Cremilde Monteiro, Cristina Micaelo, Rui Silva, Rute Granja and Warley Soares for helping in the laboratory work. Special acknowledgments are due to the diving team of the Xavisub Lda for the high-quality sediment sampling operations. The authors also thank the company Geosurveys Geophysical Consultants Ltd, for providing the flow template used in this work for processing the UHRS data and the EMSO-PT project (ref. PINFRA/22157/2016).

Funding The CHIMERA project was funded by the Agência Portuguesa do Ambiente and European Comission through the Operational Programme for Sustainability and Efficiency in the Use of Resources (POSEUR-09-2016-48-FC-000030). Emília Salgueiro and Marta Neres were funded by postdoctoral FCT fellowships: SFRH/BPD/111433/2015 and SFRH/BPD/96829/2013, respectively.

Author contributions MMH: investigation (lead), project administration (lead), resources (lead), supervision (lead), validation (lead), writing – review & editing (lead); **PB**: investigation (lead), methodology (lead), writing – review & editing (equal); **VM**: writing – original draft (lead), writing – review & editing (lead); **MR**: investigation (equal), methodology (lead), writing – review &

editing (equal); **MN**: investigation (lead), methodology (lead), writing – review & editing (equal); **MS**: investigation (equal), methodology (equal), validation (equal); **ES**: investigation (equal), methodology (equal), writing – review & editing (supporting); **TD**: investigation (equal), methodology (equal), writing – review & editing (supporting); **AIR**: investigation (equal), methodology (equal), writing – review & editing (supporting); **MTG**: investigation (lead), methodology (lead), writing – review & editing (equal); **MJG**: investigation (equal), methodology (equal), validation (lead), writing – review & editing (supporting); **EA**: investigation (equal), methodology (equal); **MS**: investigation (equal), methodology (equal); **MF**: investigation (equal), methodology (equal); **CAP**: writing – review & editing (equal); **CB**: investigation (equal), methodology (equal); **PT**: investigation (equal), methodology (equal), project administration (lead), writing – review & editing (lead).

Data availability statement The datasets generated during the current study are not publicly available due to confidentiality agreements with research collaborators. Data are however available from the authors upon reasonable request and with permission of the Núcleo de Monitorização Costeira e Risco – Agência Portuguesa do Ambiente.

References

Andrade, C., Freitas, M.C., Cachado, C., Cardoso, A., Monteiro, J., Brito, P. and Rebelo, L. 2002. Coastal zones. *In*: Santos, F.D., Forbes, K. and Moita, R. (eds) *Climate Change in Portugal: Scenarios, Impacts and Adaptation Measures*. Gradiva.

Andrade, C., Rodrigues, A. *et al.* 2015. *Gestão da Zona Costeira O Desafio da Mudança*. Sediment Working Group: Final report, http://www.apambiente.pt/_zdata/DESTAQUES/2015/GTL_Relatorio%20Final_20150416.pdf [in Portuguese].

Anthony, E.J., Cohen, O. and Sabatier, F. 2011. Chronic offshore loss of nourishment on Nice beach, French Riviera: a case of over-nourishment of a steep beach? *Coastal Engineering*, **58**, 374–383, https://doi.org/10.1016/j.coastaleng.2010.11.001

Blott, S.J. and Pye, K. 2001. Gradistat: A grain size distribution and statistics package for the analysis of unconsolidated sediments. Earth Surf. Process. *Landforms*, **26**, 1237–1248, https://doi.org/10.1002/esp.261.

Costley, C.T., Mossop, K.F., Dean, J.R., Garden, L.M., Marshall, J. and Carroll, J. 2000. Determination of mercury in environmental and biological samples using pyrolysis atomic absorption spectrometry with gold amalgamation. *Analytica Chimica Acta*, **405**, 179–183, https://doi.org/10.1016/S0003-2670(99)00742-4

Davis, R.A., Jr, Wang, P. and Silverman, B.R. 2000. Comparison of the performance of three adjacent and differently constructed beach nourishment projects on the Gulf Peninsula of Florida. *Journal of Coastal Research*, **16**, 396–407.

Dias, J.M.A. and Neal, W.J. 1990. Modal size classification of sands: an example from the Northern Portugal Continental-Shelf. *Journal of Sedimentary Petrology*, **60**, 426–437.

Dias, J.M.A. and Nittrouer, C.A. 1984. Continental-shelf sediments of northern Portugal. *Continental Shelf Research*, **3**, 147–165 https://doi.org/10.1016/0278-4343(84)90004-9

Dias, J.M.A., Boski, T., Rodrigues, A. and Magalhães, F. 2000. Coastline evolution in Portugal since the Last Glacial Maximum until present: a synthesis. *Marine Geology*, **170**, 177–186, https://doi.org/10.1016/S0025-3227(00)00073-6

Dias, J.M.A., Jouanneau, J.M. *et al.* 2002. Present day sedimentary processes on the northern Iberian shelf. *Progress in Oceanography*, **52**, 249–259, https://doi.org/10.1016/S0079-6611(02)00009-5

Dondurur, D. 2018. *Acquisition and Processing of Marine Seismic Data*. Elsevier.

Duarte, H., Wardell, N. and Monrigal, O. 2017. Advanced processing for UHR3D shallow marine seismic surveys. *Near Surface Geophysics*, **15**, 347–358, https://doi.org/10.3997/1873-0604.2017022

Ferreira, Ó. and Matias, A. 2013. Portugal. *In*: Pranzini, E. and Williams, A. (eds) *Coastal Erosion and Protection in Europe*. Routledge, 275–293.

Flemming, B.W. and Delafontaine, M.T. 2000. Mass physical properties of muddy intertidal sediments: some applications, misapplications and non-applications. *Continental Shelf Research*, **20**, 1179–1197, https://doi.org/10.1016/S0278-4343(00)00018-2

Folk, R.L. and Ward, W.C. 1957. A study in the significance of grain-size parameters. *Journal of Sedimentary Petrology*, **27**, 3–26, https://doi.org/10.1306/74D70646-2B21-11D7-8648000102C1865D

Friedman, G.M. 1961. Distinction between dune, beach and river sands from their textural characteristics. *Journal of Sedimentary Petrology*, **31**, 514–529.

Hamm, L., Capobianco, M., Dette, H.H., Lechuga, A., Spanhoff, R. and Stive, M.J.F. 2002. A summary of European experience with shore nourishment. *Coastal Engineering*, **47**, 237–264, https://doi.org/10.1016/S0378-3839(02)00127-8

ICES 2016. *Effects of extraction of marine sediments on the marine environment 2005–2011*. ICES Cooperative Research Report No. **330**.

Mougenot, D. 1988. *Geologie de la Marge Portugaise*. PhD Thesis, Université Pierre et Marie Curie Paris VI, France.

Nabighian, M. 1972. The analytic signal of two-dimensional magnetic bodies with polygonal cross-section: its properties and use for automated anomaly interpretation. *Geophysics*, **37**, 507–517, https://doi.org/10.1190/1.1440276

Neres, M., Brito, P. *et al.* 2019. Combining magnetic and seismic reflection data for characterization of inner shelf sand nourishment areas. *Geophysical Research Abstracts*, **21**, EGU2019-4339.

Oliveira, I.M., Valle, A.F. and Miranda, F. 1982. Littoral problems in the Portuguese west coast. *Cape Town, Republic of South Africa, Coastal Eng. 1982 Proceedings, III*, 1951–1969.

Payton, C.E. 1977. Seismic stratigraphy: applications to hydrocarbon exploration. *Memoir of the American Association of Petroleum Geologist*, **26**, 516.

Pinto, C.A., Silveira, T.M. and Teixeira, S.B. 2020. Beach nourishment practice in mainland Portugal (1950–

2017): Overview and retrospective. *Ocean Coast. Manag.* 192, 105211, https://doi.org/10.1016/j.oce coaman.2020.105211

Rodrigues, A., Magalhães, F. and Dias, J.A. 1991. Evolution of the North Portuguese coast in the last 18,000 years. *Quaternary International*, **9**, 67–74. https://doi.org/10.1016/1040-6182(91)90065-V

Taborda, R. and Andrade, C. 2014. *Morfodinâmica do estuário exterior do Tejo e intervenção na região da Caparica – v1*. Contribution to the Littoral Working Group. **Anex I**, 161–179.

Taborda, R., Freire de Andrade, C. *et al.* 2014. Modelo de circulação sedimentar litoral no arco Caparica-Espichel. *Comunicações Geológicas*, **101**, 641–644.

Thompson, D.T. 1982. EULDPH: a new technique for making computer-assisted depth estimates from magnetic data. *Geophysics*, **47**, 31–37, https://doi.org/10.1190/1.1441278

Udden, J.A. 1914. Mechanical composition of clastic sediments. *Bulletin of the Geological Society of America*, **25**, 655–744, https://doi.org/10.1130/GSAB-25-655

US AID 2009. *Adaptation to Coastal Climate Change: A guidebook for development planners*. US Agency for International Development (US AID).

US EPA 1996. Method 3050B: Acid Digestion of Sediments, Sludges, and Soils, Revision 2. US EPA, Washington, DC.

Van Lancker, V., Francken, F. *et al.* 2017. Building a 4D voxel-based decision support system for a sustainable management of marine geological resources. *In*: Diviacco, P., Leadbetter, A. and Glaves, H. (eds) *Oceanographic and Marine Cross-Domain Data Management for Sustainable Development*. IGI Global, 224–252.

Vera-Cruz, D. 1972. Artificial nourishment of Copacabana beach. *Proceedings 13th Coastal Engineering Conference*. New York: ASCE, 141–163.

Vicente, C. and Clímaco, M. 2012. Trecho de costa do Douro ao cabo Mondego: caracterização geral do processo erosivo. LNEC, relatório 253/2012 – DHA/NEC [abstract in English], http://repositorio.lnec.pt:8080/xmlui/handle/123456789/1004065

Wentworth, C. 1922. A scale of grade and class terms for clastic sediments. *The Journal of Geology*, **30**: 377–392, https://doi.org/10.1086/622910

Collating European data on geological events in submerged areas: examples of correlation and interpretation from Italian seas

Loredana Battaglini, Silvana D'Angelo and Andrea Fiorentino*

Istituto Superiore per la Protezione e la Ricerca Ambientale, Department for the Geological Survey of Italy, Via V. Brancati 48-00144, Roma

*Correspondence: andrea.fiorentino@isprambiente.it

Abstract: The European Marine Observation and Data Network (EMODnet) Project provides freely available data on European seas. The main purpose of EMODnet is to overcome the fragmentation and dishomogeneity of the available data, providing access to a harmonized and interoperable database. The EMODnet Geology Lot includes information at multiple scales on the seabed and its substrate (http://www.emodnet-geology.eu/). The dataset on 'Geological events and probabilities' collects information on landslides, earthquakes, volcanic structures, active tectonics, tsunamis and fluid emissions. The Geological Survey of Italy, which coordinates the collation of 'Geological events and probabilities' data, provided guidelines to compile layers complemented by comprehensive and detailed patterns of attributes for each feature in order to characterize each type of geological event. Occurrences of events are often associated with each other, particularly in tectonically active areas. Geological events affect both submerged and coastal environments. Data gathered by EMODnet Geology provide a good basis for further studies, contributing to the outlining of different tectonic settings and providing support to the use of marine resources, as well as to the management of marine-coastal areas particularly regarding the identification and assessment of geological and environmental hazards.

The European Marine Observation and Data Network (EMODnet) Project (Vallius *et al.* 2020) provides freely available harmonized data in European seas, represented as digital maps accessible through a central Portal (http://www.emodnet.eu/). It is subdivided into thematic lots (each of them provided with a dedicated portal) concerning Bathymetry, Biology, Geology, Chemistry, Physics, Seabed Habitats and Human Activities. The Geology lot collates existing information from national and international projects and accurately validated literature data at multiple scales within different Work Packages (WPs) on seafloor sediments, bedrock geology and geomorphology, coastal processes, geological events and mineral resources. WP6, 'Geological events and probabilities', coordinated by the Geological Survey of Italy, ISPRA, considers submarine landslides, earthquakes, volcanic structures, active tectonics, tsunamis and fluid emissions (Fig. 1). Layers can be viewed online (https://www.emodnet-geology.eu/map-viewer/) or downloaded and opened in a GIS application, which is the best way to explore their contents

of the available data, providing access to a harmonized and interoperable database. This goal is achieved by compiling several layers complemented by comprehensive and detailed patterns of attributes for each feature, in order to represent the diverse characteristics of each geological event occurrence. The WP6 coordinator elaborated and distributed, in cooperation with Project Partners, guidelines to compile shapefiles and attribute tables aiming at the identification of parameters that should be used to characterize events. The construction and structure of the geological events database are described in Battaglini *et al.* (2020). Particular attention has been devoted to the definition of attribute tables in order to achieve the best degree of harmonization and standardization according to the European INSPIRE Directive. Instructions indicate the format and properties for each field of the attribute tables, providing lists of terms allowed for coded domains. Each occurrence reported is complemented by the appropriate reference, whereas additional information can be provided if available. Harmonized data can be browsed through online, while original data can be recovered through references.

EMODnet Geology: collating data on 'Geological events and probabilities'

The main purpose of each EMODnet Geology WP is to overcome the fragmentation and dishomogeneity

Tectonics

Active tectonics is one of the issues included into WP6 which had not been foreseen by the project

From: Asch, K., Kitazato, H. and Vallius, H. (eds) 2022. *From Continental Shelf to Slope: Mapping the Oceanic Realm.* Geological Society, London, Special Publications, **505**, 155–167.
First published online January 21, 2021, https://doi.org/10.1144/SP505-2019-96

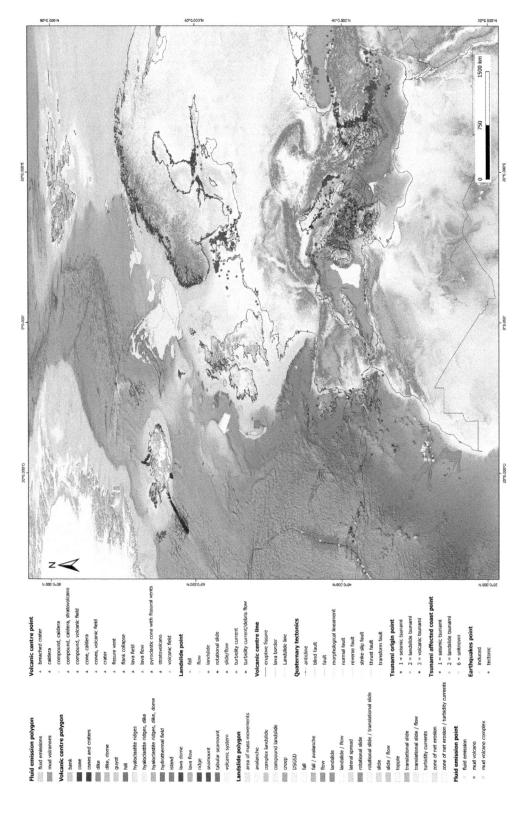

Fig. 1. GIS visualization of geological events occurrences in European seas (underlying Digital Terrain Model (DTM) from EMODnet Bathymetry).

tender. However, in agreement with the partners, it was decided to collate data on Quaternary tectonics, since it is closely connected to geological events.

More than 15 000 records have been entered into the database. They concentrate on the most active portions of European seas, from the mid-Atlantic ridge of Iceland and Portugal to the Mediterranean and Black seas. Occurrences should be complemented by information on the type and time interval of activity, specifying whether they crop out or are buried.

Earthquakes

Concerning earthquakes, since standardized updated data have already been gathered by the Seismic Portal (http://www.seismicportal.eu) of the European-Mediterranean Seismological Centre, it was decided to rely on the existing website by providing a link on the EMODnet Geology Portal. However, the European-Mediterranean Seismological Centre applies a strict automated protocol to select data to be displayed, through which not all events considered locally relevant by EMODnet Geology Partners are selected. Consequently, an earthquake layer has been created as a WP6 product in order to allow partners to provide additional harmonized data which add up to almost 3000 entries

Details to be listed in the attribute table, where they are known, are location, depth, date and time, magnitude and magnitude type, source, deformational style, strike, dip and rake.

Submerged volcanic structures

More than 1000 occurrences of volcanic centres are mainly located close to plate boundaries, along the mid-Atlantic ridge and in the Tyrrhenian and Aegean seas.

Occurrences can be characterized by morphological and activity types, age of activity, chemical composition (TAS or magmatic series), eruption frequency, Volcanic Explosivity Index and the age of its reference eruption, height, volcanic district and the presence of fluid emissions.

Submerged fluid emissions

Fluid emissions can be of volcanic and hydrothermal origin located in oceanic spreading centres, island arcs or intraplate volcanic areas, or can derive from chemical–physical alterations of organic material trapped in sedimentary layers. Emissions of volcanic origin have been included within the volcanic structures layers, whereas emissions of non-volcanic origin have been gathered within separate layers also including mud-volcanoes.

Seepage may be of microbial, thermogenic or abiogenic origin. Generally fluids are enriched in methane and can be associated with characteristic biotic communities. Fluid emissions have been classified according to the process that generates them or based on the associated materials of geological or biological derivation (e.g. salt domes, mud volcanoes, brine pools, bacterial mats, autogenous carbonates). Morphological types have also been considered since features are related to erosional or depositional processes as well as to the composition and escaping mechanisms of the fluid.

Almost 3500 records have been collated within fluid emission layers in EMODnet Geology WP6. Extended areas of fluid emissions have been identified in the Barents Sea, the North Sea, the Black Sea and offshore the Atlantic southern coast of Portugal and Spain.

Submerged landslides

Submarine landslides have been characterized by type of movement, material involved (lithology), volume, thickness, slope, age and source area. Approximately 3000 records have been entered into the landslides layers. They are common in tectonically active areas as well as in areas with high sedimentation rates, where sediment accumulation increases instability. Nevertheless, large landslides are present in more stable areas such as the northern European seas, where they affect submerged glacial deposits.

A study to tentatively identify landslides predisposing factors and triggering processes is being carried out, in the frame of ISPRA activities addressed at EMODnet Geology WP6, to assess probabilities. It aims at the assessment, by means of a mathematical model, of areas of potential instability where future events might take place. Preliminary results indicate that slope and depth are the highest loading factors (Innocenti et al. 2020).

Tsunamis

Tsunami data were gathered into two layers, one reporting the locations of events that originated tsunamis and another one reporting the stretches of coast affected by tsunamis, this latter to be compiled only in case of unknown origin. Attributes listed for tsunamis are date, type, cause, run-up, intensity and affected coast. Where events originating tsunamis had been entered in one of the other EMODnet Geology WP6 layers, links to those layers were provided in the attribute table as well. More than 500 tsunami records (including both origins and affected coasts) have been entered into the EMODnet Geology database. Tsunamis originated by volcanic activity are less common than those originated by earthquakes.

However, volcanic collapses, violent submarine eruptions or even explosive eruptions on land (when pyroclastic flows fall down and plunge into the sea) may move large volumes of water. Tsunamis originated by landslides can be strongly destructive in the vicinity of the origin area, even when they are of low energy.

Geological events in Italian seas

Digital cartography allows the visualization of selected attributes of features reported therein. Figure 1 shows all of the layers included in EMODnet Geology WP6. Each type of feature is represented by a different attribute: type of source event for tsunamis; type of element for tectonics; morphological type for volcanoes; and type for fluid emissions and landslides. Occurrences of events are often associated with each other, particularly in tectonically active areas where clusters are evident (mid-Atlantic ridge and Canary Islands, Mediterranean Sea). By taking a closer look at such areas, correlation among different types of events may be highlighted, as illustrated below for Italian seas.

Tectonics is the main driver of the complex geological evolution of Italian seas and outlines features characterizing their articulated morphobathymetry (cf. Fiorentino *et al.* 2020). More than 4000 tectonic occurrences have been reported by a systematic collation of data carried out in cooperation with colleagues from other Italian public Institutions (Institute of Marine Sciences of the National Research Council, National Institute of Oceanography and Applied Geophysics and the universities of Genova, Palermo and Trieste). Generalization of features had to be performed at different scales of representation foreseen by different phases of the project (1:250 000 in phase II and 1:100 000 in phase III). Many seismogenic faults are located in marine areas and have caused strong earthquakes and associate tsunamis throughout Italian history. Earthquake hypocentres in marine areas mainly concentrate in the southeastern Tyrrhenian Sea (https://ingv. maps.arcgis.com/apps/webappviewer/index.html? id=86a8b115cbbe4b698f7296fcbd745633). Tsunami origins are also more common in the southeastern Tyrrhenian Sea and surrounding regions (Calabria and Sicily).

Submarine landslides can be induced by volcanic eruptions or by instability generated by fluid emissions, since gases expelled through the seabed lead to a deterioration of the mechanical properties of deposits. Submarine landslides are common in Italian seas (more than 1000 occurrences). Data obtained by elaborating the Camerlenghi *et al.* (2010) database have been complemented by information gathered within the Italian Geological

Mapping Project database and within the MAGIC Project (DPC-CNR 2007–2013).

The compilation of the Italian delivery concerning volcanic structures was carried out working in close cooperation with RomaTRE University. On the basis of a detailed analysis of high-resolution morphobathymetry and validated literature, a total of 76 volcanic seamounts have been identified, 18 of which emerge above sea level as well-known volcanic islands (Giordano *et al.* 2019). Seamounts have been classified according to their morphological type as stratovolcano, fissural edifice, simple cone and composite edifice. They have been organized, according to their location, age and chemical composition, into seven Volcanic Seamount Sectors: Ligurian, Corsica–Sardinian, Etruscan, Neapolitan, Central Tyrrhenian, Aeolian – E Tyrrhenian and Sicily Channel (Pensa *et al.* 2019). Each Volcanic Seamount Sector can be associated with a different geodynamic setting, from the Miocene Ligurian structures to the presently active Aeolian Islands, which mark the progressive southeastward opening of the Tyrrhenian back-arc basin.

Fluid emissions are also common in Italian seas as both volcanic hydrothermal vents and gases originated by physical–chemical alterations of organic material, especially within deltaic palaeoenvironments where sediment input is abundant during low stands of the sea-level.

Correlation and interpretation

A few examples of correlation between geological events will be briefly discussed. A geomorphological features layer, produced by the Geological Survey of Italy for EMODnet Geology Sea-floor Geology at 1:750 000 scale, is represented in the figures together with geological event layers to help show their correlation.

The Ligurian Sea continental slope is dissected by canyons driven by faults (Fig. 2), which drain material from the narrow continental shelf and steep slope to the deeper basin, by means of mass wasting processes. The deeper central part of the Ligurian Sea was originated by Oligo-Miocene rifting, as attested by the presence of the dated seamounts.

The northern Tyrrhenian Sea (Fig. 3) is characterized by north–south-trending structures alternating ridges and depressions aligned with faults. Basins are locally limited northward and southward too by faults oriented in an anti-Apenninic direction, which also control canyons cutting into the eastern Sardinia slope. Landslides have been identified in the Corse basin, whereas a few volcanic seamounts are located in the more articulated southeastern portion of the area.

The central-southern Tyrrhenian basin (Fig. 4) develops from a bathyal plain reaching almost

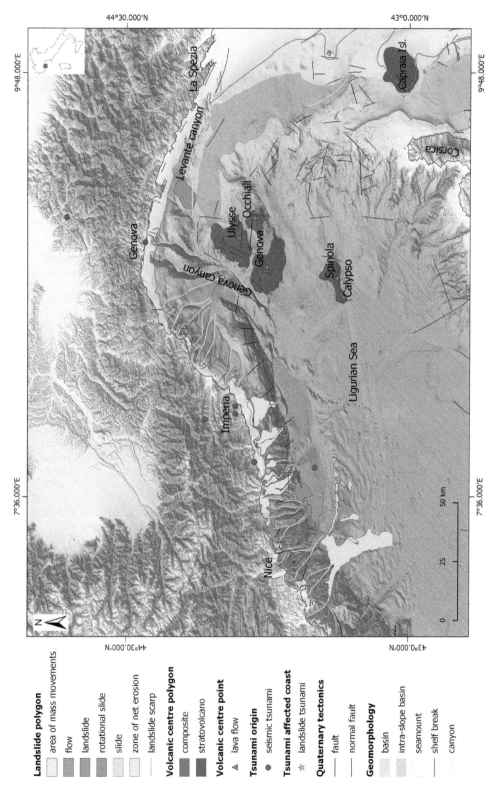

Fig. 2. GIS visualization of geological events and geomorphology in the Ligurian Sea (underlying DTM from EMODnet Bathymetry).

Landslide polygon
- flow
- landslide
- turbidity currents

Volcanic centre
- fissural
- composite

Volcanic centre point
- ▲ lava flow

Quaternary tectonics
- —— fault
- —— normal fault

Geomorphology
- basin
- intra-slope basin
- seamount
- —— shelf break
- —— canyon
- —— ridge

Fig. 3. GIS visualization of geological events and geomorphology in the northern Tyrrhenian Sea (underlying DTM from EMODnet Bathymetry).

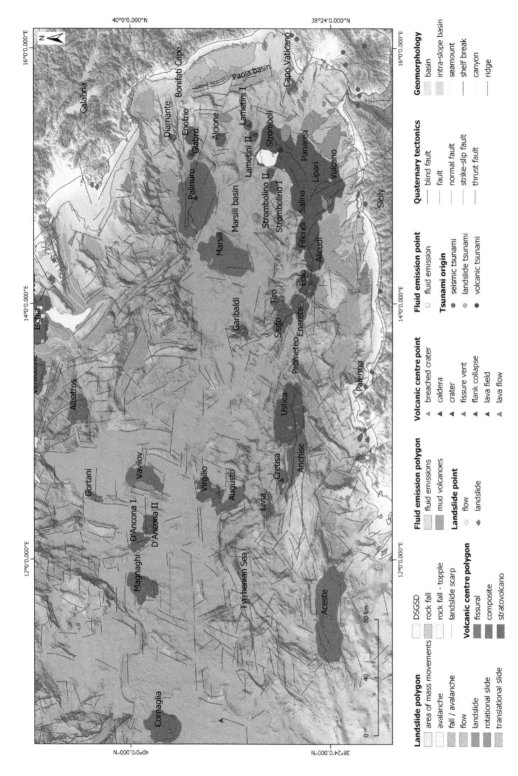

Fig. 4. GIS visualization of geological events and geomorphology in the southern Tyrrhenian Sea (underlying DTM from EMODnet Bathymetry).

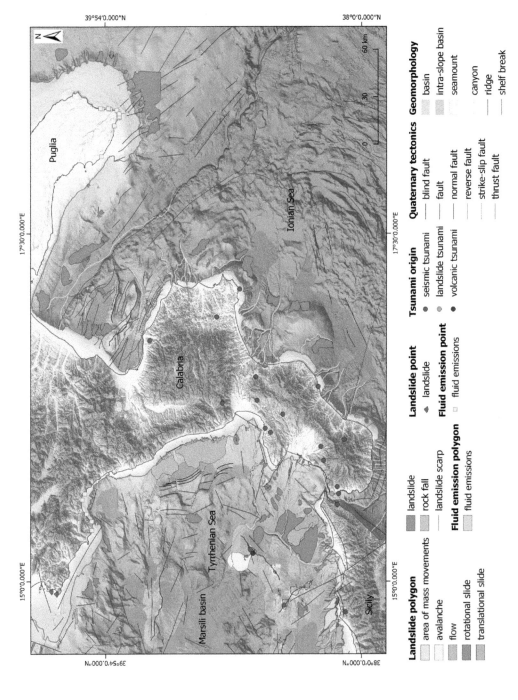

Fig. 5. GIS visualization of geological events and geomorphology in the Tyrrhenian and Ionian sides of Calabria (underlying DTM from EMODnet Bathymetry).

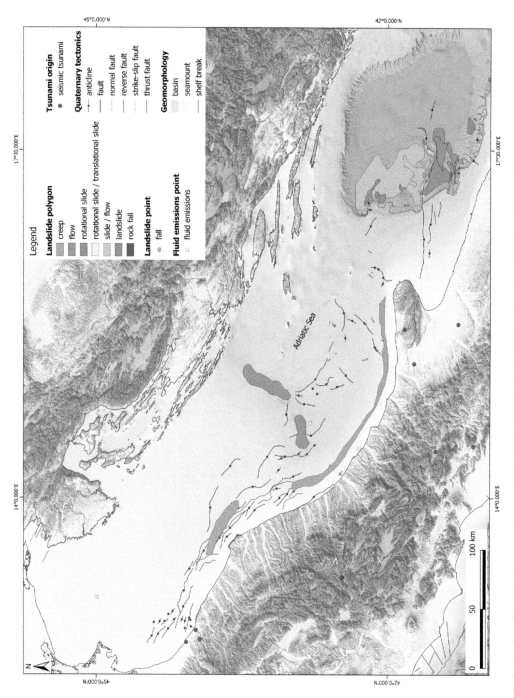

Fig. 6. GIS visualization of geological events and geomorphology in the Adriatic Sea (underlying DTM from EMODnet Bathymetry).

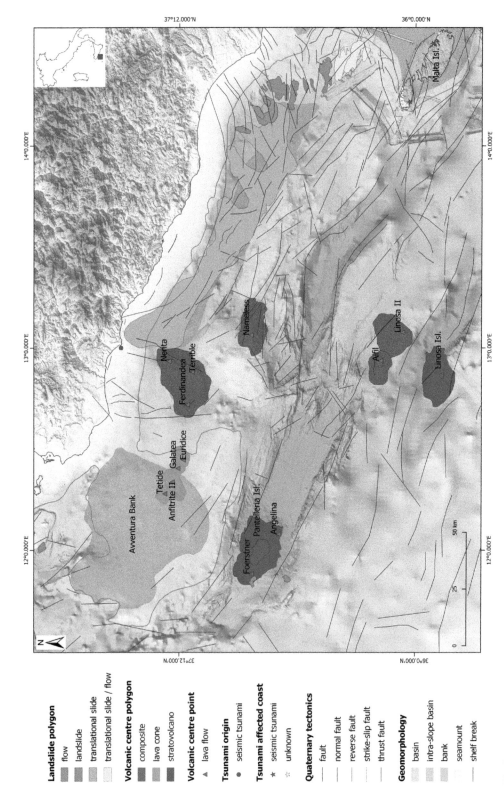

Fig. 7. GIS visualization of geological events and geomorphology in the Sicily Channel (underlying DTM from EMODnet Bathymetry).

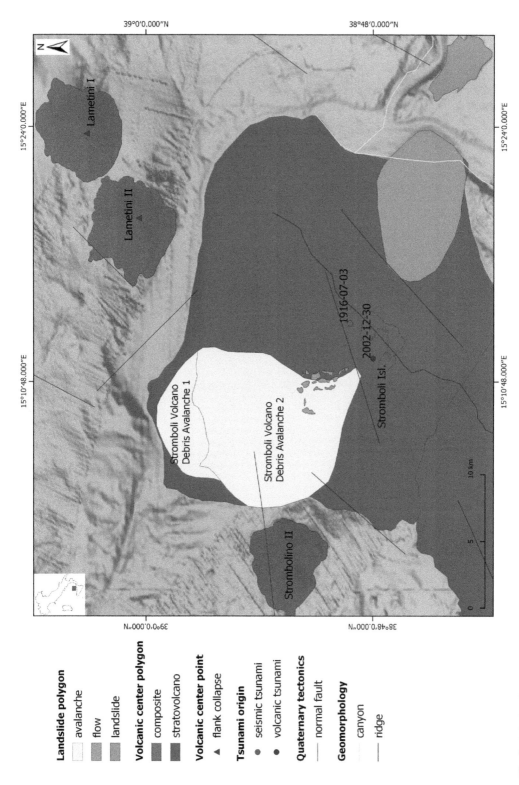

Fig. 8. GIS visualization of geological events and geomorphology of the Stromboli Island (underlying DTM from EMODnet Bathymetry).

4000 m in depth, spread by several volcanic sea-mounts of recent formation that could involve flank collapses. Landslides are common on the continental slope, especially around volcanic islands where debris avalanche may distribute material far away from its origin. Additional landslides affect steep slopes of volcanic seamounts. Numerous non-volcanic fluid emissions surround the area, whereas widespread volcanic vents characterize many sea-mounts (Marsili, Palinuro, Panarea and Vulcano; Monecke *et al.* 2019).

The uprising of the Calabrian Apennines has induced, on both its Tyrrhenian and Ionian sides, frequent and considerable mass movements (Fig. 5). These may affect the entire slope as well as the tectonically controlled canyons systems.

Geological events are less common in the Adriatic Sea (Fig. 6), which partly constitutes the foreland of the Apenninic chain. Landslides here concentrate on the slope which is located in its southern portion, whereas extended creeps parallel the coastline in its central portion and debris flows affect the northern slope of the Mid-Adriatic Depression.

The Sicily Channel (Fig. 7) is characterized by NW–SE-trending horst and graben structures. Extension is here associated with substantial Na-alkaline magmatism (Pensa *et al.* 2019), generating several volcanic seamounts and culminating in the formation of Pantelleria and Linosa islands. Landslides extensively affect the continental slope off the southwestern Sicily coast.

One of the most suitable examples of correlation between geological events is represented by the 2002–03 eruption of the Stromboli volcano. An effusive phase was followed, a few days later, by the fall of a crater ridge, causing a landslide of about 16 million m^3, originating a tsunami that hit the coasts of Stromboli Island and also reached other Aeolian islands as well as the Calabria and Sicily coasts (Fig. 8). Monitoring by scientists and the Civil Protection Department is necessary in such settings.

Conclusions

Geological events and their mutual relationships influence both submerged and coastal environments. The data inventory realized within EMODnet Geology provides a good basis for further studies. Each feature stored in GIS maps can be considered as part of a single category (i.e. landslides, volcanoes, tectonics, etc.) or in relation to other categories.

The inventory allows derivative maps to be elaborated by selecting attributes considered relevant for specific purposes, fostering additional interpretations concerning different aspects. The analysis of data contained therein can contribute to provide evidence of different tectonic settings and can be a great support in the management of marine resources, as well as in planning human activities in marine-coastal areas, particularly for the identification and assessment of geological and environmental hazards. The interaction between different geological processes, which frequently controls other natural and anthropic variables, is often difficult to estimate and predict. Submarine events could cause unquantified risks to society, either directly or through disturbances of marine biological ecosystems, and the loss of biodiversity.

The easily accessible visualization of data concerning submarine environments, available through the EMODnet Geology Portal, represents a means of communicating 'the unseen', providing a tool to deliver important information about geo-hazards and to plan specific surveys for risk assessment.

Acknowledgements We would like to thank all reviewers for their fruitful comments which helped us a lot to improve our manuscript, especially in its first draft. We also want to thank the EMODnet Bathymetry Consortium (2018): EMODnet Digital Bathymetry (DTM) for the DTM displayed in the figures.

Author contributions **LB**: writing – original draft (equal); **SD**: writing – original draft (equal); **AF**: writing – original draft (equal).

Funding This work was funded by the European Maritime and Fisheries Fund (EASME/EMFF/2016/1.3.1.2 – Lot 1/SI2.750862_EMODnet – Geology 3).

Data availability The datasets generated during and/or analysed during the current study are available in the EMODnet Geology Portal, https://www.emodnet-geology.eu/.

References

Battaglini, L., D'Angelo, S. and Fiorentino, A. 2020. Mapping geological events in submerged areas. *In*: Moses, C. (ed.) *Mapping the Geology and Topography of the European Seas (EMODnet).* Quarterly Journal of Engineering Geology and Hydrogeology.

Camerlenghi, A., Urgeles, R. and Fantoni, L. 2010. *A Database on Submarine Landslides of the Mediterranean Sea. Submarine Mass Movements and their Consequences.* Springer, Dordrecht.

DPC-CNR. 2007–2013. Progetto MaGIC Marine Geohazards along the Italian Coasts. Dipartimento della Protezione Civile – Consiglio Nazionale delle Ricerche.

Fiorentino, A., Battaglini, L. and D'Angelo, S. 2020. EMODnet collation of geological events data provides evidences of their mutual relationships and connection with underlying geology: a few examples from Italian seas. *In*: Moses, C. (ed.) *Mapping the Geology and Topography of the European Seas (EMODnet)*. Quarterly Journal of Engineering Geology and Hydrogeology.

Giordano, G., Pensa, A., Fiorentino, A., Vita, L. and D'Angelo, S. 2019. Map of Italian submarine volcanic structures. *In*: D'Angelo, S., Fiorentino, A., Giordano, G., Pensa, A., Pinton, A. and Vita, L. (eds) *Atlas of Italian Submarine Volcanic Structures*. Memorie Descrittive della Carta Geologica d'Italia, **104**.

Innocenti, C., Battaglini, L., D'Angelo, S. and Fiorentino, A. 2020. Submarine landslides: mapping the susceptibility in European seas. *In*: Moses, C. (ed.) *Mapping the Geology and Topography of the European Seas (EMODnet)*. Quarterly Journal of Engineering Geology and Hydrogeology.

Monecke, T., Petersen, S., Augustin, N. and Hannington, M. 2019. Seafloor hydrothermal systems and associated mineral deposits of the Tyrrhenian Sea. *In*: D'Angelo, S., Fiorentino, A., Giordano, G., Pensa, A., Pinton, A.

and Vita, L. (eds) *Atlas of Italian Submarine Volcanic Structures*. Memorie Descrittive della Carta Geologica d'Italia, **104**, 41–74.

Pensa, A., Pinton, A., Vita, L., Bonamico, A., De Benedetti, A.A. and Giordano, G. 2019. ATLAS of Italian submarine volcanic structures. *In*: D'Angelo, S., Fiorentino, A., Giordano, G., Pensa, A., Pinton, A. and Vita, L. (eds) *Atlas of Italian Submarine Volcanic Structures*. Memorie Descrittive della Carta Geologica d'Italia, **104**, 77–184.

Vallius, H.T.V., Kotilainen, A.T., Asch, K.C., Fiorentino, A., Judge, M., Stewart, H.A. and Pjetursson, B. 2020. Discovering Europe's seabed geology: the EMODnet concept of uniform collection and harmonization of marine data. *In*: Asch, K., Kitazato, H. and Vallius, H. (eds) *From Continental Shelf to Slope: Mapping the Oceanic Realm*. Geological Society, London, Special Publications, **505**, https://doi.org/10.1144/SP505-2019-208.

Faulting within the San Juan–southern Gulf Islands Archipelagos, upper plate deformation of the Cascadia subduction complex

H. Gary Greene[1]* and J. Vaughn Barrie[2]

[1]Moss Landing Marine Labs, Moss Landing, CA and Tombolo Mapping Lab, Orcas Island, WA

[2]Geological Survey of Canada, Institute of Ocean Sciences, Sidney, B.C., Canada

*Correspondence: greene@mlml.calstate.edu

Abstract: The San Juan–southern Gulf Islands Archipelago of Washington State, USA and western Canada is located on the upper plate of the Cascadia subduction zone, in the forearc between the trench and volcanic arc. Onland and island investigations show many faults within the region that primarily represent old, inactive faults associated with transport, subduction and accretion of tectonostratigraphic terranes. However, until recently little geologic investigation and mapping have been done in the offshore. From these narrow straits, channels and sounds we have collected and interpreted high-resolution multibeam echosounder bathymetric data, 3.5 kHz sub-bottom and Huntec seismic-reflection profiles, and piston-cores to identify and date recently active faults. Previous studies by us focused on the earlier recognized active Devils Mountain fault zone that bounds the southern part of the Archipelago and the recently reported newly mapped active Skipjack Island fault zone that bounds the northern part. These transcurrent fault zones appear to be deforming and rotating the Archipelago. We concentrate on the unique deformation occurring within the seaways to determine the relationship and styles of faulting associated with these active bounding fault zones and relate the fault geometry and kinematics to one other subduction complex, the New Hebrides island arc of Vanuatu.

The islands of the San Juan Archipelago are composed of rocks of varying ages and types ranging from Mesozoic to Cenozoic volcanic, intrusive and metamorphic basement rocks that have been subjected to subduction and tectonic transport processes (Johnson *et al.* 1986; Brown 1987; Brandon *et al.* 1988; Bergh 2002) to Tertiary sedimentary bedrock units that have been deposited in a forearc basin (Johnson 1982; Johnson *et al.* 1986; Kelsey *et al.* 2012) of which comprise the northern San Juan and southern Gulf Islands Archipelago. Structural and glacial imprinting of these rocks are apparent on the islands and have been well mapped (Easterbrook 1992, 1994; Bergh 2002; Barrie *et al.* 2009). However, until recently little geologic mapping has been done beneath the waters of the Archipelago with the exception of projecting structures mapped on the islands into the marine environment and mapping seafloor faults in the southwestern part of the San Juan Archipelago (Brandon *et al.* 1988; Tilden 2004; Greene and Barrie 2011; Barrie and Greene 2015; Greene *et al.* 2018).

Tectonic setting

The San Juan–southern Gulf Islands Archipelago (Fig. 1) lies within a forearc tectonic setting established between the surface trace (trench) of the Juan de Fuca Plate and Cascadia subduction zone and the volcanic arc present on the North American Plate, and located on the the upper plate. As such the Archipelago is prone to deformation from the downgoing slab and subject to seismicity along the decollement as well as along faults within the brittle upper plate. The forearc islands of the Archipelago are located *c.* 250 km east of the trench, *c.* 150 km west of the volcanic arc, *c.* 150 km from the locked-transition boundary, and *c.* 50 km east of the eastern margin of the transition or active deformation zone (Flück *et al.* 1997). However, Parsons *et al.* (1998) place the deformation zone at 120 km offshore. Wang *et al.* (2003) report that the locked portion of the deformation zone is located entirely offshore beneath the continental slope. Geologic studies are sparse in association with the Cascadia convergent margin and structures with no surface expression or poor exposure have been overlooked (Miller *et al.* 2001).

The regional tectonic setting of northwestern Washington is one that has evolved from the oblique convergence of the Juan de Fuca Plate with the North American Plate where the Cascadia subduction zone defines this active plate boundary. Currently, the Juan de Fuca Plate is moving northeastward in reference to fixed North America and is being subducted at a rate of *c.* 42 to 45 mm/yr (Murray and Lisowski

From: Asch, K., Kitazato, H. and Vallius, H. (eds) 2022. *From Continental Shelf to Slope: Mapping the Oceanic Realm.* Geological Society, London, Special Publications, **505**, 169–191.
First published online July 15, 2020, https://doi.org/10.1144/SP505-2019-125

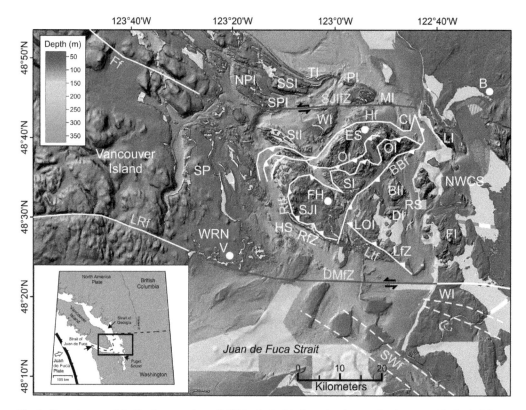

Fig. 1. Tectonic map of the San Juan–southern Gulf Islands Archipelago showing previously mapped faults (solid lines are well-defined major faults, dashed lines are inferred faults); white lines with yellow labels are faults constructed from mapping on the San Juan Islands, after Brandon (1989) and other islands (Vancouver and Whidbey), red lines represent seafloor bounding faults of the Archipelago including the recently mapped Skipjack Island fault zone (SJIfZ; Greene *et al.* 2018), the previously mapped Devils Mountain fault zone (DMfZ; Johnson *et al.* 2001; Barrie and Greene 2018). These along with islands and towns overlie multibeam echosounder bathymetric and LiDAR imagery. SWf, South Whidbey fault zone; WI, Whidbey Island and Whidbey fault zone; LfZ, Lopez fault zone (complex) or Ltf, Lopez thrust fault; Ff, Fulford fault and LRf, Leech River fault of Vancouver Island; BBf, Buck Bay fault; Rtf, Rosario thrust fault or RfZ, Rosario fault zone; Ot, Orcas thrust; Hf, Haro fault; WRN, Wrangalia terraine; NWCS Northwest Cascades–San Juan Islands tectonic terranes; FI, Fidalgo Island; DI, Decatur Island; LOI, Lopez Island; SJI, San Juan Island; V, Victoria on Vancouver Island; FH, Friday Harbor; SI, Shaw Island; BlI, Blakely Island; B, Bellingham; LI, Lummi Island; CI, Clark Island; OI, Orcas Island; ES, Eastsound; WI, Waldron Island; StI, Stewart Island; SPI, South Pender Island; SSI, Saturna Island; TI, Tumbo Island; PI, Patos Island; MI, Matia Island; HS, Haro Strait; RS, Rosario Strait. Modified after Greene *et al.* (2018).

2000; Wilson 2002; DeMets *et al.* 2010). Based on GPS velocity data collected within the Cascadia subduction zone region, the northwestern Washington/southwestern British Columbia region is presently being shortened in a N–S direction at *c.* 3 to 3.5 mm/yr, with estimates of long-term shortening at 5 to 6 mm/yr (Mazzotti *et al.* 2002; McCaffrey *et al.* 2013) or as much as 6 to 7 mm/yr. (Murray and Lisowski 2000). Dextral translation of the Sierra Nevada block of California appears to be the cause of this shortening (Wells *et al.* 1998). Convergence of this block with the Oregon forearc near the Mendocino triple junction has resulted in clockwise rotation

of the Oregon forearc, which in turn drives the Washington forearc northward against the Canadian Coast Mountains (Wells *et al.* 1998). In addition, the Cascadia margin has been described as complex, responding kinematically to the interactions of two superimposed forces at the North American–Juan de Fuca–Pacific plates boundary, consisting of oblique subduction and entrainment of the Sierra Nevada and Oregon Coast Range blocks (Miller *et al.* 2001). These authors point out that the movement of forearc tectonic blocks are associated with movement along well-known faults south of Cascadia, such as the San Andreas, Ma'acama and Bartlett Springs fault zones in central California, which may connect through Oregon with the Straight Creek Fault of Brandon *et al.* (1988) in Washington State. Based on GPS observations, Miller *et al.* (2001) divided the Cascadia convergent margin into three coastal kinematic domains. From north to south these domains consist of: (1) a relatively stable Canadian forearc, due largely to subduction coupling, (2) northwestward migrating Cascadia forearc, and (3) Pacific plate entrainment coupled with clockwise or northward translation of the southern Cascadia forearc (Snavely and Wells 1996; McCaffrey *et al.* 2000; Savage *et al.* 2000; Wells and Simpson 2001).

In the general vicinity of our study area, forearc deformation varies markedly from north to south. Along the Canadian segment the forearc inboard of the northern Juan de Fuca Plate is deforming nearly orthogonal to the strike of the deformation front and evidence of forearc migration parallel to the subduction zone is absent (Miller *et al.* 2001). Along the central Cascadia margin northwestward transport of the forearc predominates with oblique subduction of the Juan de Fuca Plate.

At depth beneath the immediate location of the San Juan–southern Gulf Islands Archipelago interpretation of seismicity data (Brudzinski and Allen 2007) and air gun seismic-reflection profiles (Parsons *et al.* 1998) indicates that the subducting slab flattens from *c.* 12° in the west to *c.* 7° in the east, is less steeply dipping than down-going slabs to the north and south, and is positioned just north of the bend in the subduction zone (Flück *et al.* 1997). Depth to the shallow dipping slab lies between 40 and 50 km in our area of study and is shown to curve at depth around the core of the eastern Olympic Mountains, extending into Vancouver Island beneath the general trends of the western Devils Mountain fault zone and Leech River fault (Brandon *et al.* 1988), mimicking the curved plate where episodic tremor and slip (ETS) have been reported to occur at intervals of *c.* 15 months (Brudzinski and Allen 2007). Brudzinski and Allen (2007) divide the Cascadia forearc into three broad ETS zones: (1) Vancouver Island or Wrangellia zone (ETS of 14 $\pm$ 2 mo.); (2) Central Oregon or Siletzia zone (19 $\pm$ 4 mo.); and (3) California or Klamath zone (14 $\pm$ 2 mo.). These authors initially interpreted the Cascadia forearc basin as the manifestation of megathrust asperities, yet they state that perhaps upper plate properties may play a major role in ETS.

Seismicity

Recently recorded earthquakes of moderate to large magnitudes (>M4) are rare in our area of study, although it lies within the northern extent of the seismically active Puget Lowland. Earthquakes that occur in the region consist of both crustal (located in the upper 37 km of the crust) and subduction earthquakes that occur at deeper depths along the subducting slab (Hyndman *et al.* 2003; McCaffrey *et al.* 2013). Crustal earthquakes of low to moderate magnitudes in the greater Puget Sound region occur primarily as a result of N–S compressive stresses along western Washington. Based on palaeoseismologic studies at least five to seven Holocene ground-rupturing earthquakes have occurred within the Puget lowlands (Bucknam *et al.* 1992; Sherrod *et al.* 2000; Nelson *et al.* 2003; Johnson *et al.* 2004; Kelsey *et al.* 2004). Also, evidence exists that earthquake-associated deformation of the seafloor or associated landsliding has produced tsunamis in the region of our study (Atwater and Moore 1992; Williams and Hutchinson 2000; Bourgeois and Johnson 2001; Williams *et al.* 2005).

Glaciation

The most recent large-scale ice-advance, the Fraser Glaciation, began around 25 to 30 ka ago. Thick, well-sorted sand deposits (Quadra Sand) were laid down in front of, and possibly along, the ice progressing south-eastwards through the Strait of Georgia and across the San Juan–southern Gulf Islands Archipelago (Clague 1976). These deposits were over-ridden and reworked by advancing ice. After advancing through the Gulf and San Juan Islands region, the ice split into two tongues: one flowing southwards into Puget Sound; and the other making a sharp right-hand turn near Victoria and on through the Strait of Juan de Fuca towards the open Pacific Ocean. The dual tongues, the southern Puget Lowland of Washington State and the western Strait of Juan de Fuca, both reached their maximum extent at about the same time, about 14 000 ^{14}C ka BP (Porter and Swanson 1998; Hewitt and Mosher 2001).

Following the Last Glacial Maximum (LGM), ice retreat through the Puget Sound basin was rapid with Juan de Fuca Strait being deglaciated by 13 600 ^{14}C ka BP (Mosher and Hewitt 2004; Barrie and Greene 2018), and the Strait of Georgia ice-free before 12 000 ^{14}C ka BP (Barrie and Conway 2002; Hetherington and Barrie 2004). During deglaciation, widespread ice-stagnation and down-wasting throughout the basin resulted in deposition of diamicton (locally 30–60 m thick), overlain by ice-proximal and ice-distal glaciomarine sediments (Barrie and Conway 2002; Guilbault *et al.* 2003). As ice-retreated from the San Juan-Gulf Islands area, the region was isostatically depressed. There was a rapid fall in sea level from a +75 m high stand before 12 000 ^{14}C ka BP to lower than the present shoreline by 11 400 ^{14}C ka BP (Clague and James 2002; James *et al.* 2009). The ultimate post-glacial sea-level low stand in the region was most likely −30 $\pm$ 5 m at approximately 9800 ^{14}C ka BP (James *et al.* 2009).

Previous work

Faults within the San Juan Archipelago have been previously mapped from exposures on the islands as older seismically inactive structures (Fig. 1). Two of these most prominent fault zones mapped on the islands, the NW–SE trending Lopez fault zone (LfZ), or Lopez Structural Complex, the Lopez thrust of Brandon *et al.* (1988), and the Rosario fault zone (RfZ) juxtapose a mixture of deformed igneous, sedimentary and metamorphic rocks that were emplaced during the Late Cretaceous terrane accretion of the San Juan thrust system. Based on the detailed microscopic structure analyses on rocks within these two fault zones Bergh (2002) concluded that the structural fabrics are related to two Late Cretaceous deformation events; (1) SW–NE shortening that occurred during the initial subduction-related accretion of the nappes; and (2) a combined orogen-parallel strike-slip motion and NW-directed thrusting associated with oblique convergence. Tilden (2004) concluded that oblique left-lateral convergence along the E–W trending Devils Mountain fault resulted in the elevation of a NW–SE trending bedrock high offshore of southwestern San Juan Island. Brandon *et al.* (1988) also show two other thrusts, the Buck Bay fault (BBf) and Orcas thrust (Ot), that occurred internally within the San Juan thrust system and a fault named the Haro fault (Hf) by Johnson (1982, his fig. 5) that separates external Cretaceous to early Tertiary units to the north from the San Juan thrust system of rocks in the south (Fig. 1).

The RfZ was traced from the southern tip of San Juan Island north as thrust or front of a nappe up and over northern San Juan Island (Brandon 1980), and eastward across Orcas Island (Fig. 1). This 1 km wide, gently dipping fault zone juxtaposes strongly foliated mudstone mélange containing ribbon chert, schist and metavolcanic rocks derived from the footwall composed of the Deadman Bay terrane, a rock sequence of early Permian to early Triassic volcanic pillow basalts and breccias interbedded with massive limestone and ribbon chert, and metasandstone from the headwall Constitution Formation.

The BBf, is a low angle thrust that juxtaposes the LfZ structural complex and Decatur terrane with the Constitution Formation and older terranes (Fig. 1). A small segment of this fault is mapped on Orcas Island by Brandon (1980). However, much of its length is inferred to extend offshore to the southwest based on the need for a fault to explain the proposed truncation of the Lopez Structural complex in southern San Juan Channel. This fault is thought to be part of what has been named the Mid-Cretaceous San Juan Thrust system (Brandon 1980).

The northern part of the San Juan Archipelago was defined by Johnson (1986) to be separated structurally from the southern Gulf Islands by the Hf, a thrust fault that separates upper Cretaceous to lower Tertiary sedimentary rocks in the north from older metamorphic rocks in the south. This fault was mapped across the northernmost part of Orcas Island and proposed to extend offshore to the west where it is shown to lie between Spieden and Stuart islands, and to the east where it is proposed to swing south around Clark and Barns islands (Fig. 1). This fault zone was considered to be inactive as no seismic events have been interpreted to be associated with it.

Objectives

The purpose of our study is to map faults in the marine environment through the interpretation of bathymetric and geophysical data that could be used to refine the structure within the San Juan–southern Gulf Islands Archipelago, and to better understand the ongoing tectonic processes. The Archipelago is bounded on the south by the active Devils Mountain left-lateral transcurrent fault zone (Johnson *et al.* 2001; Barrie and Greene 2018) and on the north by the recently mapped Skipjack Island left-lateral transcurrent fault zone (Greene *et al.* 2018), suggesting that active deformation may be occurring within the Archipelago between the two fault zones (Fig. 1). Therefore, the objective of our study was to map the fault geometry and identify any evidence of recent fault motion.

We focus on three major areas within and along the margins of the San Juan Archipelago: (1) the southern Georgia Strait–Lummi Island area in the vicinity of the recently mapped eastern extent of the Skipjack Island fault zone (SJIfZ); (2) the Southern Lopez Island area in the vicinity of the previously mapped LfZ complex and Devils Mountain fault zone (DMfZ); and (3) the Haro Strait area that separates Vancouver Island, BC Canada from the western margin of the Archipelago in Washington State, USA (Fig. 1). With the exception of the southern part of the Haro Strait area mapped by Tilden (2004), faults within these areas have not previously been mapped in detail. Our mapping provides insights into the recency and kinematics of the regional tectonic processes.

Data and methods

A suite of marine geophysical data are used to map the faults within the San Juan–southern Gulf Islands Archipelago. These data consist of wide swathe multibeam echosounder (MBES) images, Light Detection And Ranging (LiDAR) images, and sub-bottom acoustic and seismic-reflection profiles. In addition, piston cores were collected in selected locales to identify types of sediment, ages of deposition and date deformational events.

Multibeam echosounder data

Wide swathe MBES bathymetry and backscatter collected in cooperation with the Geological Survey of Canada, Canadian Hydrographic Service, Center for Habitat Studies, Moss Landing Marine Labs, and Tombolo/Sea Doc Society Tombolo Mapping Lab were used in the interpretation of structures in this study (Fig. 2). From 2001 to 2008 the Canadian Coast Guard vessels *Otter Bay*, *Revisor*, *Young* and *Vector* acquired extensive high-resolution bathymetric datasets within the waterways surrounding the San Juan–southern Gulf Islands Archipelago. The MBES Simrad EM 1002 (95 kHz frequency) and EM 3000–3002 (300 kHz frequency) systems were used for deep (>80 m) and shallow (<80 m) waters with dataset resolutions at 5 and 2 m. In most areas, the tracks were positioned so as to insonify 100% of the seafloor with a 100% overlap, providing 200% coverage. Positioning was accomplished using a broadcast Differential Global Positioning System (DGPS) and MBES data were corrected for sound speed variations in the stratified water column using frequent acoustic speed casts.

In addition to bathymetric data, the MBES systems collected and recorded backscatter intensity, which can be used to determine relative sediment differences. The multibeam bathymetry and backscatter raster datasets were processed using ESRI™ ArcGIS tools.

LiDAR data

The LiDAR data were obtained from the Puget Sound LiDAR Consortium in the USA and from Bednarski and Rogers (2012) for Canada, which cover much of the island's land areas (Fig. 2). Bare-earth LiDAR imagery was interpreted in the same fashion as the offshore MBES digital elevation model (DEM) data and was incorporated into a GIS project as a seamless digital topographic–bathymetric dataset. Identification and alignment of faults are made from LiDAR imagery where linear bedrock

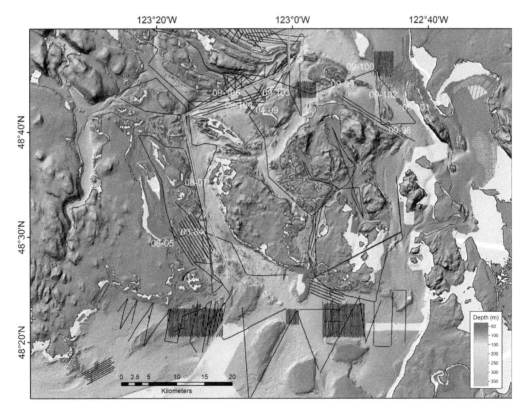

Fig. 2. Data collection map showing track lines along which seismic-reflection (Huntec, black lines) and sub-bottom (3.5 kHz CHIRP, red lines) profiles, and locations of cores (red dots) were collected within the San Juan–southern Gulf Islands Archipelagos.

scarps and shoreline morphology is generally oriented with the regional structural grain.

Seismic-reflection profile data

Two types of seismic-reflection profile data were collected: (1) Huntec boomer; and (2) 3.5 kHz sub-bottom profile data. The deep-towed Huntec DTS (2–500 kHz) seismic-reflection profiles were collected by the Geological Survey of Canada aboard the *CCGS John P. Tully* in 2000 and aboard the *CCGS Vector* in 2003, 2004, 2008, 2009 and 2013 (Fig. 2). Knudsen™ 3.5 kHz Compressed High Intensity Radiated Pulse (CHIRP) 3260 sub-bottom profile data were collected in the northern San Juan Archipelago from the *RV Tombolo* in 2016. The Huntec surveys used higher energy and lower frequency sources (16 transducers), resulting in lower (*c.* 5 m compared to *c.* 2 m for CHIRP) near-surface resolution of reflections and small-scale structures, but offer increased image depths (up to 0.5 s) into the sediment and older sedimentary rocks. The higher resolution CHIRP data provides detailed images of layering that were targeted for coring.

Interpretation of seismic-reflection profile data

The IHS Kingdom Version 8.8 software program was used to view the Huntec seismic-reflection data collected after 2013. The 64-bit copy used for this interpretation is housed at the Geological Survey of Canada, Pacific Geoscience Center. For seismic-reflection data collected prior to 2013 analogue copies only exist and the data could not be presented in Kingdom. A Knudsen software processing and visualization program was used to view and interpret the 3.5 kHz CHIRP data. Criteria for the identification of faults are: (1) the juxtaposition of acoustic characteristics such as displaced well-layered reflections; (2) truncation of layered reflections against acoustically transparent or chaotic reflections; (3) deformed reflections along a vertical or slanting plain; and (4) hyperbolic diffractions associated with truncation of layers (i.e. a probable fault plain), or a combination of one or more of the above criteria.

Core data

A Benthos piston corer was used to collect seafloor sediment samples and to determine stratigraphic ages. A total of 12 cores were collected within our area of study from the *CCGS Vector* in 2004 (Fig. 2). At the land-based sediment lab of the Geological Survey of Canada's Pacific Geoscience Center, Sidney B.C., all cores were split, described in detail and run through a Multi-Scan Core Logger (MSCL) where density, velocity and magnetic susceptibility were measured. Selected subsamples were taken where carbon datable material was present and submitted for ^{14}C radiometric dating.

Results

Examination of the multibeam bathymetric and backscatter data along with the seismic-reflection profiles shows that linear and curved seafloor features characteristic of exposed fault planes or glacial deposits such as moraines and eskers exist on the seafloor within the San Juan–southern Gulf Islands Archipelago. We focus on three areas of deformation that lie between the two major defining E–W trending transcurrent fault zones of the San Juan Archipelago, the SJIfZ and the DMfZ. In the north, near the questionably mapped eastern terminus of the SJIfZ, a series of curvilinear seafloor scarps suggest faults may be active in this NE part – the Southern Georgia Strait–Lummi Island area – of the San Juan Archipelago. In the south, in the vicinity of the DMfZ, a series of E–W trending faults known as the Lopez fault zone or complex (Brandon 1980, 1989; Brandon *et al.* 1988) are mapped along the southern margin of Lopez Island – the Southern Lopez area – and control the morphology of the bays and small islands in this area. In the western part of the San Juan Archipelago – Haro Strait area – NW–SE trending faults that sweep from near Cattle Point on southern San Juan Island towards Vancouver Island of British Columbia, Canada are mapped, which are explained and shown in detail below.

Southern Georgia Strait–Lummi Island Area

The SJIfZ, the northern structural boundary of the San Juan Archipelago, is a left-lateral strike-slip zone of faults that extends from near central Vancouver Island, possibly connecting to the Fulford fault on Vancouver Island, past southern Sucia Island, and splitting around Matia Island where it trends eastward towards Bellingham Bay (Fig. 3; Greene *et al.* 2018, their Fig. 10). The eastern offshore terminus of the fault zone is not well defined and may extend inland to connect with faults mapped onshore, such as the Vedder Mt fault, or it may be truncated along NNW curving faults described here. Evidence of possible recent activity along the SJIfZ is observed west of Skipjack Island where the fault zone bifurcates into two splays characterized as a flower structure at depth with both faults exhibiting seafloor expressions (Fig. 4). Flower structures are characteristic of strike-slip, representing wedge-shaped sediment or rock bodies as a component of convergence or divergence across different fault strands (Christie-Blick and Biddle 1985).

The MBES imagery of the Southern Georgia Strait–Lummi Island area exhibits complex seafloor

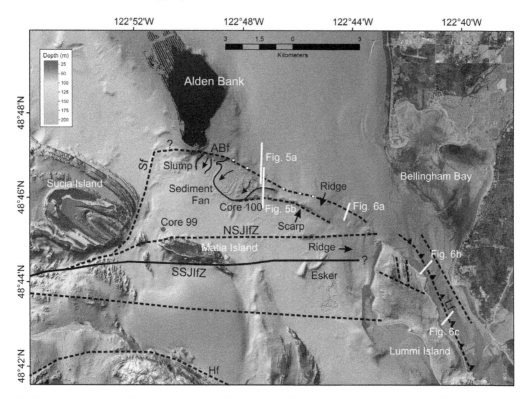

Fig. 3. Bathymetric map of the southern Georgia Strait–Lummi Island study area showing location of faults interpreted from seismic-reflection and sub-bottom profiles and correlated with seafloor morphology. NSIfZ, northern Skipjack Island fault zone; SSIfZ, southern Skipjack Island fault zone; Hf, Haro fault; Sf, Sucia fault; Lf, Lummi fault. White lines represent locations of seismic-reflection and sub-bottom profiles presented as figures. Red dots are locations of cores (Core 99 = Geological Survey of Canada Core #200909-99 and Core 100 = Geological Survey of Canada Core #200909-100). Modified after Greene *et al.* (2018).

morphology that appears to be related to faulting and glacial deposition (Fig. 3). In the bathymetry linear and arcuate NW–SE trending scarps and narrow ridges are well defined along with a zone of relatively hummocky seafloor. This is in contrast to an otherwise smooth seafloor, extending from the southern tip of Alden Bank towards the northern tip of Lummi Island. Southeast of Alden Bank, a slump, a densely rippled sediment fan, and two short subparallel ridges are imaged, south of which a short esker is visible. A more continuous narrow ridge curves southeastward towards northern Lummi Island. An E–W trending scarp just west of the northern tip of Lummi Island represents the seafloor expression of the SJIfZ as mapped by Greene *et al.* (2018) and reflects an apparent left-lateral offset of the narrow continuous ridge in the vicinity where the inferred northern strand of the SJIfZ (NSJIfZ) is mapped.

Two sets of acoustic sub-bottom profiles were used to identify and map faults within this area (Fig. 2). The first set consists of the N–S oriented Huntec seismic-reflection profile lines located

southeast of Alden bank, and the second set consists of randomly oriented 3.5 kHz and Huntec profile track lines located south and east of the first set. The Huntec profiles in this second set exhibit normal faults that align with the linear and arcuate seafloor expressions (scarps and ridges) observed in the MBES data (Figs 3 & 5).

Alden Bank–Lummi Island – a fairly continuous ridge crest is acoustically characterized in the Huntec profiles by chaotic reflections that may represent a recessional moraine (Fig. 5). East of this most easterly, and most continuous ridge, a thick (c. 20–30 m) sequence of deformed alternating well-layered to acoustically transparent reflections appear ponded against the ridge. In many of the Huntec seismic-reflection profiles the contact between the ridge reflections and the layered sequence is sharp and characteristic of a normal fault, down to the east, where folded reflections are also suggestive of faulting (Fig. 5). This contact varies in inclination (verging westward and locally ranging from near vertical to steeply dipping), but is readily correlated

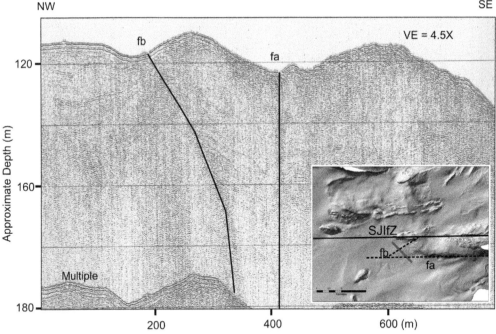

Fig. 4. Huntec seismic-reflection profile showing primary Skipjack Island fault zone and associated flower structure, seafloor expression indicates recent fault movement; (**a**) interpreted profile, and (**b**) MBES bathymetry map showing location of profile and associated faults. SJIfZ, main trace of Skipjack Island fault zone; fa, location of fault a in 'a' and fb = location of fault b in 'a'. Location shown in Figure 11.

from one line to another. Normal faults are identified displacing the well-layered acoustic unit east of this continuously mapped contact. Near-vertical normal faults are also observed in the profiles west of the continuous ridge, as are slumps and random blocks characteristic of mass wasting (Fig. 3).

Lummi Island–Bellingham Bay – the 3.5 kHz sub-bottom profiles and Huntec seismic-reflection profiles collected between Lummi Island and Bellingham Bay also image near-vertical normal faults, reverse faults, and thrusts that generally verge westward (Fig. 6). Two fairly continuous fault traces have been mapped between Lummi Island and the mainland (Fig. 3). The westernmost fault appears as a westward verging thrust that parallels the curving coastline of Lummi Island and projects northwestward towards the thrust or reverse fault that is mapped from the Huntec profiles along the eastern margin of the continuous ridge. The easternmost continuous fault is a steep eastward verging thrust fault (Fig. 6c). Between these two faults is a thick (c. 10 m) well-stratified deformed sedimentary sequence consisting of alternating well-layered and acoustically transparent reflections that are folded, faulted and eroded with normal faults locally located between the two continuous parallel faults that bound the channel (Fig. 6c).

A piston core (Geological Survey of Canada Core #200909-100, or 100 as referenced herein) located near the southern margin of the fan/slump south of Alden Bank (Fig. 3) as described in Greene *et al.* (2018) penetrated 200 cm of sand with shell and shell hash with datable material at depths of 30 cm and 122 cm that yielded a ^{14}C age of 8400 ± 40 ka BP and 9390 ± 40 ka BP. This core is the closest sample we recovered near this area of investigation, which recovered material deposited by the glaciers, materials represented by reflections that appear to be displaced by faulting.

Southern Lopez area

In the northern Strait of Juan de Fuca, south of Lopez and San Juan Islands lies the DMfZ, a W–NW-striking, 125-km-long, oblique-slip transpressional deformation zone that extends westward from the Cascade Range foothills across the eastern Strait of Juan de Fuca to Vancouver Island (Johnson *et al.* 2001; Fig. 1). Onshore and offshore geophysical and geologic data indicate left-lateral and vertical displacement of Quaternary sediments at numerous locations along the zone, indicating high seismic potential (Oldow 2000; Johnson *et al.* 2001, 2004). Possible additional segments to this fault zone

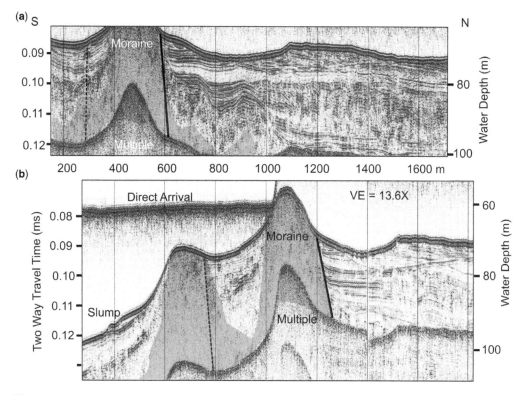

Fig. 5. Huntec seismic-reflection profiles showing glacial and recent Fraser River deposits juxtaposed with a glacial moraine indicated by purple overlay; (**a**) northern extent profile, and (**b**) southern extent profile (see Fig. 3 for profile locations).

were recognized east of Whidbey Island in Skagit Bay and south of the southeastern part of San Juan Island. Recent mapping (Barrie and Greene 2018) of the fault zone suggests that it extends onshore Vancouver Island to where it may join the Leech River fault (LRf). Based on previously reported deep-penetration seismic-reflection profiles, the dip of the DMfZ is estimated to be 61° ± 10° to the north (south verging) with the steepest dips in the east near the western shore of Whidbey Island (Johnson *et al.* 2001).

The LfZ of Brandon (1989) is located north of the DMfZ and has been mapped as a 3-km-wide system of steep to sub-vertical en echelon right-lateral strike-slip faults that control the bays and small islands of southern Lopez Island. Bergh (2002) speculated that the high-angle LfZ was located above a thrust detachment represented by the RfZ that allowed for orogen-parallel translation.

The MBES image in and around southern Lopez Island offshore generally exhibits a smooth seafloor with no or little expression of faulting or other disturbances within the harbours and bays (Fig. 7). However, a more irregular seafloor in the south suggests deformation has occurred there recently. The

LiDAR images onshore show distinct effects of past faulting as evidenced in the general E–W lineaments along the edges of rock outcrops and the shapes of the bays and offshore rock exposures. We used the offshore bathymetric images of hard rock outcrops in contact with the smooth bay sediment and 3.5 kHz sub-bottom profiles to identify and map faults. These data were used to extend the faults mapped onshore and to refine the fault geometry in the area.

The 3.5 kHz sub-bottom profiles were collected simultaneously with the MBES data and consequently the track lines along which the data were collected are sinuous, due to the priority in collecting detailed MBES data at 200% coverage (Fig. 2). Interpretation of the sub-bottom profiles indicates that beneath a fairly flat and smooth seafloor active deformation in the form of faulting and folding exists. A series of NE–SW trending normal, reverse and thrust faults are mapped along with a couple of short bifurcating and E–W trending conjugate faults.

Onshore to the north of the bays of southern Lopez Island is a surface expression in the LiDAR image of a fairly lengthy E–W trending fault that extends offshore both to the east and west (Fig. 7).

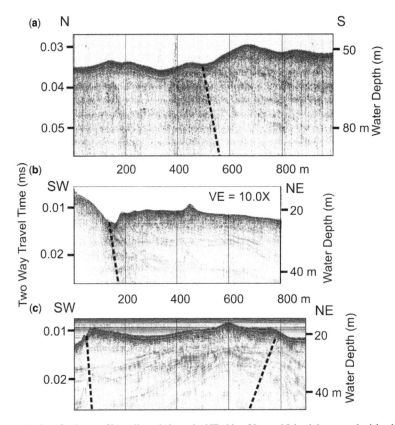

Fig. 6. Huntec seismic-reflection profiles collected along the NE side of Lummi Island, between the island and the mainland south of Bellingham Bay; (**a**) interpreted steeply dipping reverse fault north of the northern end of Lummi Island, (**b**) interpreted reverse fault off northeastern Lummi Island, and (**c**) the southern most profile showing interpreted steeply dipping reverse fault along Lummi Island and a west dipping thrust fault near the mainland (see Fig. 3 for locations).

This fault marks the northern margin of a rock outcrop in the east that is also faulted along its southern margin by a short NW–SE trending fault. The eastern projection of this lengthy fault appears to connect with a fault mapped in Rosario Strait (Fig. 8a) from a Huntec seismic-reflection profile where a fairly broad zone of acoustically transparent reflections truncates folded, well-layered reflections.

The faults mapped in the offshore of southern Lopez Island present a complex geometry but can be divided into three duplex fault zones (a series of faults defined by two continuous bounding faults), here named from north to south: (1) the Mackaye Harbor duplex (Duplex 1), which appears to fit with Brandon (1989) thrust fault in this location; (2) Johns Point duplex (Duplex 2); and (3) Iceberg Point duplex (Duplex 3). Each of the fault duplexes are composed of two separate fairly lengthy (>4 km long) faults offset (anastomosing) from each other in the central part of the study area (Fig. 7) and may be structurally connected by short conjugate faults.

Duplex 1 – Mackaye Harbor duplex is c. 0.5 km wide, NW–SE oriented, and consists of two diverging parallel faults that broaden in separation to the west where they are interconnected locally with a short sub-parallel conjugate fault (Fig. 7). This fault zone is composed of vertical to near-vertical normal faults (Fig. 8b) that juxtapose dipping and deformed high amplitude reflections against chaotic and acoustically transparent reflections. The northern strand of this duplex zone appears to control the straight bedrock northern margin of the Mackaye Harbor and extends onshore in the east where a surface expression of the fault is well exhibited. In the west, the northern strand can be extended across Richardson Bay and across land as suggested by a subtle scarp observed in the LiDAR image. To the east of Richardson Bay, in northwestern Mackaye Harbor, a fault strand bifurcates from the main strand of the northernmost duplex fault and continues to the west, converging towards and then continuing subparallel to the southernmost fault in the duplex.

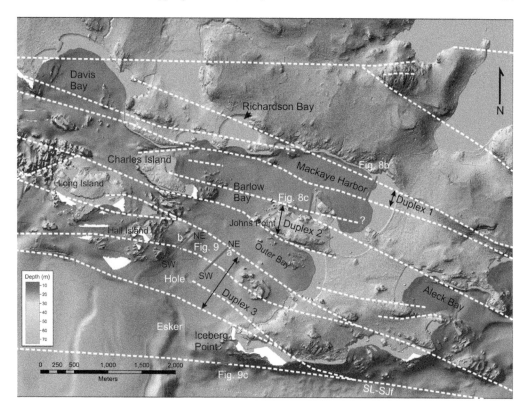

Fig. 7. Bathymetric map of the southern Lopez Island study area showing locations of faults interpreted from sub-bottom profiles, seafloor morphology, and LiDAR images. SL-SJf = South Lopez-San Juan fault. Red lines indicate 3.5 kHz sub-bottom profile locations presented as figures and used in the construction of fault geometry. Duplex 1 = Mackaye Harbor duplex, Duplex 2 = Johns Point duplex, and Duplex 3 = Iceberg Point duplex.

This relationship is well expressed in the LiDAR imagery (Fig. 7). The southernmost fault of this duplex is a continuous fault mapped as extending westward from the eastern shore of the Mackaye Harbor to cross the rocky headlands between Richardson Bay and Davis Bay in the west where it is projected to extend offshore towards San Juan Island (Fig. 7). This fault was crossed multiple times along a track line that parallels the trace of the fault and appears as a near-vertical normal fault that displaces deformed, well-layered reflections and extends to an eroded acoustic surface overlain by a <5 m thick acoustically transparent layer.

Duplex 2 – The central set of faults, Johns Point duplex (Duplex 2), is well expressed in LiDAR through the linear alignment of rock outcrops in Aleck Bay, Johns Point peninsula and Charles Island, and is narrower (c. 400 m wide) than the Mackaye Harbor duplex. The northernmost fault in this duplex appears as a near-vertical normal fault in the 3.5 kHz sub-bottom profiles and displaces well-layered, folded reflections that extend to an

eroded surface at the seafloor or are locally overlain by a thin (<1 m thick) acoustical transparent unit (Figs 7 & 8c). This fault is a continuous NW–SE trending structure extending from Aleck Bay through Barlow Bay to Charles Island. It offsets submerged rock outcrops along the eastern extension of Charles Island, and is over 3 km long. The southernmost fault in the duplex is defined in the LiDAR as converging eastward towards, and sub-paralleling the northern fault where it is well defined along, and controlling, the southern shorelines of Johns Point peninsula and Charles Island. Subsurface expression of this fault is observed in the 3.5 kHz sub-bottom profiles where a north dipping reverse fault offsets weakly (low amplitude) reflections within a syncline. The fault extends to the surface of a high amplitude subsurface layer that is overlain by an acoustically transparent layer.

Duplex 3 – The southernmost fault duplex, Iceberg Point duplex (Duplex 3), is the widest zone of the three being approximately 1 km wide and trending more northerly than the other two duplexes

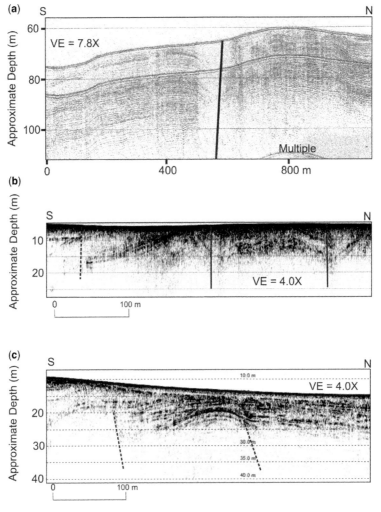

Fig. 8. Interpreted faults from seismic-reflection and sub-bottom profiles collected in the vicinity of Southern Lopez Island and within the previously mapped Lopez fault complex; (**a**) Huntec profile collected in Rosario Strait showing deformation along an acoustically transparent zone, (**b**) 3.5 kHz sub-bottom profile showing interpreted normal faults along the northern bounding fault of the Mackaye Harbor fault duplex (Duplex 1 in Fig. 7), (**c**) northern boundary fault of the Johns Point fault duplex (Duplex 2 in Fig. 7).

(Fig. 7). This duplex is comprised of the two primary (margin faults) and two intermediate faults. The two intermediate faults range in lengths from 5 km to 2 km, while the two margin faults are longer (>8 km long).

The north margin fault of this duplex is well imaged in LiDAR through the alignment of rock outcrops that form the central spine of the southern headland peninsula of Aleck Bay and the central eastern shore of Outer Bay (Fig. 7). A southeastern projection of the fault can be made based on the submerged rock exposures south of Aleck Bay where the fault appears to converge with a due E–W

trending fault we map and here named the 'South Lopez–San Juan fault' (SL–SJf) along the southern outer edge of Lopez Island (Fig. 7).

To the west this fault extends through and offsets rock outcrops exposed on the seafloor north of Hall Island and south of Charles Island. This fault is characterized in a 3.5 kHz sub-bottom profile as a near-vertical normal fault that juxtaposes a very high amplitude gently dipping acoustic reflection against a steeply dipping acoustically chaotic and transparent unit with the fault displacing reflections beneath an eroded surface and covered with a thin (c. 2 m) acoustically transparent unit.

The northern intermediate fault is paired in central Outer Bay with short (*c.* 1 km) closely spaced (*c.* 35 m apart) parallel, steeply dipping westward normal faults (Fig. 9a, b). The southern intermediate fault is a splay of a due E–W trending fault we map and here named the 'South Lopez–San Juan fault' (SL–SJf) along the southern outer edge of Lopez Island (Figs 7 & 10). The second interior fault is a splay of the southernmost Iceberg Point fault duplex and is defined onshore by the linearity exhibited in a rocky point projecting out into the bay just east of Iceberg Point. Westward extension of this fault is based on fault picks from the 3.5 kHz sub-bottom profiles that show the fault to be a vertical to south dipping reverse fault that deforms well-layered high-amplitude reflections that form a syncline and is overlain with a thick (*c.* 20 m) prism of acoustical transparent material. This fault appears to terminate in a depression that resembles a pockmark and is surrounded with sediment that appears to have been blown from a hole in the seafloor (Fig. 7).

The southernmost fault of the Iceberg Point fault duplex is well defined along the southern shoreline

of Iceberg Point where the coastline appears sheared and controlled by faulting in the LiDAR imagery (Fig. 7). This fault forms the southern edge of Iceberg Point and based on offshore bathymetry and 3.5 kHz sub-bottom profile data, it appears to extend westward as a horsetail splay from the northward verging reverse SL–SJf across the upper terminous of an esker and along the southern edge of Lopez Island shelf that supports Long and Charles islands (Figs 7, 9c & 10). Based on interpretations of a 3.5 kHz sub-bottom profile data, the southernmost fault of the Iceberg fault duplex is a near-vertical normal fault that truncates and juxtaposes well-layered high-amplitude reflections on the north with an acoustically chaotic to transparent unit on the south (Fig. 7). Several inferred normal faults are observed to the north of this fault, which displace the high amplitude well-layered reflections and extend to near the seafloor.

Cattle Pass – The seafloor in the vicinity of Cattle Pass, near Cattle Point of southern San Juan Island is defined by a very linear, near vertical, generally a NNW–SSE oriented rock scarp, initially mapped in

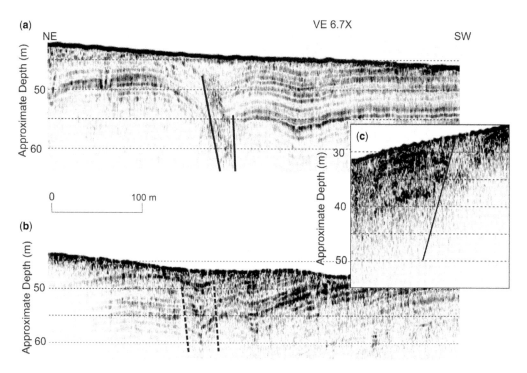

Fig. 9. Interpreted faults from 3.5 kHz sub-bottom profiles in the south part of the southern Lopez Island study area; (**a**) sub-bottom profile showing closely spaced dual faults within the Iceberg Point fault duplex in central Outer Bay (Duplex 3 in Fig. 7), (**b**) sub-bottom profile showing the correlatable dual faults east of Hall Island within Iceberg Point fault duplex (fault Duplex 3 in Fig. 7), and (**c**) sub-bottom profile across the South Lopez–San Juan fault, a south verging reverse (thrust) fault that marks the southern boundary of the Lopez fault complex (see Fig. 3 for profile track lines).

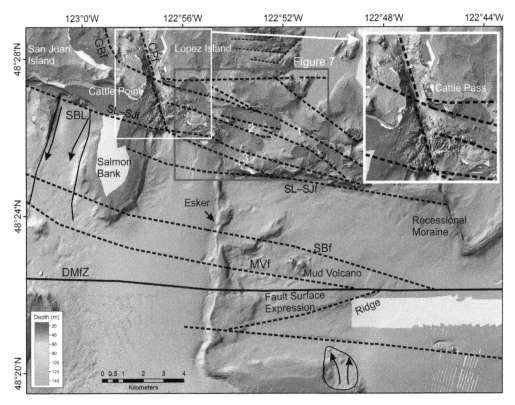

Fig. 10. Fault map constructed from MBES bathymetry, seismic-reflection profiles and LiDAR imagery with expanded MBES bathymetric map of the seafloor between southern Lopez Island and the Devils Mountain fault zone showing wrench fault-like geometry. Faults within the red-lined box are generalized because of scale, showing those faults that are most continuous; see Figure 7 for numbers and locations of faults within the Lopez fault complex. White-lined box is exploded view of the Cattle Pass area. White numbers indicate location of duplexes (1 = Duplex 1 or Mackate Harbor duplex, 2 = Duplex 2 or Johns Point duplex, 3 = Duplex 3 or Iceberg Point duplex). GBf, Griffin Bay fault; CPf, Cattle Pass fault; DMfZ, Devils Mountain fault zone; Mvf, Mud volcano fault; SBf, Salmon Bank fault; SL–SJf, South Lopez–San Juan fault; SBL, Salmon Bank landslide.

detail by Tilden (2004, her fig. 5.4, p. 93) as a fault, although it is most likely heavily modified from glaciation. We extend two NW–SE trending faults mapped along the most northerly shoreline and onshore of southern Lopez Island, the Mackaye Harbor and Johnson Point fault duplexes (Duplex 1 and Duplex 2 on Fig. 7), across Cattle Pass to San Juan Island that appear to right laterally offset the older fault (Fig. 10), which we here call the Cattle Pass fault (CPf). The most northerly fault may curve northward to extend through Griffin Bay, the Griffin Bay fault (GBf), and trends into the Friday Harbor area. Iceberg Point fault duplex (Duplex 3 in Fig. 7) appears to project into the Cattle Pass area, extending beneath the proximal margin of Salmon Bank (Fig. 10), in the general vicinity of a fault initially mapped by Tilden (2004).

None of the cores we collected in the San Juan Archipelago were located within or near the Lopez

fault complex. However, the deformation of the shallow surficial sediment we observed in the sub-bottom profiles suggest that folding and faulting of the deposits has occurred recently, since the last glacial retreat.

Haro Strait area

The Haro Strait area encompasses the western part of the San Juan Archipelago extending from the Cattle Point and Speiden Channel (western San Juan Island area) on the east to Vancouver Island on the west where faults are newly mapped between the DMfZ and SJIfZ (Fig. 11). In this area a series of NW–SE to E–W oriented fault traces have been mapped. It is a very difficult area to map as the glaciers have severely altered the bedrock and locally left thick deposits of unconsolidated material, modified from the strong tidal currents that sweep through Haro

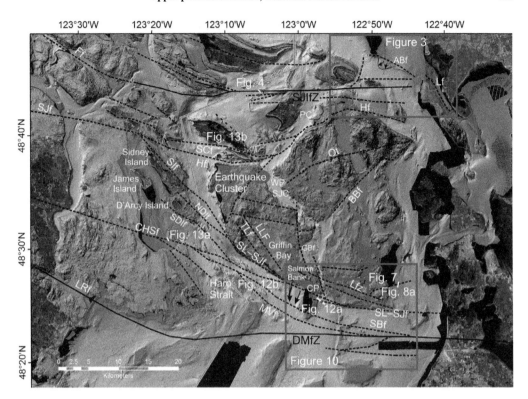

Fig. 11. Fault map of the entire Archipelago investigated, constructed from MBES bathymetry, seismic-reflection profiles and LiDAR imagery with bathymetric map of the Haro Strait study area showing a series of lengthy fault strands radiating from the Devils Mountain fault zone and associated faults mapped in the southern Lopez Island area. These faults compose two generally oriented fault zones consisting of a NW–SE and a NNW–SSE trend. The NNW–SSE trend appears to merge or be truncated by the E–W trending faults in the north. Solid lines represent well-defined major fault trends and dashed lines represent inferred faults. Red dots on fault lines indicate where fault was picked from a seismic-reflection profile. White lines show location of seismic-reflection profiles presented as figures and used to construct the fault geometry and estimate fault activity. Grey line is international boundary between the USA and Canada. DMfZ, Devils Mountain fault zone; SJIfZ, Skipjack Island fault zone; Ff, Fulford fault; SJf, San Juan fault; LRf, Leech River fault; MVf, Mud Volcano fault; SBf, Salmon Bank fault; GBf, Griffen Bay fault; SL–SJf, South Lopez–San Juan fault; CHSf, Central Haro Strait fault; NDIf, North D'Arcy Island fault; CPf, Cattle Pass fault; SIf, Sidney Island fault; Hf, Haro fault; SCf, Speiden Channel fault; SIfZ, Skipjack Island fault zone; BBf, Buck Bay fault; Lf, Lummi fault; ABf, Alden Bank fault; TLf, Trout Lake fault; LLf, Lawson Lake fault; WP, Wasp Passage; PC, Presidents Channel; CP, Cattle Pass. 'Earthquake Cluster' represents location of seismic events that initiated 5 April 2019. Red line boxes show locations of map figures.

Strait. However, using the MBES imagery and seismic-reflection profiles we were able to identify and map faults whose geometry makes reasonable geologic sense.

South of the Lopez Island study area and north of the primary trace of the DMfZ four faults are observed in the bathymetry, some appear to have recently deformed the seafloor where linear cracks and crevices exist (Fig. 10). Pockmarks and a mud volcano suggest fluid seepage is taking place in the vicinity of the faults (Fig. 10). The DMfZ in this area exhibits seafloor expression along a due E–W trend and appears to have offset a well-imaged

N–S oriented esker (Barrie and Greene 2018). To the south of the primary DMfZ two converging NE–SW and E–W faults bound a subtle ridge that appears to be unstable as indicated by cracks and sediment creep features. The northernmost fault in this set runs through a linear crack of a small, elongated mound that sits on top of the ridge (Fig. 10). The southern fault parallels the primary trace of the DMfZ.

Two parallel splay faults trend NW from the primary DMfZ trace, bound the mud volcano complex described by Barrie and Greene (2018), and appear to offset the esker (Fig. 10). The north bounding fault of the mud volcano complex is here named the 'Salmon Bank fault' (SBf), as it appears to extend beneath the southern tip of Salmon Bank. The south bounding fault is here named the 'Mud Volcano fault' (MVf) and is projected into Haro Strait.

A major fault that trends E–W, the SL–SJf, is imaged in the bathymetry and 3.5 kHz sub-bottom profiles and defines the base of the slope of southern Lopez Island (Figs 10 & 11). To the east, this fault cuts through the recessional moraine in Rosario Strait. To the west it cuts through and appears to offset the esker where an apparent east verging reverse fault cuts to the seafloor and juxtaposes well defined, apparent westward dipping, thinly layered reflections on the west against poorly defined reflections and an acoustically transparent unit on the east (Fig. 12a). To the east of the acoustically transparent unit, another near-vertical normal fault (Fig. 12a) juxtaposes this unit with well-layered thinly bedded reflections folded into an anticline and a syncline that abuts a near-vertical normal fault. These faults are projected to extend towards the central part of Salmon Bank (Fig. 10).

The primary trace of the SL–SJf extends NW–SE where it appears to trend beneath the proximal end of Salmon Bank and defines the head scarp of a landslide located along the western side of the bank, here called the 'Salmon Bank landslide' (SBL, Fig. 10). This fault appears to extend northward as a controlling structure of the southwestern coastline of San Juan Island, bifurcates into two NNW–SSE strands and crosses Haro Strait to where one strand offsets submerged rocks of northern D'Arcy Island (Fig. 11), here named the 'North D'Arcy Island fault' (NDIf). The fault appears to terminate at the southern tip of Sidney Island, although it may continue onto the island. The other fault strand, here named the 'Sidney Island fault' (SIf), again based on linear rock scarps and a seismic-reflection profile, also appears to cross Haro Strait, paralleling the NDIf. It extends along the straight coastline and rock outcrops of northeastern Sidney Island, controlling the lineament of small islands and straight edges of submerged rock outcrops north of Sidney Island. Core 08-07 (Fig. 2) recovered near the base of the SIf scarp NW of D'Arcy Island contained datable

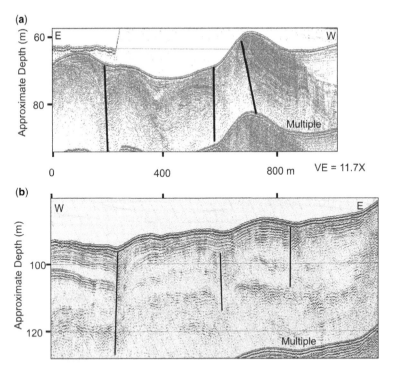

Fig. 12. Huntec seismic-reflection profiles showing faults interpreted near Salmon Bank within the Haro Strait study area: (**a**) line 2003-85 showing the South Lopez–San Juan fault (SL–SJf) east of Cattle Pass that truncates south-dipping thinly layered reflections juxtaposed with poorly defined reflections and an acoustically transparent unit, (**b**) line 2004-18-2 showing the interpreted continuation of the Salmon Bank fault (SBf) as a normal north up fault that juxtapose well-layered flat lying reflections on the south with folded weakly layered reflections on the north and two other faults that displace acoustic basement to the north; note seafloor expression of the faults (see Fig. 11 for locations).

material that yielded a ^{14}C age of 10,760 $\pm$ 60 ka BP. Short, parallel faults step northward from the SL–SJf and are mapped based on the straight edges along submerged rock outcrops of north-central San Juan Island and appear to control the coastline in this area.

Another fault, here called the 'Central Haro Strait fault' (CHSf), oriented generally NW–SE, splays from the SL–SJf just west of the Salmon Bank landslide. This fault continues across Haro Strait where it truncates a rock exposure in the central part of the strait, and based on evidence in a seismic-reflection profile (Fig. 13a) is projected westward to possibly connect with a fault-like lineament observed in the LiDAR image of northern Saanich Peninsula of Vancouver Island, north of Victoria. Two cores (08-05 and 03-84) located in the vicinity of the western extent of the CHSf (Fig. 2) contained glacial silt and sand that appear in the seismic-reflection profile to correlate with faulted reflections, but no datable material was recovered from the cores. A NNW–SSE trending strand splays northward from the central Haro Strait rock outcrop at the intersection with the CHSf and forms the western flank of D'Arcy

Island, here named the 'Southern D'Arcy Island fault' (SDIf), which appears to extend to James Island.

South of the CHSf near the coast of Vancouver Island, two generally NW–SE oriented faults, the southern one (MVf) converging on the northern one (SBf), have been mapped in this area (Fig. 11) based on linear bathymetric scarps and displaced reflections observed in seismic-reflection profiles (Fig. 12b). Farther east and south of the CHSf, two sub-parallel faults (SBf and MVf) splay from the DMfZ south of Lopez Island with the more lengthy and northern SBf extending through the southern tip of Salmon Bank and connecting with the CHSf on the Vancouver Island shelf. The MVf converges with the SBf in the central Haro Strait area (Fig. 11).

Using MBES and acoustical geophysical data our attempt to confirm the offshore extension of the faults mapped onshore in previous studies met with mixed results. However, with the use of LiDAR we were able to confirm the onshore locations of these faults. As reported by previous workers, in the central part of the San Juan Archipelago a series of

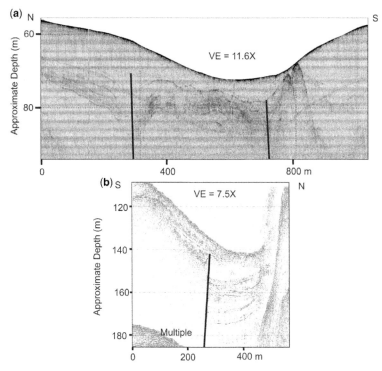

Fig. 13. Huntec seismic-reflection profiles collected in Haro Straight areas in 2004 and 2008 showing faults displacing glacial materials; (**a**) line 2008-37 showing the Central Haro Strait fault CHSf displacing well-layered reflections interpreted as glacial deposits in a swale filled with acoustically transparent material, (**b**) profile 2004–18 showing a splay of the Haro fault in Speiden Channel called the Speiden Channel fault (SCf) that displaces well-layered cross-bedded reflections from an acoustically transparent unit (see Fig. 11 for locations).

generally NE–SW trending thrust faults have been mapped on the islands and extended between islands (Bergh 2002). On Orcas Island three major thrusts were mapped (Fig. 1): in the south the Buck Bay fault (BBf); in the central part of the island the Rosario thrust fault (Rt); and in the north the Hf. On San Juan Island, the Buck Bay fault was mapped as a thrust by Bergh (2002) that extended around southern Shaw Island and curved around the northern part of the island. The Rt was mapped to extend from southwestern Orcas Island through southern Shaw Island. The Hf was shown to bifurcate and swing westward separating San Juan Island from Spieden Island by one fault, and Spieden from Stuart islands by the other fault (Bergh 2002). The offshore trend of some of these faults was confirmed locally, but due to the age of faulting (pre-glacial erosion) and the strong currents that prevent deposition of sediments in the narrow channels and sounds the likelihood of detection was reduced.

Based on LiDAR imagery we map the BBf to trend NW–SE along the northern spur of Orcas Island across the mouth of East Sound. Then using linear rock outcrops observed in the MBES data we extend the fault along the southeastern flank of Shaw Island to the San Juan Channel where it appears to terminate (Fig. 11), similar to that mapped by Brandon *et al.* (1988). The Rt is oriented more E–W than BBf and based on LiDAR imagery appears to extend across the central part of Orcas Island, crosses Wasp Passage, trending onto San Juan Island (Fig. 11). Similar to the mapping of Bergh (2002), based on MBES imagery and a seismic-reflection profile data (Fig. 13b) we map the Hf as bifurcated strands that split the islands of San Juan, Spieden and Stuart, but interpret the main strand to continue across Haro Strait. This strand appears to connect with the SIf, extending in a westerly direction across the Saanich Peninsula to Vancouver Island where we propose it connects with the San Juan fault (SJf) mapped on Vancouver Island (Fig. 11). A NW–SE splay of Hf curves across northern San Juan Island from its intersection in Presidents Channel near the eastern side of Orcas Island, extending into San Juan Island. This fault may have recently been activated as a shallow focused (*c.* 13 km deep) M2.9 earthquake (Pacific Northwest Seismic Network, psns.org/events?) occurred along the trace of this fault on 5 April 2019 (Fig. 11) followed by several days of lower magnitude aftershocks.

Examination of the LiDAR imagery of San Juan Island shows two sets of faults in the central part of the island. In the southeastern part of the island two parallel NW–SE oriented faults are mapped, the Trout Lake (TLf) and Lawson Lake faults (LLf), that appear to control the position of the lakes. In contrast to these faults are a set of two parallel faults mapped in the central part of the island that trend

more E–W and appear to bound a more rugged hilly area of the island (Fig. 1).

Discussion

The interpretation of the MBES and geophysical data collected within the San Juan Archipelago during the past two decades indicates that this part of the upper plate of the Cascadia subduction zone is tectonically deformed with considerable faulting and folding taking place. Ages of sediments obtained from ^{14}C dating methods on shells collected from cores suggest that many faults are active as they displace Quaternary sediment. This activity is along faults that lie between the two major E–W oriented transcurrent fault zones, SJIfZ and DMfZ that bound the San Juan Archipelago and likely results from stress associated with movement along these fault zones. We focus on three areas within the Archipelago – Southern Georgia Strait–Lummi Island, Southern Lopez Island, and Haro Strait areas – where faulting and folding characteristics define distinct kinematic processes.

Southern Georgia Strait–Lummi Island Area

The Southern Georgia Strait–Lummi Island area lies near the presently mapped terminus of the SJIfZ in an area of complex seafloor morphology that appears to have been formed by glacial ice retreat about 10 ka. In this area between Alden Bank, Matia Island and the northern tip of Lummi Island a dramatic change in seafloor morphology is present along a line between Alden Bank and Lummi Island. Here a change in water depth ranges from 50 m in the north where a relatively smooth seafloor exists to 100 m in the south where an irregular seafloor occurs (Fig. 3). Along this line of bathymetric demarcation are several gentle arcuate south-facing scarps, an offset subtle ridge and a distinct highly rippled fan or debris apron that appears to be the result of slumping and modern sediment transport and spillover from the north. Two E–W trending strands of the SJIfZ that bound Matia Island have been mapped in this area (Greene *et al.* 2018) and converge with the southern part of the ridge where it broadens into a bathymetric irregular and hummocky seafloor.

Seismic-reflection profiles collected across the scarps and ridge exhibit sharp acoustical contrasts between well-layered reflections in the north, representative of early Fraser River deposits based on ^{14}C ages of 8400 ± 40 ka BP to 9390 ± 40 ka BP from a core (Core 09-100, Greene *et al.* 2018). Acoustically transparent to chaotically distributed sediment from the last ice retreat (*c.* 10 ka) abuts a buried recessional moraine that formed between Alden Bank and the northern tip of Lummi Island, thus ponding Fraser River deposits behind (north) of the moraine. We interpret that the sharp contact and deformed beds along a steeply to gently dipping southward verging plane (Fig. 5) suggests faulting. Thus, we map a fairly continuous normal fault, the Alden Bank fault (ABf) in this area that may have formed prior to the construction of the moraine, and thus influenced the position of the moraine through the presence of an elevated bedrock ridge. At depth in the seismic-reflection profile (Fig. 5a) two smaller moraine-like features are observed that may represent later, shorter ice advances and recessions that were subsequently buried by glacial outwash and Fraser River-derived sediment during the final stagnation and recession of the ice and the onset of distal delta construction.

We interpret the series of arcuate south-facing scarps as en-echelon faults that trend along a general NW–SE line along the forefront of the moraine, although these could be erosional features. The seismic-reflection profiles suggest deformation is occurring along the moraine's front in association with normal movement of the ABf. In addition, local faulting is observed behind the moraine, which cut and offset ponded glacial and recent Fraser River deposits, and although these faults are short and not continuous throughout the area, they displace and deform recent (<10 ka) glacial deposits and suggest active deformation.

The ABf mapped along the backside of the moraine appears to extend towards northern Lummi Island where a similar but reverse fault (the Lummi fault or Lf) is mapped along the eastern side of the island (Fig. 6b, c). Here, the arcuate coastline of NE Lummi Island appears to be fault controlled as westward verging reverse faults have been identified here in the 3.5 kHz sub-bottom profiles. The intersection of the northern splay of the SJIfZ is difficult to resolve, although we see hints in the bathymetry that this fault zone offsets the primary reverse Lf. If this is the case, then the kinematics of faulting transfers from normal faulting (dilatational) in the north to reverse and possible thrusting (compressional) in the south, which may result from the impingement of relatively brittle block against a mainland rigid block. Thus, we suspect that the crustal deformation occurring in the South Georgia Strait–Lummi Island area is the result of shifting transtensional–transpressional forces associated with movement along the SJIfZ. The reverse faults mapped in this area suggest that the northeastern part of the San Juan Archipelago is pushing against and underthrusting the mainland. Alternatively, Christie-Blick and Biddle (1985) and references therein report that linear or curvilinear faults in map view reflect principle displacements in strike-slip fault systems, thus the ABf may represent deformation associated strike-slip motion.

Southern Lopez Island Area

The LfZ has been structurally described as 'steep to subvertical and up to 3 km wide' that in map view consist of faults that 'define a left-stepping, en echelon fault system' (Bergh 2002, p. 937). Our mapping based on interpretation of LiDAR and MBES bathymetric data support the mapping of Bergh (2002) and (Brandon et al. 1988), but illustrates better the continuity of faulting that forms the bays and waterways of the southern Lopez Island area (Fig. 7). We map three sets of duplex faults as a fault complex, all oriented sub-parallel to each other and consisting of NW–SE and NNW–SSE trending faults that are expressed as rock scarps and lineaments, both on land and on the seafloor. In addition, sub-bottom profiles collected in the area exhibit faults that displace and deform glacial sediment indicating active deformation within the fault zone. Three sets of duplex faults make up the LfZ complex and splay in a horsetail fashion from a primary E–W trending SL–SJf that marks the southern boundary of the LfZ. The northern boundary is defined as the most northerly fault of fault Duplex 1, which Bergh (2002) mapped as a steep south-verging thrust.

Local areas of transpression, exhibited in elevated ridges and islands, and transtension, exhibited in basins (bays), formed by the left-lateral motion of faults within the Lopez fault zone comprise the geomorphology of southern Lopez Island. These en echelon fault duplexes splay at angles of c. 30 degrees from the primary SL–SJf indicating a wrench fault tectonic setting. This fault zone may have developed along essential pre-existing Mid-Cretaceous structures (Brandon 1980) that significantly influenced the location and orientation of faults and folds of the strike-slip movement of the braided (horsetail) splays of the DMfZ, a structural relationship described by Christie-Blick and Biddle (1985) and references therein. These authors state that overstep, branching and braiding fault strands are fundamental features of many strike-slip fault zones or fault systems and report on many examples from the southern California region.

Another major E–W fault trace is mapped on land north of the northern fault duplex that extends eastward offshore in an en echelon fashion to connect with the fault observed in a seismic-reflection profile in Rosario Strait (Fig. 8a), and westward where it cuts across the northern part of Davis Bay and across Cattle Pass to southern San Juan Island where it may bend northward, cutting through Griffin Bay and into Friday Harbor (Fig. 11). This fault is shown to offset right-laterally an older fault scarp (CPf) that defines the seafloor of Cattle Pass (Fig. 10). Also, the extension of the northern fault of the LfZ complex extends across Cattle Pass to southern San Juan Island where it also appears to right-laterally offset the Cattle Pass fault near Cattle Point.

Between the SL–SJf and the DMfZ another area of apparent transpression and transtension exists. In this area, the primary strand of the DMfZ is a well defined E–W oriented fault with two splays (the SBf and MVf) trending more northwestward at an angle of less than 30 degrees and a splay trending southwesterly at *c*. 30 degrees from the primary fault, which merges with another E–W fault parallel-ing the primary splay of the DMfZ in the south (Fig. 10). Located between the SBf and the MVf is a mud volcano indicating that fluid flow is occurring within the DMfZ. Fluid flow along with seismic activity on these faults may have stimulated mass wasting in the area as demonstrated by the slide along the western side of Salmon Bank (Salmon Bank slide) and a small slide located on the northern side of a bank just south of the DMfZ (Fig. 10). Recent (in the past 10 ka) motion along these faults is suggested by apparent offset of the esker and moraines in the area.

Similar to the Southern Georgia Strait–Lummi Island area in the northeastern corner of the San Juan Archipelago, the Southern Lopez Island area exhibits a complex structural picture that is difficult to interpret. However, like the northeastern corner of the Archipelago, we view this area as a structural transition zone, an area where fault trends splay from E–W (DMfZ) to NW–SE trending faults (SL–SJf, SBf, HVf).

Haro Strait area

In the Haro Strait area a series of generally NW–SE and NNW–SSE trending faults are mapped from MBES data and are extended onto Vancouver Island using LiDAR data. The SBf and MVf merge into the CHSf in southern Haro Strait that continues on land crossing the Saanich Peninsula (Fig. 11). This is a major thoroughgoing fault (the CHSf) with a more northern splay trending north as the SDIf and possibly forming the northern flank of James Island. The major CHSf splays from the SL–SJf at the head of the submerged Salmon Bank slide off southern San Juan Island and continues northwestward as the SIf to where it merges with the Hf north of Sidney Island (Figs 10 & 11). The CHSf appears to be a sub-parallel strike-slip horsetail splay fault to the DMfZ with possible conjugate faults between the two repre-sented by the SBf and MVf. The SL–SJf and its splays represent the western boundary of the San Juan Archipelago. Four cores recovered in the vicinity of these faults were composed of glacial material with samples obtained from two cores that yielded ^{14}C age from shells of $10,760 \pm 60$ ka and 9640 ± 50 ka BP that appear to correlate with the upper reflections displaced by the faults.

Our interpretation of the regional fault geometry that we mapped is that clockwise rotation of the Archipelago is occurring as suggested by the splay faults of Haro Strait and strike-slip motion along the SJIfZ and DMfZ. Motion along the SJIfZ pro-duces a N to S transition from transtension to trans-pression in the northeast corner of the San Juan Archipelago with wrench faulting and associated transtension and transpression occurring between the left-lateral transcurrent DMfZ and CHSf. The mechanism for this rotation is poorly understood but may result from the impingement of the Oregon block against the Washington forearc. Internal defor-mation of the Archipelago appears to be occurring today as suggested by the cluster of earthquakes in the lower part of the upper plate that was initiated 5 April 2019 followed by several aftershocks, the apparent re-activation of an older fault splay of the Hf. Deformed sediment within the LfZ also suggests that this fault zone has been recently reactivated or may be younger than previously proposed.

Comparison of Cascadia with the young Cocos Plate that subducts beneath Mexico has been made based on the observed similar seismogenic depth limit (Tichelaar and Ruff 1993; Flück *et al.* 1997; Parsons *et al.* 1998). Flück *et al.* (1997) compare Cas-cadia with the Nankai margin of Southwestern Japan and other areas, however, a good analogy of the upper plate deformation in the central part of the zone that we discuss here can be made with the New Hebrides convergence zone of Vanuatu where the nearly orthogonal collision of the Australia–India Plate is subducting beneath the Pacific Plate (Falvey and Greene 1988). Similar fault geometry has been mapped on the islands (e.g. Malakula and Espiritu Santo islands) and offshore in the forearc and interarc areas (Greene *et al.* 1988a). In the New Hebrides sub-duction complex three major strike-slip fracture zones, the Santa Maria, Aoba and Ambrym fracture zones, have been mapped trending across the forearc perpen-dicular to the trend of the trench, similar in fault geo-metry to the Skipjack Island, Devils Mountain and Seattle fault zones mapped in the Cascadia forearc. However, the nature and dip of the subduction zone is much different from Cascadia, consisting of a subducting ridge of seamounts and the D'Entrecas-teaux fracture zone, that may lock subduction locally, and is a steeply dipping (near vertical to 60°) frag-mented down-going slab (Louat *et al.* 1988). Distinct structural rotation has occurred within the New Hebri-des forearc islands as exhibited in the rotation of Malakula Island between the Aoba right-lateral strike-slip and Ambrym left-lateral strike-slip fracture zones (Burne *et al.* 1988; Greene *et al.* 1988b), and similar to the Cascadia forearc, contraction is taking place, but within a narrower zone whereas the distance between the arc and the forearc islands is 50–100 km with the islands located *c*. 100 km west of the arc.

Conclusions

Recent mapping of faults within the San Juan–southern Gulf islands Archipelago displays a complex structural pattern that results from older Mesozoic to Cenozoic crustal accretion and modern deformation occurring from the present-day subduction and impingement of northward migrating tectonstratigraphic terranes within the Cascadia subduction complex. This new mapping based on the interpretation of MBES, LiDAR and coring data indicates that many of the faults mapped on the seafloor are extensions of previously mapped island faults that offset and displace late Pleistocene and Holocene sediment, demonstrating that active faulting is occurring within the Archipelago. These reactivated faults formed during accretion consist of the southern splay of the Haro fault and faults that comprise the Lopez fault zone. The San Juan Archipelago is structurally well defined by the north bounding Skipjack Island fault zone and the south bounding Devils Mountain fault zone, two fault zones essentially well mapped on the seafloor. The kinematics of the region appears to result in the form of clockwise rotation around a vertical axis in the centre of the Archipelago with major movement along primarily transcurrent faults with local areas undergoing transpression and transtension.

Rotation around a vertical axis postulated on structural grounds such as proposed here have been confirmed by palaeomagnetic data with such rotations tending to be pervasive in strike-slip regimes over a wide range of scales (Christie-Blick and Biddle 1985). A recent (5 April 2019), shallow (c. 13 km), small magnitude (2.9–3.2 M) earthquake and aftershocks along a reactivated fault, the Haro fault, within the central part of the Archipelago suggests that internal deformation within the Archipelago is occurring. Similar type of kinematics may be taking place within the upper plate of the New Hebrides subduction complex of Vanuatu.

Acknowledgements The Canadian Hydrographic Service – Pacific (Department of Fisheries and Oceans) is thanked for the diligent collection of multibeam bathymetry. The officers and crew of the *CCGS Vector* are acknowledged for able seamanship in collection of the geophysical and sediment sample data. Peter Neelands, Robert Kung, Kim Conway and Greg Middleton are thanked for invaluable assistance at sea and in the laboratory. We thank Royal Roads University for the use of the Geotek split core multi-sensor core logger and Randy Enkin of the Geological Survey of Canada, Pacific for logging the cores and contributing to the interpretations. Brian Todd of the Geological Survey of Canada, Atlantic assisted in the collection and interpretation of data in the Lummi Island area and we thank him for his generous contribution. We appreciate the insightful and constructive reviews provided by Kristin Rohr of the Geological Survey of Canada including her excellent guidance and Janet Watt of the US. Geological Survey. In addition, we are thankful for the patience and encouragement shown to us by the Geological Society Special Publications Coordinating Editor Kristine Asch, who we found to be very enjoyable to work with.

Funding Funding for J. Vaughn Barrie is provided by The Geological Survey of Canada, ID0E3BBG2412. Funding for H. Gary Greene is provided by Dickenson Foundation, ID0EHJBG2413.

Author contributions HGG: conceptualization (equal), data curation (equal), formal analysis (equal), funding acquisition (equal), investigation (lead), methodology (equal), project administration (lead), resources (equal), software (supporting), supervision (lead), validation (equal), visualization (supporting), writing – original draft (lead), writing – review & editing (lead); **JVB**: conceptualization (supporting), data curation (equal), formal analysis (equal), funding acquisition (equal), investigation (equal), methodology (equal), project administration (supporting), resources (equal), software (lead), supervision (supporting), validation (equal), visualization (lead), writing – original draft (supporting), writing – review & editing (supporting).

Data availability statement The datasets generated during the current study are available at the Geological Survey of Canada – Pacific (Natural Resources Canada).

References

Atwater, B.F. and Moore, A.L. 1992. A tsunami about 1000 years ago in Puget Sound, Washington. *Science*, **258**, 1614–1617, https://doi.org/10.1126/science.258.5088. 1614

Barrie, J.V. and Conway, K.W. 2002. Contrasting glacial sedimentation processes and sea-level changes in two adjacent basins on the Pacific margin of Canada. *Geological Society, London, Special Publications*, **203**, 181–194, https://doi.org/10.1144/GSL.SP.2002.203. 01.10

Barrie, J.V. and Greene, H.G. 2015. Active faulting in the northern Juan de Fuca Strait, implications for Victoria, British Columbia. *Geological Survey of Canada Current Research: 2015–2016*, https://doi.org/10.4095/296564

Barrie, J.V. and Greene, H.G. 2018. The Devil's Mountain fault zone: an active Cascadia upper plate zone of deformation, Pacific Northwest of North America. *Sedimentary Geology*, **364**, 228–241, https://doi.org/10.1016/j.sedgeo.2017.12.018

Barrie, J.V., Conway, K.W., Picard, K. and Greene, H.G. 2009. Large-scale sedimentary bedforms and sediment dynamics on a glaciated tectonic continental shelf: examples from the Pacific margin of Canada. *Continental Shelf Research*, **29**, 796–806, https://doi.org/10.1016/j.csr.2008.12.007

Bednarski, J.M. and Rogers, G.C. 2012. *LiDAR and digital aerial photography of Saanich Peninsula, selected Gulf Islands, and coastal regions from Mill Bay to*

Ladysmith, southern Vancouver Island, British Colum-bia. Geological Survey of Canada, Open-File, **7229**. https://doi.org/10.4095/291819

Bergh, S.G. 2002. Linked thrust and strike-slip faulting dur-ing Late Cretaceous terrane accretion in the San Juan thrust system, Northwest Cascade, Washington. *Geo-logical Society of America Bulletin*, **114**, 934–949, https://doi.org/10.1130/0016-7606(2002)114<0934: LTASSF>2.0.CO;2

Bourgeois, J. and Johnson, S.Y. 2001. Geologic evidence of earthquakes at the Snohomish delta, Washington, in the past 1200 years. *Geological Society of America Bulletin*, **113**, 482–494, https://doi.org/10.1130/ 0016-7606(2001)113<0482:GEOEAT>2.0.CO;2

Brandon, M.T. 1980. Structural geology of middle Creta-ceous thrust faulting on southern San Juan Island, Washington. M.S. thesis, Seattle, University of Washington.

Brandon, M.T. 1989. Geology of the San Juan-Cascade Nappes, Northwestern Cascade Range and San Juan Islands. *Washington Division of Geology and Earth Resources Information Circular*, **86**, 137–162.

Brandon, M.T., Cowan, D.S. and Vance, J.A. 1988. The Late Cretaceous San Juan thrust system, San Juan Islands, Washington. *Geological Society of America Special Paper*, **221**, 81.

Brown, E.H. 1987. Structural geology and accretionary his-tory of the northwest Cascades system, Washington and British Columbia. *Geological Society of America Bulle-tin*, **99**, 201–214, https://doi.org/10.1130/0016-7606 (1987)99<201:SGAAHO>2.0.CO;2

Brudzinski, M.R. and Allen, R.M. 2007. Segmentation in episodic tremor and slip all along Cascadia. *Geology*, **35**, 907–910, https://doi.org/10.1130/G23740A.1

Bucknam, R.C., Hemphill-Haley, E. and Leopold, E.B. 1992. Abrupt uplift within the past 1700 years at south-ern Puget Sound, Washington. *Science*, **258**, 1611–1614, https://doi.org/10.1126/science.258. 5088.1611

Burne, R.V., Collote, J.-Y. and Daniel, J. 1988. Superficial structures and stress regimes of the downgoing plate associated with the subduction-collision of the central New Hebrides Arc (Vanuatu). *Circum-Pacific Council for Energy and Mineral Resources Earth Science Series*, AAPG, Tulsa, OK, **8**, 357–376.

Christie-Blick, N. and Biddle, K.T. 1985. Deformation and basin formation along strike- slip faults. *Society of Eco-nomic Paleontologist and Mineralogists Special Publi-cations*, **37**, 1–34.

Clague, J.J. 1976. Quadra sand and its relation to the late Wisconsin glaciation of south-west British Columbia. *Canadian Journal of Earth Sciences*, **13**, 803–815.

Clague, J.J. and James, T.S. 2002. History and isostatic effects of the last ice sheet in southern British Colum-bia. *Quaternary Science Reviews*, **21**, 71–87, https:// doi.org/10.1016/S0277-3791(01)00070-1

DeMets, C., Gordon, R.G. and Argus, D.F. 2010. Geolo-gially current plate motions. *Geophysical Journal International*, **181**, 1–80, https://doi.org/10.1111/j. 1365-246X.2009.04491.x

Easterbrook, D.J. 1992. Advance and retreat of Cordilleran ice sheets in Washington, U.S.A. *Geographie Physique et Quaternaire*, **46**, 51–68, https://doi.org/10.7202/ 032888ar

Easterbrook, D.J. 1994. Chronology of pre-Late Wisconsin Pleistocene sediments in the Puget Lowland, Washing-ton. *Washington Division of Geology and Earth Resources Bulletin*, **80**, 191–206.

Falvey, D.A. and Greene, H.G. 1988. Origin and evolution of the sedimentary basins of the New Hebrides Arc. *Circum-Pacific Council for Energy and Mineral Resources Earth Science Series*, AAPG, Tulsa, OK, **8**, 413–442.

Flück, P., Hyndman, R.D. and Wang, K. 1997. Three-dimensional dislocation model for great earthquakes of the Cascadia subduction zone. *Journal of Geophys-cial Research*, **102**, 20, 539-20-550.

Greene, H.G. and Barrie, J.V. (eds) 2011. *Potential Marine Benthic Habitats of the San Juan Archipelago*. Geolog-ical Survey of Canada Map Series, 4 Quadrants, 12 sheets, scale 1:50,000.

Greene, H.G., Macfalane, A. and Wong, F.L. 1988*a*. Geol-ogy and offshore resources of Vanuatu – introduction and summary. *Circum-Pacific Council for Energy and Mineral Resources Earth Science Series*, AAPG, Tulsa, OK, **8**, 1–25.

Greene, H.G., Macfarlane,, Johnson, D.P. and Crawford, A.J. 1988*b*. Structure and tectonics of the central New Hebrides Arc. *Circum-Pacific Council for Energy and Mineral Resources Earth Science Series*, AAPG, Tulsa, OK, **8**, 377–412.

Greene, H.G., Barrie, J.V. and Todd, B.J. 2018. The Skip-jack Island fault zone: an active transcurrent structure within the upper plate of the Cascadia subduction com-plex. *Special Issue, Journal of Sedimentary Geology*, **378**, 61–79, https://doi.org/10.1016/j.sedgeo.2018. 05.005

Guilbault, J.P., Barrie, J.V., Conway, K.W., Lapointe, M. and Radi, T. 2003. Paleoenvironments associated with the deglaciation process in the Strait of Georgia off Brit-ish Columbia: microfaunal and microfloral evidence. *Quaternary Science Reviews*, **22**, 839–857, https:// doi.org/10.1016/S0277-3791(02)00252-4

Hetherington, R. and Barrie, J.V. 2004. Interaction between local tectonics and glacial unloading on the Pacific mar-gin of Canada. *Quaternary International*, **120**, 65–77, https://doi.org/10.1016/j.quaint.2004.01.007

Hewitt, A.T. and Mosher, D.C. 2001. Late Quaternary strat-igraphy and seafloor geology of eastern Juan de Fuca Strait, British Columbia and Washington. *Marine Geol-ogy*, **177**, 295–316, https://doi.org/10.1016/ S0025-3227(01)00160-8

Hyndman, R.D., Massotti, S., Wiechert, D. and Rogers, G.C. 2003. Frequency of large crustal earthquakes in Puget Sound – south Georgia Strait predicted from geo-detic and geologic deformation rates. *Journal of Geo-physical Research*, **109**, B1. https://doi.org/10.1029/ 2001JB001710

James, T., Gowan, E.J., Hutchinson, I., Clague, J.J., Barrie, J.V. and Conway, K.W. 2009. Sea-level change and paleogeographic reconstructions, southern Vancouver Island, British Columbia, Canada. *Quaternary Science Reviews*, **28**, 1200–1216, https://doi.org/10.1016/j. quascirev.2008.12.022

Johnson, S.Y. 1982. Stratigraphy, Sedimentology, and Tec-tonic Setting of the Eocene Chuckanut Formation, North Cascades, Washington. Ph.D. thesis, Seattle, University of Washington.

Johnson, S.Y., Zimmerman, R.A., Naeser, C.W. and Whetten, J.T. 1986. Fission-track dating of the tectonic development of the San Juan Islands, Washington. *Canadian Journal of Earth Sciences*, **23**, 1318–1330, https://doi.org/10.1139/e86-127

Johnson, S., Dadisman, S., Mosher, D., Blakely, R. and Childs, J. 2001. *Active tectonics of the Devil's Mountain Fault and related structures, northern Puget lowland and eastern Strait of Juan de Fuca region, Pacific Northwest*. U.S. Geological Survey Professional Paper, **1643**.

Johnson, S.Y., Nelson, A.R. *et al.* 2004. Evidence for late Holocene earthquakes on the Utsalady Point Fault, northern Puget Lowland, Washington. *Bulletin of the Seismological Society of America*, **94**, 2299–2316, https://doi.org/10.1785/0120040050

Kelsey, H.M., Sherrod, B.L., Johnson, S.Y. and Dadisman, S.V. 2004. Land-level changes from a late Holocene earthquake in the Northern Puget lowland, Washington. *Geology*, **32**, 469–472, https://doi.org/10.1130/G20361.1

Kelsey, H.M., Sherrod, B.L., Blakely, R.J. and Haugerud, R.A. 2012. Holocene faulting in the Bellingham forearc basin: upper-plate deformation at the northern end of the Cascadia subduction zone. *Journal of Geophysical Research*, **117**, B03409. https://doi.org/10.1029/2011JB008816

Louat, R., Hamburger, M. and Monzier, M., 1988. Shallow and intermediate-depth seismicity in the New Hebrides Arc, contraints on the subduction process. *Circum-Pacific Council for Energy and Mineral Resources Earth Science Series*, AAPG, Tulsa, OK **8**, 329–356.

Mazzotti, S., Dragert, H., Hyndman, R.D., Miller, M.M. and Henton, J.A. 2002. GPS deformation in a region of high crustal seismicity, N. Cascadia forearc. *Earth and Planetary Sciences Letters*, **198**, 41–48, https://doi.org/10.1016/S0012-821X(02)00520-4

McCaffrey, R., Long, M.D., Goldfinger, C., Zwick, P.C., Nabelek, J.L., Johnson, C.K. and Smith, C. 2000. Rotation and plate locking along the southern Cascadia subduction zone. *Geophysical Research Letters*, **27**, 3117–3120, https://doi.org/10.1029/2000GL011768

McCaffrey, R., King, R.W., Payne, S.J. and Lancaster, M. 2013. Active tectonics of northwestern U.S. inferred from GPS-derived surface velocities. *Journal of Geophyscical Research, Solid Earth*, **11**, 709–723, https://doi.org/10.1029/2012JB009473

Miller, M.M., Johnson, D.J., Rubin, C.M., Dragert, H., Wang, K., Qamar, A. and Goldfinger, C. 2001. GPS-determination of along-strike variations in Cascadia margin kinematics: implications for relative plate motion, subduction zone coupling, and permanent deformation. *Tectonics*, **20**, 161–176, https://doi.org/10.1029/2000TC001224

Mosher, D.C. and Hewitt, A.T. 2004. Late Quaternary deglaciation and sea-level history of eastern Juan de Fuca Strait. *Quaternary International*, **121**, 23–39, https://doi.org/10.1016/j.quaint.2004.01.021

Murray, M.H. and Lisowski, M. 2000. Strain accumulation along the Cascadia subduction zone. *Geophysical Research Letters*, **27**, 3631–3634, https://doi.org/10.1029/1999GL011127

Nelson, A.R., Johnson, S.Y. *et al.* 2003. Late Holocene earthquake on the Toe Jam Hill fault, Seattle fault zone, Bainbridge Island, Washington. *Geological Society of America Bulletin*, **115**, 1386–1403, https://doi.org/10.1130/B25262.1

Oldow, J.S. 2000. *Fault characteristics and assessment of slip on the Devils Mountain and North Whidbey Island fault systems, northwestern Washington*. U.S. Geological Survey Final Technical Report, 10.

Parsons, T., Trehu, A.M. *et al.* 1998. A new view into the Cascadia subduction zone and volcanic arc: implications for earthquake hazards along the Washington margin. *Geology*, **26**, 199–202, https://doi.org/10.1130/0091-7613(1998)026<0199:ANVITC>2.3.CO;2

Porter, S.C. and Swanson, T.W. 1998. Radiocarbon age constraints on rates of advance and retreat of the Puget lobe of the Cordilleran ice sheet during the last glaciation. *Quaternary Science Research*, **50**, 205–213.

Savage, J.C., Svarc, J.L., Prescott, W.H. and Murray, M.H. 2000. Deformation across the forearc of the Cascadia subduction zone at Cape Blanco, Oregon. *Journal of Geophysical Research*, **105**, 3095–3102, https://doi.org/10.1029/1999JB900392

Sherrod, B.l., Bucknam, R.C. and Leopold, E.B. 2000. Holocene relative sea-level changes along the Seattle Fault at Restoration Point, Washington. *Quaternary Research*, **54**, 384–393.

Snavely, P.D. and Wells, R.E. 1996. Cenozoic evolution of the continental margin of Oregon and Washington. *U.S. Geological Survey Professional Paper*, **1560**, 161–182.

Tichelaar, B.W. and Ruff, L.J. 1993. Depth of seismic coupling along subduction zones. *Journal of Geophysical Reseach Solid Earth*, **98**, B2, 2017–2037, https://doi.org/10.1029/92JB02045

Tilden, J.E. 2004. Marine Geology and Potential Rockfish Habitat in the Southwestern San Juan Islands, WA. M.Sc. thesis, Moss Landing Marine Labs, California State University, Monterey Bay.

Wang, K., Wells, R., Mazzotti, S., Hyndman, R.D. and Sagiya, T. 2003. A revised dislocation model of interseismic deformation of the Cascadia subduction zone. *Journal of Geophysical Research*, **108**, 2026, https://doi.org/10.1029/2001JB001227

Wells, R.E. and Simpson, R.W. 2001. Northward migration of the Cascadia forearc in the northwestern U.S. and implications for subduction deformation. *Earth Planet Space*, **53**, 275–283, https://doi.org/10.1186/BF03352384

Wells, R.E., Weaver, C.S. and Blakely, R.J. 1998. Fore-arc migration in Cascadia and its neo-tectonic significance. *Geology*, **26**, 759–762, https://doi.org/10.1130/0091-7613(1998)026<0759:FAMICA>2.3.CO;2

Wilson, D.S. 2002. The Juan de Fuca plate and slab: Isochron structure and Cenozoic plate motions. U.S. Geological Survey, *Open-File Report 02-328*, 9–12.

Williams, H. and Hutchinson, I. 2000. Stratigraphic and microfossil evidence for late Holocene tsunamis at Swanton Marsh, Whidbey Island, Washington. *Quaternary Research*, **54**, 218–227, https://doi.org/10.1006/qres.2000.2162

Williams, H.F.L., Hutchinson, I. and Nelson, A.R. 2005. Multiple sources for late-Holocene tsunamis at Discovery Bay, Washington State, USA. *The Holocene*, **15**, 60–73, https://doi.org/10.1191/0956683605hl784rp

Geological mapping of coastal and offshore Japan (by GSJ-AIST): collecting and utilizing the geological information

Kohsaku Arai

Geological Survey of Japan, AIST, Central 7, Higashi 1-1-1, Tsukuba, 305-8567 Japan

   0000-0001-9683-0671

ko-arai@aist.go.jp

Abstract: Devastating earthquakes and tsunamis capable of causing catastrophic damage to human societies and economies have frequently occurred on and around the Japanese islands. Because Japan is a long and narrow island arc in the Pacific Ocean located at the junction of four plates (the Eurasian and Philippine Sea Plates in the SW and the North America and Pacific Plates in the NE), it has probably suffered the highest number of earthquake and tsunami events anywhere in the world. Hence, geological and geoinformation investigations are supremely important, not only for understanding the geological development of Japan but also for forecasting the risks associated with geohazards and securing the safety of human lives and infrastructures. For these reasons, the Geological Survey of Japan, National Institute of Advanced Industrial Science and Technology (GSJ-AIST) has been conducting marine geological surveys since 1974 and, as of the end of 2019, produced a Marine Geology Map Series consisting of 90 geological maps. In addition, a coastal zone research project that aims to connect marine, coastal and land area geoinformation seamlessly was launched in 2008 to survey areas that have not yet been investigated, and six maps of geologically distinct areas have already been published based on that collected data.

Map series of marine geology

The Geological Survey of Japan, National Institute of Advanced Industrial Science and Technology (GSJ-AIST) has been conducting marine geological surveys as part of a programme known as the 'Marine Geological Investigations on the Continental Shelves and Slopes around Japan' since 1974. Based on the results of those surveys, as of the end of 2019, a Marine Geology Map Series consisting of 90 geological maps has been compiled and published (Fig. 1), with eight 1:1000 000 maps being created in addition to the more detailed 1:200 000 maps.

The 1:200 000 maps are separated into sedimentological and geological maps, and the geological maps include magnetic and gravity anomaly data. Survey efficiency was maximized by performing nocturnal geophysical observations and daytime sediment sampling. The aim was to obtain 'uniform' and 'high-density' mapping data for all areas around the Japanese islands. Furthermore, since there were no major variations in the items observed and the data collection methods that have been and are being used to systematically and comprehensively collected this uniform data, the results obtained are of sufficient quality to allow accurate geological interpretations.

Seismic reflection surveys are among the most basic methods used for better understanding marine geological structures. In such surveys, geological structures are normally examined using dense two-dimensional (2D) grid data collected at 2 mile intervals across the geological structure, and at 4 mile intervals parallel to the structure. In this study, simultaneously with our seismic reflection surveys, we collected bathymetric data using a multibeam echo sounder, shallow sedimentary structure data using a sub-bottom profiler (SBP) and gravity and magnetic anomaly survey data.

The sedimentological maps were created based on sediment samples collected from the seafloor surface, primarily using grab and/or core samplers, and show the seafloor deposits as determined from the grain sizes and compositions of the collected sediments. The grab sampler was also equipped with a conductivity depth profiler, a turbidimeter, a water sampler and a submarine camera, which allowed the collection of a wide variety of submarine environment information at the sampling points (Fig. 2). As a result, vast quantities of data, not only on the seafloor but also on the surrounding water conditions, could be collected in a single operation of the sampler, and the resulting sedimentological maps were created through comprehensive interpretations of the observed sedimentology and geochemistry of the marine environment. Since the sampling points were located at the crossing points of the seismic survey lines, we were also able use the SBP data to gain an understanding of the sedimentation processes used to form surface sediments by identifying vertical changes in the sedimentary processes.

From: Asch, K., Kitazato, H. and Vallius, H. (eds) 2022. *From Continental Shelf to Slope: Mapping the Oceanic Realm.* Geological Society, London, Special Publications, **505**, 193–201.
First published online July 22, 2020, https://doi.org/10.1144/SP505-2019-95

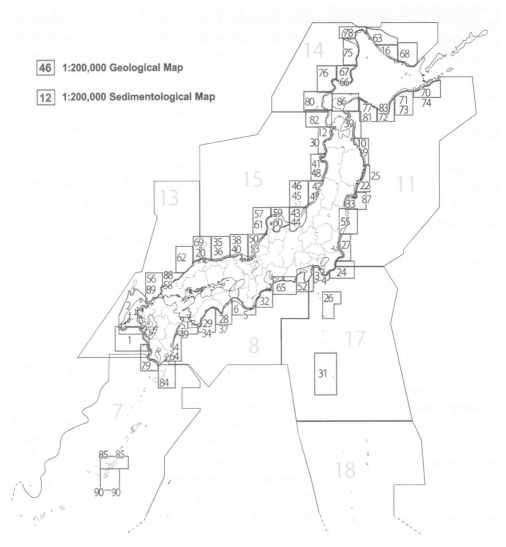

Fig. 1. Marine Geological Map Series of areas surrounding Japan published by GSJ-AIST.

The marine geological maps are images of the area below the seafloor based on seismic reflection profiles (Fig. 3) and the age of the sampled sediments. These maps, which show the geological structure and stratification of sediments, were created through integrated interpretations based on structural geology, seismic stratigraphy, geophysics, sedimentology and stratigraphy, as well as other observations.

The offshore surveys of the four main islands of Japan (Honshu, Hokkaido, Kyushu and Shikoku) were completed in FY 2006, and the various data that were collected during those surveys have since been published as a database and as marine geological maps, and thus have become important

intellectual infrastructure information. The marine geological maps, for example, show the presence of active marine faults and their geological activities and are thus a vital part of efforts aimed at safeguarding human lives and infrastructure. Meanwhile, the sedimentological maps are being used to advance understanding of the distribution of the sediment deposits (such as sand and gravel) that can potentially be used as aggregate construction materials.

Geological survey database

Based on the survey data collected from around the Japanese islands, the GSJ-AIST has published its

Fig. 2. Grab sampler equipped with a conductivity depth profiler, turbidimeter, water sampler and submarine camera.

'Database of Offshore Geologic Structure' to show a portion of the seismic reflection profiles used for the mapping project. These can be seen on the GSJ-AIST website at https://gbank.gsj.jp/marine seisdb/index_E.html (Fig. 4). Since seismic reflection profiles are among the most critical and fundamental types of geological structure data, SBPs are among the useful types of acoustic survey systems for obtaining continuous of geological profile images. SBPs operate by transmitting a sound pulse (commonly 3.5 kHz) from a research vessel and then collecting reflections from the sea bottom and sub-bottom. Such profilers have higher below-sea-bottom resolution levels than seismic reflection surveys conducted using air-gun systems, in spite of their shallower survey depths, primarily owing to the higher-frequency sound sources used. Hence, SBP data are useful for investigating recent sedimentary processes and/or fault activities, and a database of information collected using a 3.5 kHz SBP has also been published on the GSJ-AIST website at https://gbank.gsj.jp/sbp_db/pages/cover-E.html

The various data types collected during the offshore surveys have been published in the form of a database and as marine geological maps. As a result, they can be used as preparation data for drilling and/or coring proposals for scientific purposes. Elsewhere, geological structure data is used to evaluate active faults and active deformation structures extending from offshore seas to populated coastlines, and thus can be effectively used in constructing safe cities that are resistant to geological disasters.

Furthermore, since geological structures are thought to control the areas where minerals and energy resources are dispersed, these processes help researchers to organize geological information that can be used to identify potential areas of exploration for mineral deposits. In particular, the marine geological surveys conducted on uniform grids by AIST can help identify the locations of potential sediment deposits in previously overlooked areas.

Current project on Okinawa since 2008

Following the completion of the surveys around the nation's four major islands, GSJ-AIST marine geological surveys of the region surrounding Japan's southwestern prefecture of Okinawa (hereafter, the Okinawa Project) were started in FY (fiscal year) 2008. In this project, geological information around the islands of the Ryukyu Arc, very little of which had been collected previously, was collected and compiled to make geological maps. In previous conventional geological studies of the Ryukyu Arc, investigations were mostly limited to onshore field surveys of small islands, so numerous points of the tectonic history of the entire region remain unresolved. Because of this, the collection of marine geological data obtained via the Okinawa Project is expected to facilitate a large number of new findings.

The Ryukyu Arc extends for over 1200 km along the eastern coast of Asia from Kyushu to Taiwan. The major islands of the Ryukyu Arc are considered to be the forearc highs and, along with the associated Ryukyu Trench, are products of subduction of the Philippine Sea Plate (PSP) beneath the Eurasian Plate. The PSP is subducting in a NW direction beneath the Eurasian Plate at a convergence rate of 4–9 cm a^{-1} (Seno *et al.* 1993). The Okinawa Trough is a backarc basin located next to the Ryukyu Arc that was formed in the late Miocene (Gungor *et al.* 2012) or the late Pliocene–early Pleistocene (Park *et al.* 1998; Sibuet *et al.* 1998; Shinjo 1999). As such, the formation of the Okinawa Trough was a key geological event associated with the complex tectonism and resulting changes in the topographic configuration of the Ryukyu Arc.

The GSJ-AIST Okinawa Project was started around Okinawa-jima, which is the most densely populated island in the Ryukyus. Since its inception, the project has resulted in 1:200 000 Marine Geology Maps published as the 'Vicinity of Northern Okinawa-jima Island' in 2015 and the 'Vicinity of Southern Okinawa-jima Island' in 2018 (Fig. 5).

To accomplish this, more than 12 000 km of high-resolution seismic profile data were acquired during three cruises (GH08, 28 July to 29 August 2008; GH09, 16 July to 17 August 2009; GH10, 27 October to 25 November 2010) aboard the *R/V Hakurei-Maru No. 2* operated by the Japan Oil, Gas and Metals National Corporation, primarily with air-gun systems using a 16-channel digital

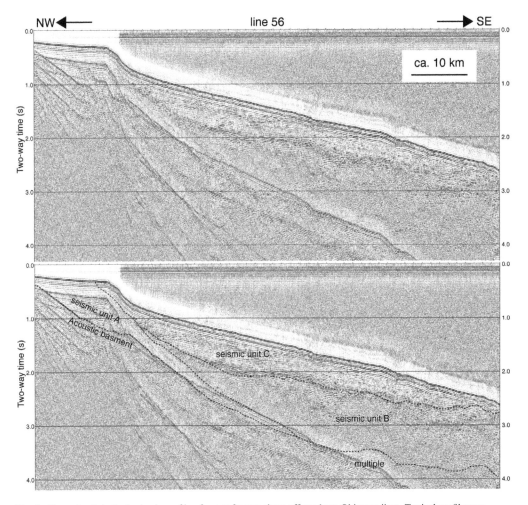

Fig. 3. Example of stacked seismic profile of upper forearc slope off northern Okinawa-jima. Typical profiles are shown across the upper forearc slope in the central Ryukyu Island arc (Arai *et al.* 2018*a*).

streamer cable (Arai *et al.* 2018*a*), as part of efforts to investigate present-day geological deformations occurring off the central Ryukyu Arc (Fig. 6).

Direct observations of geological structures on an overriding plate are important because forearc slope deformation is related to large earthquakes and tsunami generation. Through these surveys, it became clear that several faults have developed in the shallow areas close to the land in the Ryukyu Arc and its upper fore-arc slope (Fig. 5). Since the activities of these faults could result in tsunamis, it will be necessary to investigate their detailed distributions and related activities further. Indeed, from the standpoint of disaster preparedness and mitigation, a complete and unified geological survey of the entire Ryukyu Arc is the most crucial mission of the Okinawa Project.

The land and sea links on geology (seamless coastal zone geoinformation)

The Noto Peninsula earthquakes in March (Noto Hanto Earthquake) and off Niigata in July (Niigata-ken Chuetsu-Oki Earthquake) 2007 caused severe damage to the coastal industries and livelihoods of numerous people in central Japan. However, the active faults responsible had not been imaged before the earthquakes occurred, because the coastal area remains generally underexplored and is typically presented as an investigation gap (Fig. 7). Here, it should be noted that investigations of such coastal areas have long been hampered because conventional research vessels are too large to conduct operations in such nearshore shallow coastal waters and because most coastal plain

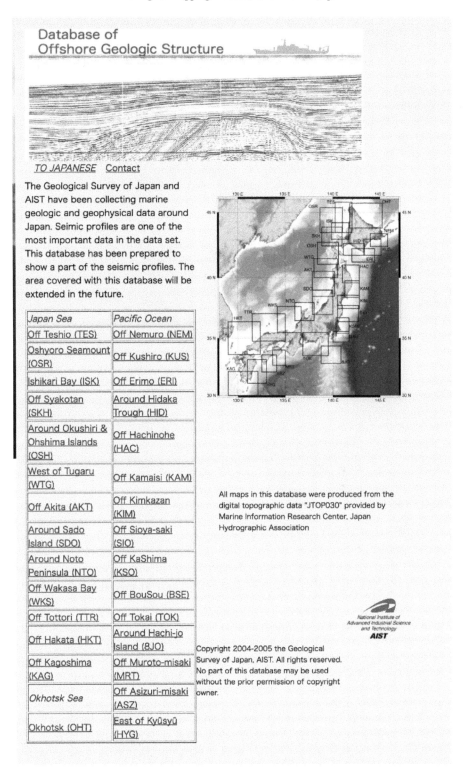

Database of
Offshore Geologic Structure

TO JAPANESE Contact

The Geological Survey of Japan and AIST have been collecting marine geologic and geophysical data around Japan. Seimic profiles are one of the most important data in the data set. This database has been prepared to show a part of the seismic profiles. The area covered with this database will be extended in the future.

Japan Sea	Pacific Ocean
Off Teshio (TES)	Off Nemuro (NEM)
Oshyoro Seamount (OSR)	Off Kushiro (KUS)
Ishikari Bay (ISK)	Off Erimo (ERI)
Off Syakotan (SKH)	Around Hidaka Trough (HID)
Around Okushiri & Ohshima Islands (OSH)	Off Hachinohe (HAC)
West of Tugaru (WTG)	Off Kamaisi (KAM)
Off Akita (AKT)	Off Kimkazan (KIM)
Around Sado Island (SDO)	Off Sioya-saki (SIO)
Around Noto Peninsula (NTO)	Off KaShima (KSO)
Off Wakasa Bay (WKS)	Off BouSou (BSE)
Off Tottori (TTR)	Off Tokai (TOK)
Off Hakata (HKT)	Around Hachi-jo Island (8JO)
Off Kagoshima (KAG)	Off Muroto-misaki (MRT)
Okhotsk Sea	Off Asizuri-misaki (ASZ)
Okhotsk (OHT)	East of Kyûsyû (HYG)

All maps in this database were produced from the digital topographic data "JTOP030" provided by Marine Information Research Center, Japan Hydrographic Association

National Institute of
Advanced Industrial Science
and Technology
AIST

Fig. 4. Database of offshore geological structure linked in https://gbank.gsj.jp/marineseisdb/index_E.html.

Fig. 5. Marine geology maps around Okinawa-jima (compiled from Arai *et al.* 2015 and Arai *et al.* 2018*b*).

areas are heavily urbanized, which restricts the amount of terrestrial domain fieldwork that can be carried out.

This investigation gap between terrestrial and marine domains impedes the recognition of active fault distribution and must be eliminated. Accordingly, since 2008, the GSJ-AIST, which also plays a vital role in understanding the national land use of Japan, has been engaged in a project known as the 'Geology and Active Fault Survey of the Coastal Area' aimed at supplementing coastal zone geological information. The eventual goal of this project is to obtain sufficient information to produce seamless

geological maps for the densely populated coastal areas where active faults pose potential threats, while also contributing to earthquake disaster risk information related to the important infrastructure and industrial sites in the same coastal areas.

In the first five-year phase (2008–13), multidisciplinary marine geology and geophysics surveys (high-resolution reflection seismic profiles along with gravity and magnetic anomalies, etc.) were conducted in the marine domains. Concurrently, terrestrial domain outcrop and drilling investigations were conducted in the coastal areas off the Noto Peninsula (Ishikawa Prefecture, central Japan) in 2008,

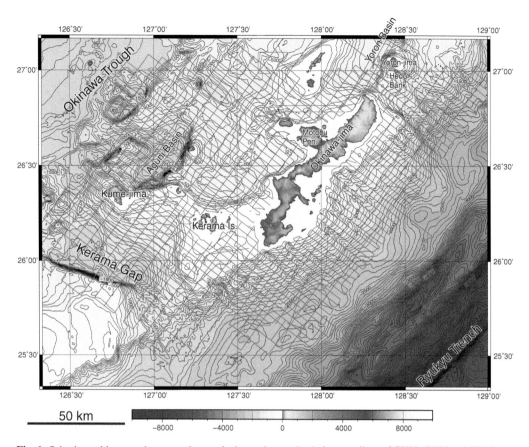

Fig. 6. Seismic, multi-narrow-beam swath sonar bathymetric, geophysical survey lines of GH08, GH09 and GH10 cruises (Arai *et al.* 2018*a*).

the Echigo Plain (Niigata Prefecture, central Japan) in 2009, Hakata Bay (Fukuoka Prefecture, SW Japan) in 2010, Yufutsu Plain (Hokkaido Prefecture, northern Japan) in 2011–12 and Suruga Bay (Shizuoka Prefecture, central Japan) in 2013.

These new results were then integrated with fundamental data collected in previous investigations and the maps that resulted provide a series of seamless geological observations that connect marine, coastal and land areas using the latest and most comprehensive survey techniques to fill in the gaps in previously existing mapping data (Fig. 7). The succeeding three-year phase of the coastal area investigation that began in 2014 dealt with the southern part of the Kanto Plain (Boso Peninsula) in central Japan and was aimed at obtaining a better understanding of geological factors, especially around the metropolitan areas.

Bound by the Sagami Trough to the SW and the Japan Trench to the east, the complicated tectonic structure of the Boso Peninsula exists in a region where both the PSP and Pacific plates subduct

westward beneath the Tokyo Metropolitan Area. In that phase of our investigations, high-resolution seismic profiling surveys, sediment sampling, geophysical surveys and the compilation of existing data on the Boso Peninsula were carried out, thus providing fruitful geological information and detailed knowledge regarding the distribution and continuity of the geological structures along the coastal area east of the peninsula. These data and results have since been summarized and are published as *The Eastern Coastal Zone of the Boso Peninsula*, following the Seamless Geological Map of Coastal Zone series (e.g. Furuyama *et al.* 2019; Ozaki *et al.* 2019).

As our results have already provided critical information regarding the distribution and continuity of subsurface geological structures, it is hoped that they will be effectively utilized for geohazard mitigation and urban planning purposes, including such uses as references when formulating infrastructure and industrial site construction plans along Japan's coastal areas.

Fig. 7. The 1:200 000 scale Seamless Land–Marine Geological Map (right) derived from the 1:200 000 geological map (left). Comprehensive survey techniques were used to cover the investigation gap. All figures are extracted from published offshore maps (Okamura 2002 and 2007) and seamless maps (Inoue and Okamura 2010 and Inoue *et al.* 2010) of areas around the Noto Peninsula.

Summary and prospects

In this paper, the author reported on surveys conducted by GSJ-AIST as part of efforts to create submarine geological maps of the seas around Japan, including nearshore shallow coastal waters off densely populated areas where active faults pose significant potential threats to human lives and infrastructure. The resulting maps not only present necessary information related to the areas around the islands but can also be useful as fundamental data for active fault evaluations, crustal movement analyses, submarine resource developments, seabed utilization efforts and research material for scholars studying these sea areas and their related environments.

The various data that were collected during these offshore surveys have since been published in database form as well as marine geological maps, and have thus become important intellectual infrastructure information. We hope that the datasets presented in our results will assist others in their research efforts and that the valuable geoinformation collected in our surveys will be used productively in various fields.

Funding This work was carried out as a part of the "Geological Mapping Project around Japan" by the GSJ-AIST.

Author contributions KA: writing – original draft (lead).

Data availability statement Please contact author for data requests.

References

Arai, K., Sato, T. and Inoue, T. 2015. *Geological map of the vicinity of northern Okinawa-jima island*. Marine Geology Map Series 85 (CD). Geological Survey of Japan, AIST [in Japanese with English abstract].

Arai, K., Inoue, T. and Sato, T. 2018a. High-density surveys conducted to reveal active deformations of the upper forearc slope along the Ryukyu Trench, western Pacific, Japan. *Progress in Earth and Planetary Science*, **5**, 45. https://doi.org/10.1186/s40645-018-0199-0

Arai, K., Inoue, T. and Sato, T. 2018b. *Geological map of the vicinity of southern Okinawa-jima Island*. Marine

Geology Map Series 90 (CD). Geological Survey of Japan, AIST [in Japanese with English abstract].

Furuyama, S., Sato, T. and Arai, K. 2019. *1:200,000 Marine geological map along the coastal zone of eastern part of the Boso Peninsula*. Seamless Geoinformation of Coastal Zone 'Eastern Coastal Zone of Boso Peninsula' S-6, https://www.gsj.jp/researches/project/coastal-geology/results/s-6.html

Gungor, A., Lee, G.H., Kim, H.J., Han, H.C., Kang, M.H., Kim, J. and Sunwoo, D. 2012. Structural characteristics of the northern Okinawa Trough and adjacent areas from regional seismic reflection data: geologic and tectonic implications. *Tectonophysics*, **522–523**, 198–207, https://doi.org/10.1016/j.tecto.2011.11.027

Inoue, T. and Okamura, Y. 2010. *1:200,000 Marine geological map around the northern part of Noto Peninsula*. Seamless geoinformation of coastal zone 'Northern coastal Zone of Noto Peninsula' S-1 (DVD). Geological Survey of Japan, AIST (in Japanese with English abstract).

Inoue, T., Ozaki, M. and Okamura, Y. 2010. *1:200,000 Seamless geological map of the northern part of Noto Peninsula*. Seamless geoinformation of coastal zone 'Northern Coastal Zone of Noto Peninsula' S-1 (DVD). Geological Survey of Japan, AIST [in Japanese with English abstract].

Okamura, Y. 2002. *Geological map east of Noto Peninsula*. Marine Geology Map Series 59 (CD). Geological Survey of Japan, AIST [in Japanese with English abstract].

Okamura, Y. 2007. *Geological map west of Noto Peninsula*. Marine Geology Map Series 61 (CD). Geological Survey of Japan, AIST [in Japanese with English abstract].

Ozaki, M., Furuyama, S., Sato, T. and Arai, K. 2019. *1:200,000 Marine and land geological map of the eastern coastal zone of the Boso Peninsula and its explanation, especially with Quaternary crustal deformation*. Seamless Geoinformation of Coastal Zone 'Eastern Coastal Zone of Boso Peninsula' S-6.

Park, J.O., Tokuyama, H., Shinohara, M., Suyehiro, K. and Taira, A. 1998. Seismic record of tectonic evolution and backarc rifting in the southern Ryukyu island arc system. *Tectonophysics*, **294**, 21–42, https://doi.org/10.1016/S0040-1951(98)00150-4

Seno, T., Stein, S. and Gripp, A.E. 1993. A model for the motion of the Philippine Sea Plate consistent with NUVEL-1 and geological data. *Journal of Geophysical Research*, **89**, 17941–17948, https://doi.org/10.1029/93JB00782

Shinjo, R. 1999. Geochemistry of high Mg andesites and the tectonic evolution of the Okinawa Trough–Ryukyu arc system. *Chem. Geol.*, **157**, 69–88, https://doi.org/10.1016/S0009-2541(98)00199-5

Sibuet, J.C., Deffontaines, B., Hsu, S.K., Thareau, N., Le Formal, J.P., Liu, C.S. and Party, A. 1998. Okinawa Trough backarc basin: early tectonic and magnetic evolution. *Journal of Geophysical Research*, **103**, 30245–30267, https://doi.org/10.1029/98JB01823

Geological controls on dispersal and deposition of river flood sediments on the Hidaka shelf, Northern Japan

Ken Ikehara[1]*, Hajime Katayama[1], Tsumoru Sagayama[2] and Tomohisa Irino[3]

[1]Research Institute of Geology and Geoinformation, Geological Survey of Japan, National Institute of Advanced Industrial Science and Technology (AIST), Tsukuba Central 7, 1-1-1 Higashi, Tsukuba, Ibaraki 305-8567, Japan

[2]NPO Hokkaido Research Center of Geology, 18-12 Bunkyodai-higashi, Ebetsu, Hokkaido 069-0834, Japan

[3]Faculty of Environmental Earth Science, Hokkaido University, N10-W5 Kitaku, Sapporo, Hokkaido 060-0810, Japan

KI, 0000-0003-3906-4303; TI, 0000-0001-6941-770X
*Correspondence: k-ikehara@aist.go.jp

Abstract: The distribution and characteristics of marine surface sediments are a basic marine geological information. Large river floods are a frequent natural hazard that transport substantial terrigenous sediments into the marine environment. In August 2003, TY ETAU (0310) caused heavy rainfall in the southern coast of Hokkaido, north Japan, where some mountainous rivers in the Hidaka region flooded. Two deposition modes for the 2003 flood sediments can be identified by comparing the pre- and post-flood surface sediment distribution. Shore-normal shallow depressions off the mouth of the Saru and Atsubetsu rivers served as channels for the discharged floodwater preventing dispersion and maintaining the necessary water density to transport the materials as density bottom currents. This action also promoted long-distance transport of flood materials across the continental shelf. Absence of depression on the inner shelf off the mouth of the Niikappu and Shizunai rivers may have dispersed floodwaters near the river mouth and deposited the flood materials close to the shore. Marine geological mapping suggests that the differences in submarine topography (the presence or absence of shallow depressions) are closely related to the regional geological structure. Thus, submarine geology is a controlling factor of the seafloor environments influenced by the river flood.

Understanding the features of the seafloor, including marine geomorphology, is often the goal of a seafloor mapping program (Thorsnes *et al.* 2018). Distribution and characteristics of marine surface sediments are key parameters in the study of marine geology. Marine geological maps have been published in many countries including the UK, Norway, Korea, Australia and Japan. Typically, a marine sedimentological map or marine surface geological map describes surface sediment distribution (e.g. Ikehara 2000). For areas around the Japanese Islands, the Geological Survey of Japan, National Institute of Advanced Industrial Science and Technology (AIST) have been publishing marine geological and sedimentological maps (Arai *et al.* 2013). The Japanese Islands are located along active plate boundaries, where large earthquakes, tsunamis and volcanic eruptions have occurred. This high level of tectonic activity has resulted in the young, mountainous landscape of Japan. Each year in the Japanese Islands, heavy typhoons and monsoon precipitation cause large landslides and floods with

severe damage owing to its middle latitude location that faces the Pacific Ocean. All these events are considered natural hazards or 'geohazards' (Gares *et al.* 1994; Solheim *et al.* 2005). Some such geohazards agitate the seafloor, supplying several new materials to the ocean that in turn form event deposits. However, the exact evaluation of a geohazard's influence on the seafloor is sometimes difficult because of a lack of seafloor data on pre-geohazard events. When such seafloor data are available, then the marine sedimentological maps or marine surface geological maps can be used to compare the pre- and post-geohazard conditions on seafloors.

In August 2003, TY ETAU (0310) occurred in central Hokkaido, north Japan. The typhoon brought heavy rainfall to the southern coast of Hokkaido, particularly in the upper reaches of the Saru and Atsubetsu rivers. Many landslides occurred on the hill slopes, and substantial terrigenous materials were delivered and deposited in the ocean; this significantly influenced the seafloor environment and flooding was expected as a consequence. Because

From: Asch, K., Kitazato, H. and Vallius, H. (eds) 2022. *From Continental Shelf to Slope: Mapping the Oceanic Realm.* Geological Society, London, Special Publications, **505**, 203–215.
First published online July 14, 2020, https://doi.org/10.1144/SP505-2019-114

flood-induced density currents can transport consid-erable discharged materials over long distances (Mulder and Syvitski 1995; Mulder *et al.* 2001, 2003; Nakajima 2006), both shallow (coastal) marine environments and deep-water marine envi-ronments can be affected by such flooding. How-ever, the exact mode of flood sediment transport and the long-term influence on the seafloor environ-ment is unknown.

In this study, we examine the temporal changes observed in marine surface sediments following the 2003 typhoon and flood. We compare the pre-events, immediate post-event, and 2 years post-event surface sediment datasets. Although the pre- and post-event surface sediment datasets have been published (Suga *et al.* 1997; Yamashita 2004; Katayama *et al.* 2007), no detailed discussion on the depositional pro-cesses of the flood event has been conducted. Based on the bathymetrical and sedimentological analyses, two modes of dispersal are considered and continen-tal shelf depositional patterns are discussed. Further-more, the influence of the topographical and geological controls on the dispersal and depositional modes is examined. The results demonstrate the importance of pre-event surface sediment datasets and sedimentological maps.

Physiographic setting and the 2003 Hokkaido Hidaka flood event

Hidaka Trough is a foreland basin that extends from the Ishikari lowland. It lies at the intersection of the NE–SW collision of the Kuril arc with the northeast-ern Japan arc (Fig. 1a; Kimura 1986, 1994) and forms the Hidaka Mountains. The trough is a 120 km wide and 150 km long sedimentary basin. At the northern Hidaka region, the shelf becomes narrower south-eastward: *c.* 21 km wide at the Mu-kawa River, *c.* 20 km wide at the Saru River mouth, *c.* 17 km wide at the Atsubetsu River mouth, *c.* 11 km wide at the Shizunai River and *c.* 10 km (narrowest) wide at Mit-suishi. The shelf then becomes wider at *c.* 13 km off Urakawa. Then, the direction of the shelf break changes from NW–SE to S–N. Heading southward, the width of the shelf increases until it is *c.* 27 km wide east of Cape Erimo, which is the southernmost tip of the Hidaka region (Fig. 1a). The water depth of the shelf break increases from north to south, ranging from *c.* 110 m at the Mu-kawa River mouth to *c.* 120 m at the Saru River and Niikappu River mouth, *c.* 130 m from Shizunai to Urakawa, and *c.* 140 m east of Cape Erimo. A wide inner shelf terrace with a water depth ranging from 10 to 30 m is found in the northern region, between the Mu-kawa River mouth and the Atsubetsu River mouth (Figs 1a, b); further, a wider outer shelf extends from the centre to the southern region. Two shallow depressions

(<5 m deeper than the surrounding terraces) crossing the shelf are found offshore of the Saru River and Atsubetsu River mouths (the off Saru River depres-sion (OSD) and the off Atsubetsu River depression (OAD). Several small mountainous rivers flow from the Hidaka Mountains; Mu-kawa, Saru, Atsu-betsu, Niikappu, Shizunai, Horobetsu and Horoman rivers from west to east (Fig. 1a). The coastal lowland occurs only along these rivers. Most of the coastline along the Hidaka region is rocky with coastal cliffs, except around the river mouths where gravelly and sandy coasts are developed.

The submarine geology offshore of the Hidaka region (Fig. 1c) has been recorded by TuZino and Inoue (2012) and TuZino *et al.* (2014). According to their maps, numerous anticlines reflecting the arc collision occur, trending shore-parallel (NW–SE) and shore-oblique (NNW–SSE). Reverse faults are also associated with some major anticlines. Older strata are exposed along the anticlines and in the area between Urakawa and Cape Erimo. The Mukawa-oki Anticline (MuA) marks the southward extension of the exposure of the middle Hidaka Trough Group (H1020 Formation to H3040 Forma-tion; older than Early–Middle Pleistocene in age) and the Tomakomai-oki Group (H4050 Formation; pre-Middle Miocene in age) on the offshore shelf between the Saru River mouth and the Atsubestu River mouth. The younger upper Hidaka Trough Group (H0010 Formation; Late Pleistocene to Holo-cene in age) is revealed at a syncline structure off-shore of the Niikappu and Shizunai river mouths.

The Hidaka region of Hokkaido, north Japan, has a cold climate with a hot–warm summer and humid-ity (classified Dfa-b Koppen classification). The annual mean precipitation at Urakawa, central Hidaka coast, during 1971–2000 was 1104.0 mm (National Astronomical Observatory of Japan 2007). There is no seasonal (monsoonal) rainfall, and the influence of tropical typhoons is small when compared with south and central Japan. River discharge in this region is at its peak during the snow-melt season (April–June). The Saru and Mu-kawa rivers are among the largest rivers in this region with lengths of 104 and 135 km, drainage areas of 1350 and 1270 km^2, annual mean water discharges of 47 and 46 m^3/s, and maximum water discharges of 760 and 1170 m^3/s, respectively (National Astro-nomical Observatory of Japan 2007). Maximum height and period of the significant wave at Tomi-hama (T in Fig. 1a, b) near the Saru River mouth was 4.0 m and 9.0 s, respectively, in August (Japan Ocean Data Center and Hydrographic Department of Maritime Safety Agency 1986). The direction of coastal currents along the Hidaka coast changes sea-sonally (Suga *et al.* 1997). Prevailing coastal current is northwestward in summer, and southeastward in winter. The direction of coastal sand movements

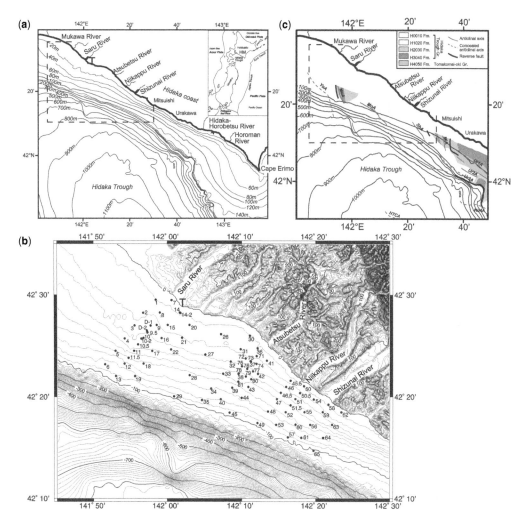

Fig. 1. Location, bathymetry and submarine geology of the studied area. (**a**) Geological setting and bathymetry around the Hidaka region (IL; Ishikari Lowland, HM; Hidaka Mountains, T; Tomihama), (**b**) bathymetry of the studied area with sampling sites of surface sediments (after Katayama *et al.* 2007) (T; Tomihama), and (**c**) simplified submarine geology offshore of the Hidaka region (YuA, Yufutsu Anticline; MuA, Mukawa-oki Anticline; MoA, Monbetsu-oki Anticline; ShA, Shizunai-oki Anticline; MiA, Mitsuishi-oki Anticline; Ur1A, Urakawa-oki No.1 Anticline; Ur2A, Urakawa-oki No.2 Anticline; Ur3A, Urakawa-oki No.3 Anticline; HTCA, Hidaka Trough Central Anticline) modified from TuZino and Inoue (2012) and TuZino *et al.* (2014). Concealed reverse faults are not shown. Rectangle in Figure 1a and 1c indicates area of Figure 1b.

along the Hidaka coast is considered to be southeast to northwest (Tanaka 1990; Suga *et al.* 1997).

On 9 and 10 August 2003, TY ETAU (0310) hit the Hidaka region and heavy rainfall occurred on the western slope of the Hidaka Mountains. The upper reaches of the Saru and Atsubetsu rivers recorded the highest rainfall (358 mm/day), and landslides occurred on the hill slopes. Swollen rivers carried suspended sediments and floating tree trunks overtopped levees and eroded riverbanks. The Saru

River channel carried at least 2.5×10^6 m^3 of sediment, and the Atsubetsu River was supplied with 0.425×10^6 m^3 (Hasegawa 2004). Maximum water discharges from the Saru and Atsubetsu rivers were estimated to be greater than 4000 and 2000–2680 m^3/s, respectively (Hasegawa 2004). Yamashita (2004) reported that yellowish grey mud with a median diameter of 10–20 µm was deposited at the Saru River mouth, over a 10-km-wide area having a water depth >30 m. The total amount of mud collected on the inner shelf during the period of 10–25 days after the flood was calculated to be 2.9×10^6 t (Yamashita 2004). Ogawa and Watanabe (2004) used water discharge rates and average suspended particle concentrations to estimate that 2.7×10^6 t of suspended particles were transported by the Saru River during the flood.

Materials and methods

Surface sediment sampling was conducted in 2005 and 2006, at 84 sites along the Hidaka coast using a Smith-McIntyre grab sampler (20 × 20 cm). Two sub-cores (maximum 7.5 cm long) were collected to examine vertical sedimentary succession and structures. By manually hammering a plastic pipe into the sediments, an additional two short cores (D-1; 12 cm long, and D-2; 18.7 cm long) were obtained offshore of the Saru River mouth by divers. Grain-size analysis was conducted on surface sediment (0–1 cm deep) using a Beckman Coulter LS-230 laser diffraction particle size analyser in the Geological Survey of Hokkaido. The grain size and distribution of surface sediment for these samples were reported by Katayama *et al.* (2007). Detailed sedimentary structures were observed via X-radiography using a Sofron STA 1005 at the Geological Survey of Japan, AIST, on a 1 cm-thick slab sample taken from a sub-core. Diatom assemblages were examined in 61 of the surface sediment samples in the Geological Survey of Hokkaido. Wet sediments (*c.* 3 g) were placed in a beaker with hydrogen peroxide to remove organic matters and hydrochloric acid to remove carbonate. After removing the acid by repeated washing with water, the samples were suspended in 200 cc water. Slides were made using 0.3 cc of suspended water, mounted using Media mount, and observed with a microscope at 1250 × magnification. The first 100 diatom frustules observed were identified and counted.

Results and discussion

Pre- and post-flood sediment distribution on the Hidaka shelf

Pre-flood surface sediment distribution on the Hidaka shelf was compiled by Suga *et al.* (1997). The surface sediment distribution (Fig. 2a) is closely related to the submarine geology described above. Although fine sand is widely distributed on the outer shelf, medium-to-coarse sand with rocky bottoms is found in between the Saru and Atsubetsu River mouths. Medium-to-coarse sands and gravels with rocky bottoms are also distributed on the shelf south of Urakawa (Suga *et al.* 1997). Medium-to-coarse sands and gravels are coincident

with exposures of old strata (middle Hidaka Trough Group and older) according to TuZino and Inoue (2012) and their shallower extension. Suga *et al.* (1997) also observed and recorded detailed grain-size distributions of surface sediments around the mouths of some major rivers. Offshore of the Saru River mouth, they collected surface sediments at 42 locations, in water depths less than 20 m in June 1993 (Fig. 3a). Although sandy sediments prevailed in this area, muddy sediments were found in front of the river mouth at water depths of 16–18 m. Sandy sediments with low mud contents were found offshore of the Shizunai River mouth in July 1993 (Fig. 3c). Mud contents at sites with water depths <30 m was <10%. The more offshore they sampled, the higher was the mud content, which is about 10–30% at a water depth of *c.* 50 m (Fig. 3c).

Post-flood surface sediment distributions were reported by Yamashita (2004), Katayama *et al.* (2007) and Noda and Katayama (2013). Yamashita (2004) conducted surface sediment samplings thrice (ranging from 10 days to 2 months after the 2003 flood) at the mouths of the Saru and Mu-kawa rivers and reported thick depositions (>20 cm) of yellowish mud near both river mouths (Fig. 2b). Katayama *et al.* (2007) also compiled surface sediment distributions offshore of the Saru, Atsubetsu, Niikappu and Shizunai river mouths based on the 2005–06 surveys. These results showed that sandy sediments prevailed in this area, and the general distribution of surface sediments (Figs 2c & 4) was almost the same as reported by Suga *et al.* (1997). However, Katayama *et al.* (2007) observed muddy sediments in two shallow depressions (OSD and ASD) and on the inner shelf off the Shizunai River mouth. The yellowish mud described by Yamashita (2004) was not found around the Saru River mouth; however, it was found in the OSD at water depths of 20–35 m and at ASD around 70 m. Noda and Katayama (2013) reported some additional surface sediment samples in this area based on the survey in 2006, with a similar distribution pattern to Katayama *et al.* (2007).

Temporal changes of shelf sediment distributions on the Hidaka shelf

The comparison of the three surface sediment distribution datasets indicated that there was large and widespread terrigenous mud deposition associated with the 2003 flood followed by reworking of the flood mud under the post-flood wave and current conditions at the Hidaka inner shelf. Higher mud content was obvious in sediments at the mouths of the major rivers including the Mu-kawa, Saru, Niikappu and Shizunai (Figs 2c, 3 & 4). The thick

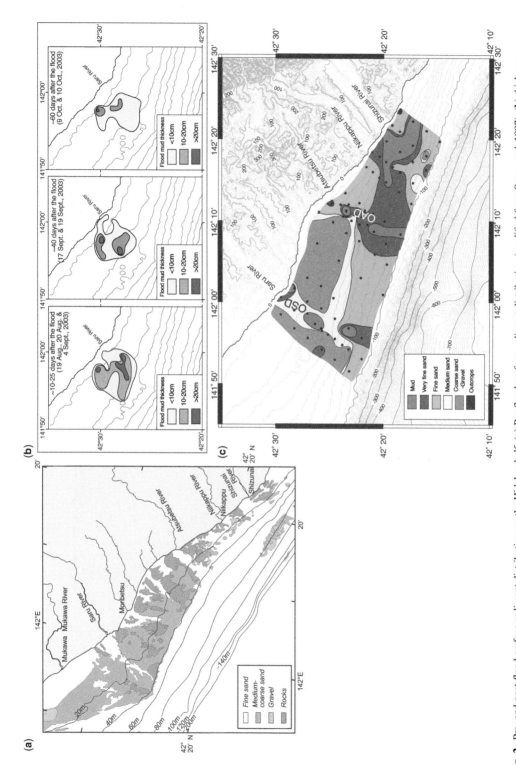

Fig. 2. Pre- and post-flood surface sediment distribution on the Hidaka shelf. (a) Pre-flood surface sediment distribution (modified from Suga *et al.* 1997), (b) thickness distribution of yellow mud supplied by the 2003 flood offshore of the Saru River mouth (data after Yamashita 2004), and (c) post-flood surface sediment distribution (after Katayama *et al.* 2007).

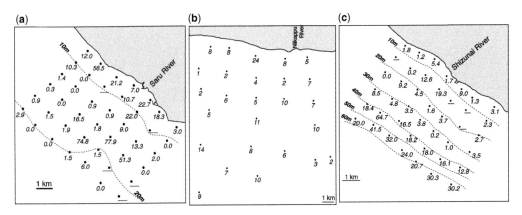

Fig. 3. Pre-flood distribution of mud contents in surface sediment offshore from the mouths of the Saru, Niikappu and Shizunai rivers. (**a**) Pre-flood mud content distribution offshore of the Saru River mouth (data after Suga *et al.* 1997), (**b**) pre-flood mud content distribution offshore of the Niikappu River mouth (data after Niikappu Fishermen's Cooperative Association 1991), and (**c**) pre-flood mud content distribution offshore of the Shizunai River mouth (data after Suga *et al.* 1997).

yellowish mud that has been deposited on the inner shelf near the mouths of the Mu-kawa and Saru rivers (Yamashita 2004) is the most characteristic change in surface sediments after the 2003 flood (Fig. 2b). However, Katayama *et al.* (2007) indicated

that there was no yellowish mud on the inner shelf near the Saru River mouth. Yamashita (2004) indicated that there was a change in the thickness of the yellowish mud offshore of the Saru River mouth within a few months of the 2003 flood

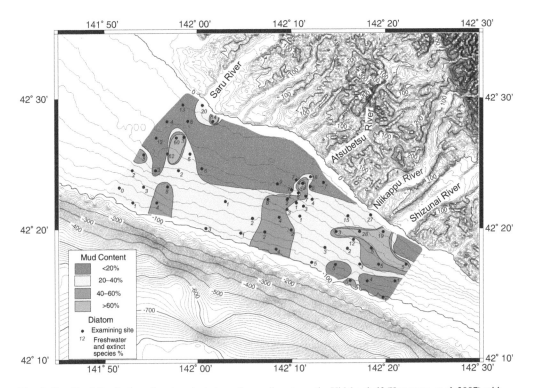

Fig. 4. Post-flood distribution of mud contents in surface sediments on the Hidaka shelf (Katayama *et al.* 2007) with concentrations of freshwater diatoms and extinct species of diatom.

(Fig. 2b). Based on the seismic reflection surveys by TuZino and Inoue (2012), old basement rocks (Tomakomai-oki and Hidaka Trough Groups) were exposed in this area (Fig. 1c) and the unconsolidated sediment layer was thought to be thin. The wide distribution of wave-cut terraces at a water depth of 10–30 m was found in this area (Figs 1a, b). Under these conditions, preservation of the yellowish flood mud is unlikely. Fan *et al.* (2004) and Sommerfield *et al.* (2007) indicated the offshore transport of shallow-water sandy sediments associated with low concentration regime or dry storm, which is characterized by a lack of buoyant river plumes with muddy particles. Because there have been no major typhoons or storms since the 2003 flood and the 2005 survey of this area, relatively large wave and coastal current action in the inner shelf area of the Hidaka region according to rough weather such as a dry storm is the most likely mechanism for reworking and rearranging the yellowish mud. Katayama *et al.* (2007) only located yellowish flood mud in the OSD and OAD (Figs 2c & 4). The topographic depression in this area generates weaker wave and current conditions that prevent the reworking of the soft yellowish mud. Muddy sediments in the OSD and OAD contain a high concentration of freshwater diatom species (such as *Achnantes lanceolate, Cymbella ventricosa, Navicula contenta* and *N. microcephala*) and extinct (Miocene) diatom species (such as *Actinocyclus ingens, Denticulopsis dimorpha, D. hyaline* and *D. hustedtii*) (Fig. 4 & Table 1). This suggests that the mud had a terrigenous origin. Stable carbon isotope is useful to determine the source of organic matters (Jasper and Gagosian 1990). The low carbon isotope values measured in organic matter found in the muddy sediments (−23.8‰ to −25.6‰) suggest the presence of terrigenous organic carbon. Similarly, the suspended materials in the Hidaka rivers (−26.0‰ to −28.8‰) (Omura *et al.* 2017) support this idea. Higher occurrence of freshwater and extinct diatoms and lower carbon isotopes of organic matters evidence cross-shelf sediment transport through the OSD (Omura *et al.* 2017) (Table 1 & Fig. 4).

Another characteristic change in surface sediment was found offshore of the Niikappu and Shizunai river mouths. Pre-flood surface sediments on the inner shelf near the Shizunai River mouth (Fig. 3c) exhibited low mud content (<10%) (Suga *et al.* 1997). The Niikappu Fishermen's Cooperative Association (1991) reported mud contents (less than 15%) of surface sediments off the Niikappu River mouth (Fig. 3b). In contrast, a wide distribution of muddy sediments was reported in a post-flood surface sediment survey (Fig. 4; Katayama *et al.* 2007). Although the sampling locations of pre- and post-flood surveys were not the same, the percentage increase in mud content between the nearby sites is calculated to be greater than 10% and sometimes at 20% or 30%. High concentrations of freshwater and extinct diatom species in the muddy sediments (Fig. 4 & Table 1) suggest that the source of muddy materials is the terrigenous. Omura *et al.* (2017) indicated low carbon isotope (δ^{13}C) values (−25.2‰) for organic matters in the inner shelf mud (Table 1) offshore from the Shizunai River mouth. An increase in δ^{13}C values when moving away from the shore (Table 1) suggests a decreasing influence of riverine materials on the shelf. The mud content increased because of both the 2003 flood discharge from the Niikappu and Shizunai rivers and the discharge by the daily precipitation on 11 September 2001 (175 mm/day) at Shizunai being nearly identical to the precipitation on 9 August 2003 (173 mm/day; by the database of Japan Meteorological Agency 2019). The occurrence of the subsurface mud layers found off the Niikappu River mouth (Fig. 5) suggests repeated deposition of mud layer on this inner shelf. However, daily precipitation greater than 100 mm/day was observed only four times between 1993 and 2003 at Shizunai (122 mm on 9 August 1995, 131 mm on 2 May 1998, 175 mm on 11 September 2001 and 173 mm on 9 August 2003). These heavy rainfalls are potential contributors to the increase in mud content of the inner-shelf surface sediments near the Niikappu and Shizunai river mouths.

Modes of flood sediment dispersal and deposition and their controlling factors

Post-flood surface sediments (Fig. 2c; Katayama *et al.* 2007) exist in two types of spatial distribution patterns. The first type is found in the shallow depressions. Surface sediments in such depressions are generally finer than those found on the surrounding terraces where medium-to-coarse sand occurs (Fig. 2c). The depressions are formed from previous paleochannels of the Saru and Atsubetsu rivers at the glacial sea-level lowstands (Suga *et al.* 1997). Offshore from the Saru and Atsubetsu rivers' shelf, basement rocks are widely exposed (Fig. 1c; TuZino and Inoue 2012). During the last glacial maximum sea-level lowstands, the river mouths progressed to the paleocoastline at around − 120 m (Shackleton 1987; Oba 1988; Siddall *et al.* 2003; Okuno *et al.* 2014). The basement rocks were eroded by river streams, which created shallow depressions. Although they were buried in 15–25 m of sediment during post-glacial transgression and Holocene sea-level highstands (Maritime Safety Agency 1982), the depressions have maintained their deepness at the OSD and OAD. Omura *et al.* (2017) reported sedimentary structures and organic geochemical characteristics of the 2003 flood mud in the OSD.

Table 1. *Grain size, occurrence of freshwater diatoms and extinct diatoms, and organic geochemical characteristics of surface sediment samples from post-flood surveys*

| Site | Water Depth (m) | Md (φ)* | Mud (%)* | Diatom | | TOC (%)† | δ13C (‰)† | TOC/TN (atomic)‡ |
				Fresh water species (%)	Extinct species (%)			
1	10	−1.83	0.1	13	0			
2	17	0.29	0.6	4	0			
3	23	−2.72	0.2	9	3			
4	34	3.21	45.3	8	0			
5	60	0.92	32.5	0	2	0.7	−23.0	9.6
6	79	2.30	25.8	0	0	0.5	−22.8	9.4
7	6	3.41	24.1	18	2	0.8	−24.5	10.0
8	16	1.54	10.1	6	2			
9	23	5.64	80.3	3	0	1.3	−23.8	10.2
9.5	28	5.64	67.8	62				
10	32	6.26	84.6		0	2.9	−25.6	13.6
10–2	33	0.79	6.0					
10.5	34	0.12	0.4					
11	42	1.24	10.2	2	1			
11.5	61	1.93	24.9					
12	74	−0.12	23.1	3	0	0.4	−22.2	8.3
13	86	2.54	27.4	1	0	1.0	−24.5	10.2
14–2	10	6.79	84.7	12	2			

| Site | Water Depth (m) | Md (φ)* | Mud (%)* | Diatom | | TOC (%)† | δ13C (‰)† | TOC/TN (atomic)‡ |
				Fresh water species (%)	Extinct species (%)			
42	34	2.83	37.5	4	0			
43	59	3.00	27.2	1	0			
44	71	3.05	31.5	1	1			
45	98	4.21	52.4	2	2			
45.5	17	NS						
46	28	2.59	24.5	14	1			
46.5	41	NS						
47	53	4.92	62.8	2	1			
48	71	2.53	22.0	2	1			
49	100	2.78	26.2	4	1			
50	10	3.31	30.6	26	1			
50.5	23	G						
51	34	5.32	67.2	27	1			
51.5	48	2.55	23.6	10	2			
52	61	3.03	34.5	1	1			
53	82	1.45	12.9	5	0			
54	19	4.96	75.2	19	0	1.6	−25.2	11.6
55	45	4.75	58.6	5	0	0.7	−24.2	10.1

Site		Md / TOC				TN	δ¹³C	C/N
15	15	G						
16	25	0.28	12.1	4	1			
17	40	1.59	25.2	1	1	0.5	−22.7	9.2
18	53	4.75	56.2	7	1	1.1	−24.6	11.3
19	81	3.77	47.6	3	1			
20	17	G						
21	21	f.s.						
22	36	1.18	12.0	6	0			
26	18	G						
27	32	G						
28	69	2.89	30.4					
29	101	2.95	34.2	1	0			
30	12	2.37	7.2	2	1			
31	19	G						
32	32	0.92	3.4	5	4			
33	53	3.62	46.6	7	1			
34	74	3.07	35.2	0	2			
35	88	2.90	29.2	1	0			
36	15	MS						
37	26	6.26	78.1	43	5			
38	44	3.59	46.9	2	0			
39	67	4.32	55.4	3	1			
40	82	3.57	43.8	0	2			
41	18	2.49	7.6	7	0			
56	61	4.01	50.4	2	3	0.8	−24.0	9.6
57	92	3.50	44.7	0	0	0.3	−22.6	7.9
58	11	2.83	6.1					
59	25	2.18	7.6	3	0	0.5	−23.4	8.6
60	70	2.93	35.6	3	1			
61	83	0.25	8.4					
62	6	NS (R?)						
63	32	5.06	60.1	5	0			
64	73	2.36	14.4	1	0			
65	106	2.33	8.9	1	0			
71	18	2.74	35.5	14	4			
72	22	−0.68	15.2	7	0			
73	25	5.44	78.3	3	1	1.0	−24.7	10.4
74	25	2.33	15.5	7	1			
75	32	1.28	19.6	7	0			
76	33	1.12	23.5	6	2			
77	33	3.22	43.6	5	2			
78	43	3.33	43.8	2	0			
79	44	3.09	38.3	0	1			
80	44	2.81	32.9	3	0			
81	56	3.20	37.0	3	0			
D-1	25	6.96	91.4	60				
D-2	29	si						

*Data after Katayama et al. (2007) except of Sites D-1 and D-2.
†Data after Omura et al. (2017).
‡Analyzed by A. Omura.
Md: si; silt, f.s.; fine sand, G; gravel, MS; mudstone, NS; no sample obtained, R?; rocky?
TOC: Total organic carbon content, TN: Total nitrogen content.

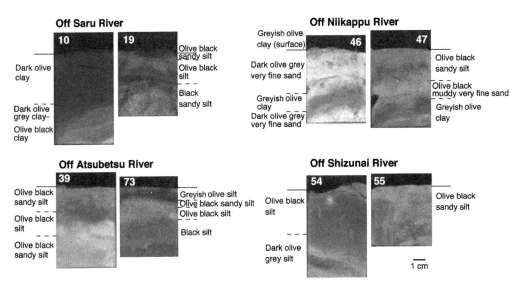

Fig. 5. Simplified sediment description and soft-X radiographs of post-flood surface sediments offshore of Saru (Sts. 10 and 19), Atsubetsu (Sts. 39 and 73), Niikappu (Sts. 46 and 47) and Shizunai (Sts. 54 and 55) rivers.

According to their interpretation, there was an upward coarsening and then an upward fining structure in a core that reflected changes in the water discharge (increasing then decreasing) from the Saru River during the flood. Such a change in sedimentary structures in relation to the river discharge has been discussed by the previous studies; the structures were considered to be hyperpycnites (e.g. Mulder *et al.* 2001; Saitoh *et al.* 2005). Although most hyperpycnites described by Mulder *et al.* (2001) were coarser than those found in the OSD, muddy hyperpycnites exhibit similar change in sedimentary structures (Bourget *et al.* 2010, 2011; Mulder and Chapron 2011). Furthermore, Ikehara *et al.* (2020) showed upward coarsening and then upward fining structures in sandy and muddy hyperpycnites on the slope off the Kumano River, although significant variation in sediment thickness, grain size and sedimentary structures were recognized for three historical flood deposits from the same river. Therefore, the bottom currents with large amounts of flood-sourced suspended particles were confined within the OSD and maintained their density and flow, moving offshore through the OSD and beyond. Shelf-normal distribution, where finer-grained surface sediments are found in the OSD than in the surrounding terraces (Fig. 2c), is likely to result from these processes. Thin layers with similar sedimentary structures but a higher concentration of freshwater and extinct diatoms (Fig. 4) were collected at the outer shelf extension of the OSD (St. 19, Fig. 5). Further offshore transport of the 2003 flood materials was inferred from the organic geochemical analysis

of the upper slope surface sediment collected from a water depth of 538 m off the OSD (Omura *et al.* 2017). A similar fine-grained surface sediment distribution with a shelf-normal trend was found in the ASD and further offshore (Figs 2c, 4 & 5). Upward coarsening sedimentary structures were observed in the surface sediments (Fig. 5) similar to those found in the OSD. The similarity in the bathymetric features, sediment distribution and sedimentary structures suggests that the same processes occurred in the ASD and the OSD during the 2003 flood. Therefore, shallow depressions, which were originally formed as paleo-river channels during the glacial sea-level lowstands, have an important role in the long-distance transport of the flood materials across the shelf.

However, no current-related sedimentary structures were found in muddy sediments on the inner shelf offshore of the Niikappu and Shizunai rivers (Fig. 5). The only structure found off the Niikappu River mouth was an alternation between mud and fine sand (Fig. 5). Similar alternated sands and muds were found on the Eel River shelf off northern California (Hill *et al.* 2007). Flood-related sediments were widely distributed on the Eel shelf and upper slope (Fan *et al.* 2004; Hill *et al.* 2007; Nittrouer *et al.* 2007). Sommerfield *et al.* (2007) proposed the repeated occurrence of dry and wet storms for the formation of mud and sand alternation on the flood-influenced shelf. The concentration of freshwater and extinct diatoms in surface sediment decreased with increasing distance from the river mouths (Fig. 4). This remarkable observation may suggest a rapid deposition of riverine suspended materials on the inner shelf near the river mouths. There is no obvious depression in the inner shelf connected to the modern Shizunai and Niikappu rivers. The absence of an inner shelf depression off the Shizunai and Niikappu rivers is most likely controlled by the geological structure. TuZino and Inoue (2012) indicated that they found no basement rock exposure offshore of the Shizunai and Niikappu river mouths but only the distributions of a young sedimentary unit (Fig. 1c). Under this condition, the paleo-river channels could be shallow, barely maintaining their depth owing to sediment burials, waves and currents erosion during the post-glacial transgression and Holocene sea-level highstands. In this situation, the discharged floodwater containing the suspended materials from the rivers not confined in an inner shelf depression is widely dispersed near the river mouths. Widely dispersed suspended water reduced its density by mixing with seawater, and deposited riverine mud near the river mouths. Although the vertical grain-size trend at the Shizunai River mouth (St. 54, Fig. 5) exhibited an upward fining grading structure with coarse-grained layers suggesting that a density flow occurred near the river mouth, the diffused water rapidly reduced its density, thus, depositing the suspended flood materials on the inner shelf near the river mouths. Based on the post-flood yellow mud distribution (Yamashita 2004; Fig. 2b), rapid deposition of the suspended flood materials also occurred near the Saru and Mu-kawa river mouths. Considering the offshore transport by the gravity flows through the shallow depression (OSD), the total amount of flood-related yellow mud discharged from the Saru River was larger than 2.9×10^6 t deposited on the inner shelf (Yamashita 2004). However, thick yellow mud disappeared near the Saru River mouth (Katayama *et al.* 2007). Further studying the controlling factors of the post-flood sediment remobilization along the Hidaka shelf is necessary. Furthermore, although the change in clay mineral compositions between the northern (OSD and OAD) area and the southern (Niikappu and Shizunai river) area is another possible control on different dispersal patterns due to different chemical (ionic) roles of the clay minerals (Sudo 1953), we require clay mineral analysis to discuss this effect.

Conclusions

Based on the post-flood surface sediment surveys, the 2003 Hokkaido Hidaka flood sediments occurred in two spatial distribution patterns on the Hidaka shelf: (1) in the river-connected shallow depressions offshore of the Saru and Atsubetsu rivers; and (2) on the inner shelf near the river mouths offshore from the Shizunai and Niikappu rivers. Comparison of pre- and post-flood surface sediment distributions shows an increase in the mud content. The shore-normal shallow depressions work as a channel for the discharged floodwater containing the suspended materials. The depression may have prevented the dispersion of floodwater and maintained the water density causing the flood materials to move as density bottom currents. Under these conditions, sedimentary influence from the flood expands in an offshore direction. However, in locations where there is no such depression, the floodwater disperses near the river mouth and deposits the flood-sourced sediments on the inner shelf. Under these conditions, the influence of the flood is restricted to the inner shelf. This interpretation is supported by the pre- and post-flood surface sediment mapping data. Sedimentological mapping on the shelf is useful for evaluating the influence of geohazard events on seafloor environments. Furthermore, marine geological mapping can reveal shallow depressions in the submarine topography because their presence is highly related to the geological structures. Submarine geology is a controlling factor in seafloor environments. Therefore, systematic collection of both marine geological and sedimentological data and publication of these results as marine geological maps should be a standard process and results should be made public property. The data are not only used for evaluating marine environmental changes from large floods but also for evaluating other geohazard events such as large earthquakes, tsunamis and anthropogenic activities such as marine mining, dumping and pollution.

Acknowledgements We greatly appreciated the Hidaka Fishermen's Cooperative Association's kind support during our near-shore field surveys in 2005–06. We also thank Mr Kazuya Suga, Dr Taqumi TuZino and Dr Takahiko Inoue for their assistance with field surveys and sediment analysis. This paper includes some of the results from the KAKENHI project '2003 Hokkaido Hidaka flood: sedimentary processes in the open ocean and influence to sea bottom environments' (No. 17340151) and is the result of a cooperative study between the Geological Survey of Japan, AIST and the Geological Survey of Hokkaido. Submarine geological data were obtained by the Geological Survey of Japan, AIST project 'Marine geological and geophysical studies on the collision zone of Kuril and northeast Japan arcs'. We also express our thanks to Prof. Hiroshi Kitazato for giving us an opportunity to write this paper.

Funding Funding was provided by the Japan Society for the Promotion of Science (JSPS).

Author contributions KI: investigation (equal), project administration (lead), writing – original draft (lead); **HK**: investigation (equal), project administration (supporting), writing – original draft (supporting); **TS**: investigation (equal), project administration (supporting), writing – original draft (supporting); **TI**: investigation (equal), writing – original draft (supporting).

Data availability statement All data generated or analysed during this study are included in this published article. If additional data is required, please contact the corresponding author.

References

Arai, K., Shimoda, G. and Ikehara, K. 2013. Marine geological mapping project in the Okinawa area – Geoinformation for the development of submarine mineral resources. *Synthesiology*, **6**, 162–169. [in Japanese with English abstract], https://doi.org/10.5571/synth.6.162

Bourget, J., Zaragosi, S. *et al.* 2010. Hyperpycnal-fed turbidite lobe architecture and recent sedimentary processes: a case study from the Al Batha turbidite system, Oman margin. *Sedimentary Geology*, **229**, 144–159, https://doi.org/10.1016/j.sedgeo.2009.03.009

Bourget, J., Zaragosi, S. *et al.* 2011. Turbidite system architecture and sedimentary processes along topographically complex slopes: the Makran convergent margin. *Sedimentology*, **58**, 376–406, https://doi.org/10.1111/j.1365-3091.2010.01168.x

Fan, S., Swift, D.J.P., Traykovski, P., Bentley, S., Borgeld, J.C., Reed, C.W. and Niedoroda, A.W. 2004. River flooding, storm resuspension, and event stratigraphy on the northern California shelf: observations compared with simulations. *Marine Geology*, **210**, 17–41, https://doi.org/10.1016/j.margeo.2004.05.024

Gares, P.A., Sherman, D.J. and Nordstrom, K.F. 1994. Geomorphology and natural hazards. *Geomorphology*, **10**, 1–18, https://doi.org/10.1016/0169-555X(94)90004-3

Hasegawa, K. 2004. Introduction. *In*: Hasegawa, K. (ed.) *Report of the 2003 Hokkaido flood disaster by typhoon 200310*. Committee on Hydroscience and Hydraulic Engineering, Japan Society of Civil Engineering, Tokyo, 1–7. [in Japanese].

Hill, P.S., Fox, J.M. *et al.* 2007. Sediment delivery to the seabed on continental margins. *International Association of Sedimentologists, Special Publication*, **37**, 49–99.

Ikehara, K. 2000. Sedimentological map and marine surface sediment study. *Chishitsu News*, **549**, 50–53. [in Japanese].

Ikehara, K., Usami, K. and Irino, T. 2020. Variations in sediment lithology of submarine flood deposits on the slope off Kumano River, Japan. *Geological Society, London, Special Publications*, **501**, https://doi.org/10.1144/SP501-2019-53

Japan Meteorological Agency 2019. Past meteorological database. http://www.data.jma.go.jp/obd/stats/etrn/view/annually_a.php?prec_no=22&block_no=0140&year=1995&month=&day=&view=

Japan Ocean Data Center and Hydrographic Department of Maritime Safety Agency 1986. *Nihon kinaki harou tokei zuhyou (Statistics Charts of Coastal Wave around the Japanese Islands)*. Japan Ocean Data Center and Hydrographic Department of Maritime Safety Agency, Tokyo.

Jasper, J.P. and Gagosian, R.B. 1990. The source and deposition of organic matter in the Late Quaternary Pigmy basin, Gulf of Mexico. *Geochimica Cosmochimica Acta*, **54**, 1117–1132, https://doi.org/10.1016/0016-7037(90)90443-O

Katayama, H., Ikehara, K., Sagayama, T., Suga, K., Irino, T., TuZino, T. and Inoue, T. 2007. Distribution of surface sediments after the 2003 flood on the shelf off Hidaka, southern Hokkaido. *Bulletin of Geological Survey of Japan*, **58**, 189–199. [in Japanese with English abstract], https://doi.org/10.9795/bullgsj.58.189

Kimura, G. 1986. Oblique subduction and collision: Fore-arc tectonics of the Kuril Arc. *Geology*, **14**, 404–407, https://doi.org/10.1130/0091-7613(1986)14<404:OSACFT>2.0.CO;2

Kimura, G. 1994. The latest Cretaceous-Early Paleogene rapid growth of accretionary complex and exhumation of high pressure series metamorphic rocks in northwestern Pacific margin. *Journal of Geophysical Research*, **99**, 22147–22164, https://doi.org/10.1029/94JB00959

Maritime Safety Agency 1982. *Eastern Part of Tomakomai*. Basic Map of the Sea in Coastal Waters (1:50,000), No. 6374-5. Maritime Safety Agency, Tokyo. [in Japanese].

Mulder, T. and Chapron, E. 2011. Flood deposits in continental and marine environments: character and significance. *American Association of Petroleum Geologists Studies in Geology*, **61**, 1–30.

Mulder, T. and Syvitski, J.P.M. 1995. Turbidity currents generated at river mouths during exceptional discharges to the world oceans. *Journal of Geology*, **103**, 285–299, https://doi.org/10.1086/629747

Mulder, T., Migeon, S., Savoye, B. and Faugeres, J.-C. 2001. Inversely graded turbidite sequences in the deep Mediterranean: a record of deposits from flood-generated turbidity currents? *Geo-Marine Letters*, **21**, 86–93, https://doi.org/10.1007/s003670100071

Mulder, T., Syvitski, J.P.M., Migeon, S., Faugeres, J.-C. and Savoye, B. 2003. Marine hyperpycnal flows: initiation, behaviour and related deposits. A review. *Marine and Petroleum Geology*, **20**, 861–882, https://doi.org/10.1016/j.marpetgeo.2003.01.003

Nakajima, T. 2006. Hyperpycnites deposited 700 km away from river mouths in the central Japan Sea. *Journal of Sedimentary Research*, **76**, 60–73, https://doi.org/10.2110/jsr.2006.13

National Astronomical Observatory of Japan 2007. *2008 Chronological Scientific Tables*. Maruzen, Tokyo. [in Japanese].

Niikappu Fishermen's Cooperative Association 1991. *Base map of Niikappu Fishery Area*. Niikappu Fishermen's Cooperative Association, Niikappu. [in Japanese].

Nittrouer, C.A., Austin, J.A., Jr, Field, M.E., Kravitz, J.H., Syvitski, J.P.M. and Wiberg, P.L. 2007. Writing a Rosetta stone: insights into continental-margin sedimentary processes and strata. *International Association of Sedimentologists, Special Publication*, **37**, 1–48.

Noda, A. and Katayama, H. 2013. *Sedimentological Map of Hidaka Trough*. Marine Geology Map Series, No. 81 (CD), Geological Survey of Japan, AIST, Tsukuba. [in Japanese with English abstract].

Oba, T. 1988. Comment for sea level change. *The Quaternary Research (Daiyonki-kenkyu)*, **26**, 243–250. [in Japanese with English abstract], https://doi.org/10.4116/jaqua.26.3_243

Ogawa, N. and Watanabe, Y. 2004. Behavior of SS in the Saru river basin. *In*: Hasegawa, K. (ed.) *Report of the 2003 Hokkaido flood disaster by typhoon 200310*. Committee on Hydroscience and Hydraulic Engineering, Japan Society of Civil Engineering, Tokyo, 125–137. [in Japanese].

Okuno, J., Nakada, M., Ishii, M. and Miura, H. 2014. Vertical tectonic crustal movements along the Japanese coastlines inferred from late Quaternary and recent relative sea-level changes. *Quaternary Science Reviews*, **91**, 42–61, https://doi.org/10.1016/j.quascirev.2014.03.010

Omura, A., Ikehara, K., Katayama, H., Irino, T. and Sagayama, T. 2017. Characteristics of shallow marine flood sediments by organic carbon analyses, examples from shelf sediments off Hidaka, southern Hokkaido, after the 2003 typhoon no. 10. *Journal of Geological Society of Japan*, **123**, 321–333. [in Japanese with English abstract], https://doi.org/10.5575/geosoc.2017.0002

Saitoh, Y., Tamura, T. and Masuda, F. 2005. Characteristics of hyperpycnal flow and its deposits as an innovative factor for the turbidite paradigm. *Journal of Geography (Chigaku-zasshi)*, **114**, 687–704. [in Japanese with English abstract], https://doi.org/10.5026/jgeography.114.5_687

Shackleton, N.J. 1987. Oxygen isotopes, ice volume and sea level. *Quaternary Science Reviews*, **6**, 183–190, https://doi.org/10.1016/0277-3791(87)90003-5

Siddall, M., Rohling, E.J., Almogi-Labin, A., Hemleben, Ch., Meischner, D., Schmeizer, I. and Smeed, D.A. 2003. Sea-level fluctuations during the last glacial cycle. *Nature*, **423**, 853–858, https://doi.org/10.1038/nature01690

Solheim, A., Bhasin, R. *et al.* 2005. International Centre for Geohazards (ICG): assessment, prevention and mitigation of geohazards. *Norwegian Journal of Geology*, **85**, 45–62.

Sommerfield, C.K., Ogston, A.S. *et al.* 2007. Oceanic dispersal and accumulation of river sediment. *International Association of Sedimentologists, Special Publication*, **37**, 157–212.

Sudo, T. 1953. *Clay minerals. Iwanami-zensho*, **178**, Iwanami Shoten, Tokyo. [in Japanese].

Suga, K., Sagayama, T. and Higaki, N. 1997. Environment of submarine geology in the coastal area of Hokkaido -1-: West part of the Pacific Ocean. Geological Survey of Hokkaido Special Report, No. **28**. [in Japanese with English abstract].

Tanaka, N. 1990. Littoral drift along the Japanese coast. *In*: Coastal Oceanography Research Committee, Oceanographic Society of Japan (ed.) *Coastal Oceanography of Japanese Islands, Supplementary Volume,* Tokai University Press, Tokyo, 359–378.

Thorsnes, T., Bjarnadottir, L.R. *et al.* 2018. National programmes: geomotphological mapping at multiple scales for multiple purposes. *In*: Micallef, A., Krastel, S. and Savini, A. (eds) *Submarine Geomorphology*. Springer, 535–552.

TuZino, T. and Inoue, T. 2012. *Geological Map of Hidaka Trough. Marine Geological Map Series, No. 77 (CD),* Geological Survey of Japan, AIST, Tsukuba. [in Japanese with English abstract].

TuZino, T., Inoue, T. and Arai, K. 2014. *Geological Map offshore of Cape Erimo*. Marine Geological Map Series, No. 83 (CD), Geological Survey of Japan, AIST, Tsukuba. [in Japanese with English abstract].

Yamashita, T. 2004. Behavior of fine-grained sediment in the coastal area. *In*: Hasegawa, K. (ed.) *Report of the 2003 Hokkaido flood disaster by typhoon 200310*. Committee on Hydroscience and Hydraulic Engineering, Japan Society of Civil Engineering, Tokyo, 160–167. [in Japanese].

Tectonic evolution in the Early to Middle Pleistocene off the east coast of the Boso Peninsula, Japan

Seishiro Furuyama[1]*, Tomoyuki Sato[2], Kohsaku Arai[2] and Masanori Ozaki[2]

[1]Marine Resources and Energy, Tokyo University of Marine Science and Technology, 4-5-7 Konan, Minato-ku, Tokyo 108-8477, Japan

[2]Research Institute of Geology and Geoinformation, Geological Survey of Japan, National Institute of Advanced Industrial Science and Technology, Tsukuba, Ibaraki 305-8567, Japan

SF, 0000-0002-5723-2027; TS, 0000-0002-9310-2107; KA, 0000-0001-9683-0671; MO, 0000-0001-6457-6402
*Correspondence: sfuruy1@kaiyodai.ac.jp

Abstract: The Kanto Basin developed, starting c. 3 Ma, influenced by the subduction of the Philippine Sea plate and the Pacific plate. Sediments in this basin have influenced the geomorphology of the Kanto region. In the Boso Peninsula, located at the eastern edge of the Kanto Basin, uplift continues in what is called the Kashima–Boso uplift zone. Although the development of this uplift after the Late Pleistocene is well understood, there are few data from the Early to Middle Pleistocene. In this study, we investigated the offshore shelf area east of the Boso Peninsula using a high-resolution seismic reflection survey, and report new information on the geological structure and uplift processes in the area from the Early to Middle Pleistocene. We identified the Kujukuri-oki anticline and the Kujukuri-oki normal fault zone. The Kujukuri-oki anticline, more than 47 km long, is north–south striking and deforms the Kujukuri-oki Group. There are numerous normal faults with displacements of less than tens of metres spread widely in the survey area (Kujukuri-oki normal fault zone). These findings reveal that the Kujukuri-oki anticline uplifted during the end of the Early Pleistocene and attenuated during the Middle Pleistocene. This anticline comprised the axis of the Kashima–Boso uplift zone at the Boso Peninsula from the Early to Middle Pleistocene and the Boso Peninsula is located at the western limb of this anticline.

Quaternary sediments distributed widely in the Kanto region in central Japan characterize the geology in this region, and the development of tectonic and sedimentary processes of those strata have been studied actively (e.g. Sugai *et al.* 2013; Kazaoka *et al.* 2015). The depositional basin (Kanto Basin, Fig. 1a) originated from the palaeoforearc basin and developed remarkably after the Pleistocene (e.g. Kaizuka 1987). This basin has subsided westward from c. 3 Ma (Nakazato and Sato 2001; Suzuki *et al.* 2011) and its eastern part also uplifted from around 1 Ma (Kashima–Boso uplift zone; Kaizuka 1987; Fig. 1b). Although Kaizuka (1987) estimated the timing of this uplift based on previous studies of changes in the thickness of the Kazusa Group (Kawai 1961) and of sedimentary and palaeontological facies (Naruse 1959), the substance of the uplift is unclear. Additionally, the eastern part of the Kanto Basin located on the Boso Peninsula has been affected by the subduction of the Philippine Sea plate and the Pacific plate (e.g. Seno and Takano 1989). In particular, the change in the direction of subduction of the Philippine Sea plate from north to NW between 3 and 0.5 Ma shifted the stress state in the Boso Peninsula (e.g. Yamaji 2000; Takahashi 2006). The relationship between the stress state and uplift is also unclear.

This study introduces high-resolution seismic profiles recorded with a boomer energy source to investigate the wider shelf area offshore east of the Boso Peninsula (35°12' to 35°42' N, 140°00' to 141°11' E Fig. 1b) considered in Furuyama *et al.* (2019). We provide the geological structure in the coastal area as well as offering new insights into the subsidence of the uplift in the eastern part of the Kanto Basin from the Early to Middle Pleistocene.

Geological settings

The Boso Peninsula is on the Pacific Ocean side of the Kanto region (Fig. 1a, b). In the offshore region east of the Boso Peninsula, the Philippine Sea and Pacific tectonic plates subduct NW and westwards under the Eurasian and North American plates, respectively (Fig. 1a). The complicated plate motions in this region have influenced the development of the subduction–accretionary system on the Boso Peninsula and in offshore areas to its east during the Paleogene to Quaternary (e.g. Yamamoto *et al.* 2017). The slope and outer ridge deposits of the system compose the southern part of the Boso Peninsula, whereas the forearc deposits of the system characterize the geology of the middle part of the peninsula. In the area

From: Asch, K., Kitazato, H. and Vallius, H. (eds) 2022. *From Continental Shelf to Slope: Mapping the Oceanic Realm.* Geological Society, London, Special Publications, **505**, 217–227.
First published online November 20, 2020, https://doi.org/10.1144/SP505-2019-116

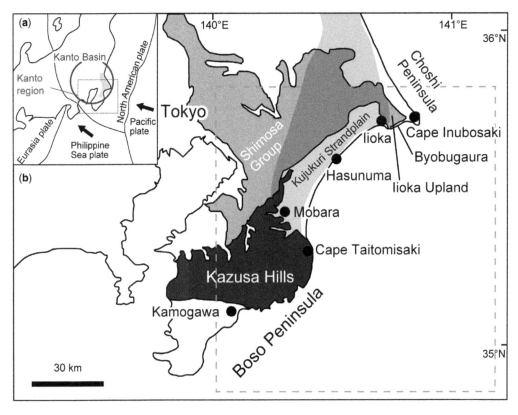

Fig. 1. (**a**) Kanto Basin (blue circle) and plate boundaries around the Kanto region. The red rectangle shows the area covered by (**b**). Red shading shows the Kashima–Boso uplift zone (Kaizuka 1987). (**b**) Topographic map of the Boso Peninsula. Red shading shows the Kashima–Boso uplift zone (Kaizuka 1987). The green rectangle represents the survey area.

of the Choshi Peninsula in the northernmost onshore part of the survey area, the Middle to Late Pleistocene and pre-Neogene sedimentary rocks crop out (Takahashi *et al.* 2003). Southwest of Cape Inubosaki, the Byobugaura sea cliffs on the Pacific side of the Iioka Upland extend about 10 km along the coast (Fig. 1b). The lower to middle Pleistocene Inubo Group crops out in the cliffs and correlates with the Kazusa Group in the middle of the Boso Peninsula (Fujioka and Kameo 2004). Southwest of the Byobugaura sea cliffs, the arcuate Kujukuri Strandplain (60 km × 10 km) extends NE–SW from Iioka to Mobara. The strandplain has expanded seaward as sediments have been deposited on the beach during the past 6000 years (Moriwaki 1979). The Kazusa Hills extend from Mobara to Kamogawa in the central part of the Boso Peninsula and are composed of Awa Group (Middle Miocene to Late Pliocene) and Kazusa Group (Early to Middle Pleistocene) rocks (Figs 2 & 3). Both groups were deposited in the fore-arc basin of the accretionary system (Ogawa *et al.* 1985). Numerous north–south-trending normal faults

developed from the Awa Group to the Kokumoto Formation of the upper Kazusa Group (Fig. 3; Yamaji 2000). These faults have not been recognized in the Shimosa Group overlying the Kazusa Group (Ishiwada *et al.* 1971).

The offshore area from Cape Inubosaki to Cape Taitomisaki is a broad shelf that extends up to 42 km from the coastline (Fig. 3) and dips SE, reaching about 200 m depth at its seaward edge. This shelf is deepest off Hasunuma. Between Choshi and Cape Taitomisaki, the seafloor beyond the shelf is almost flat. At the shelf edge off Hasunuma, the Katakai submarine canyon opens to the SE. The shelf narrows abruptly from 30 to 7 km from just south of Cape Taitomisaki to the area off Kamogawa where the Kamogawa submarine canyon opens to the SE.

Offshore to the east of the Boso Peninsula, Tanahashi and Honza (1983) and Okuda and Miyazaki (1986) examined the stratigraphy and geological structure at subsea depths of 1–1.5 km using seismic reflection profiles recorded with an airgun energy source. The Maritime Safety Agency (1986, 2000)

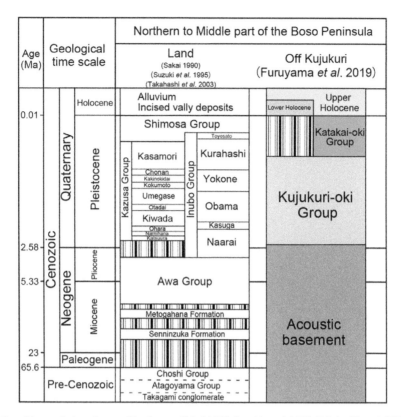

Fig. 2. Stratigraphic correlation of sea and land areas (Sakai 1990; Suzuki *et al.* 1995; Takahashi *et al.* 2003; Furuyama *et al.* 2019).

reported on the shallow subsea geological structure of the shelf area immediately east of the peninsula on the basis of a single-channel seismic reflection survey recorded with sonoprobe and sparker energy sources. These studies focused on geological mapping and there is little information about the tectonic development of the shelf area.

Furuyama *et al.* (2019) used high-resolution seismic reflection data acquired in 2014 and 2015 (Fig. 3; also see the Method section) to examine the stratigraphy and structure beneath the shelf between Cape Inubosaki and Kamogawa. They divided the strata overlying the acoustic basement into the Kujukuri-oki Group, the Katakai-oki Group and the lower and upper Holocene in ascending order (Fig. 2). The seismic reflection pattern of the Kujukuri-oki Group, mainly discussed in this study, clearly shows continuous stratification broken by anticlines, synclines and small-scale faults (Figs 4 & 5). This group mostly dips NW though some part of this group are dipped by faults and folds toward SW. The thickness of the Kujukuri-oki Group identified from seismic profiles is over 533 ms (400 m assuming a sound velocity of 1500 m s^{-1}). The thickness of this group becomes

gradually thinner northward. The unconformity between the acoustic basement and the Kujukuri-oki Group is recognized by an onlap pattern off Cape Inubosaki (Figs 4 & 5). The upper limit of this group is widely eroded and covered by the upper strata. Furuyama *et al.* (2019) estimated the depositional ages of those strata from previously published stratigraphic correlations, analyses of calcareous nanoplankton and other microfossils (Tanahashi and Honza 1983; Nishida 1984; Nishida *et al.* 2016, 2019), and correlation with cores from the Kujukuri Strandplain (Komatsubara 2019). They concluded the sedimentary ages of the Kujukuri-oki and Katakai-oki Groups to be Lower to Middle Pleistocene and Upper Pleistocene, respectively. They correlate the Kujukuri-oki Group with the Kazusa Group and Inubo Group on land (Fig. 2). We have adopted their stratigraphic subdivision and correlation here.

Methods

We acquired high-resolution seismic profiles using a boomer source (300 J AA301; Applied Acoustic

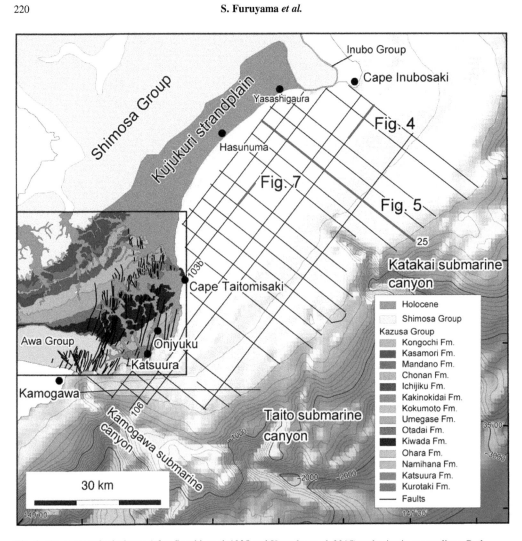

Fig. 3. Onshore geological map (after Suzuki *et al.* 1995 and Kazaoka *et al.* 2015) and seismic survey lines. Red lines represent the locations of the following figures. The square contains a detailed geological map of the Kazusa Group (after Kazaoka *et al.* 2015).

Engineering Ltd, Great Yarmouth, UK) at shot intervals of 3.125 m and a 24-channel streamer with a 3.125 m hydrophone spacing (Geometrics Inc., CA, USA) during surveys conducted from 27 August to 4 October 2014 and from 30 June to 27 July 2015. The record length was 3.0 s at a sample rate of 0.125 s. Because sea conditions varied considerably during the surveys, some survey lines were recorded several times and later merged based on post-stacked common mid-point positions. Merging can result in different degrees of noise cancellation, so noise levels may differ among the merged profiles. Seismic data were recorded in SEG-D format by GeoEel (Geometrics Inc.) and processed with Seismic Processing Workshop software

(Parallel Geoscience Co., NV, USA). The pre-stack seismic processing sequence included bandpass filtering (100–2000 Hz), muting, gain normalization, velocity analysis and normal moveout after aligning the bottom by velocity analysis, and deconvolution. After 12-fold stacking, horizontal trace summing, deconvolution, bandpass filtering and gain normalization, the data were output in SEG-Y format.

Results

This study focuses on two dominant geological structures – the Kujukuri-oki anticline and the Kujukuri-oki normal fault zone (Fig. 6) – recognized

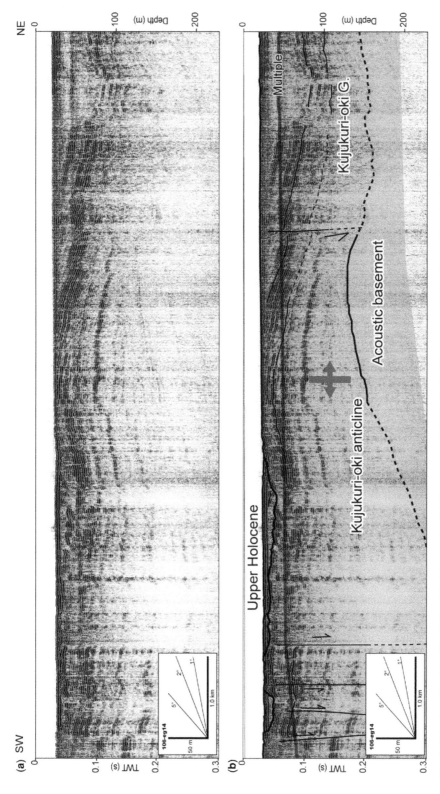

Fig. 4. Seismic section (**a**) of line 106 and its interpretation (**b**) (Furuyama *et al.* 2019). Dashed lines represent inferred parts. The seismic reflections are time profiles and the depth of reflections is calculated using 1500 m s^{-1}. The location of this profile is represented in Figure 3.

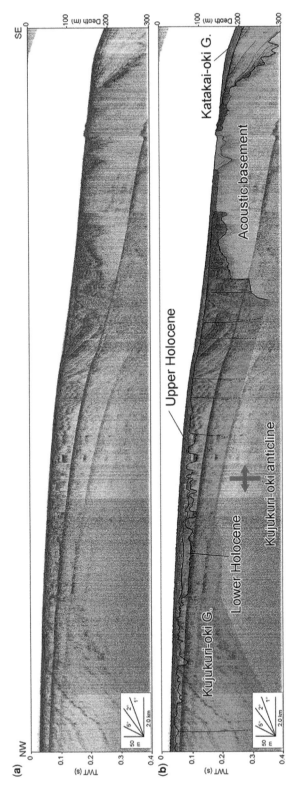

Fig. 5. Seismic section (**a**) of line 25 and its interpretation (**b**) (Furuyama *et al.* 2019). Dashed lines represent inferred parts. The seismic reflections are time profiles and the depth of reflections is calculated using 1500 m s^{-1}. The location of this profile is represented in Figure 3.

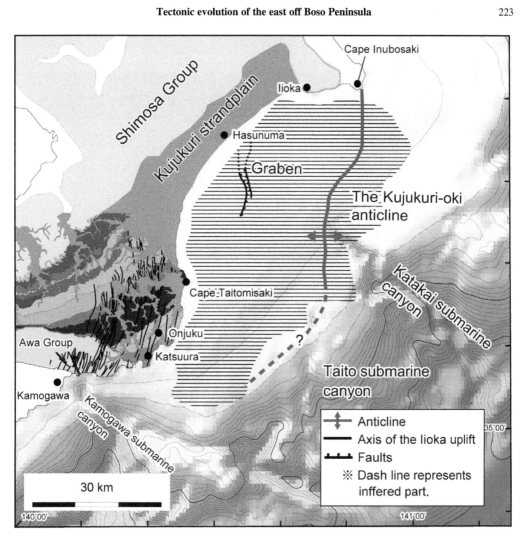

Fig. 6. Structural map of the eastern Boso Peninsula. The horizontal hatching shows the area where numerous normal faults are common in the Kujukuri-oki Group. The geological map of the land area refers to Suzuki *et al.* (1995) and Kazaoka *et al.* (2015).

in the Lower to Middle Pleistocene Kujukuri-oki Group (Fig. 2).

Kujukuri-oki anticline

We recognized an approximately north–south-striking major anticline more than 47 km long extending from just south of Cape Inubosaki to the shelf edge about 20 km south of the Katakai submarine canyon. We defined this feature (the Kujukuri-oki anticline) based on the dip of the Kujukuri-oki Group (Figs 4 & 5). The Kujukuri-oki Group dips at about 2–20° to the NW or NNW on the western limb of the Kujukuri-oki anticline and about 3–10° or more to the SE on its eastern

limb (Fig. 5). The amount of uplift of the Kujukuri-oki anticline appears on our seismic profiles to be about 150 ms two-way-travel (*c.* 113 m; Fig. 5). However, erosion of the crest of the anticline and horst and graben faulting around its axis makes accurate estimation of the amount of uplift difficult.

Fault zone developing in the survey area

Normal faults with displacements of less than tens of metres are widely developed in the Kujukuri-oki Group (Figs 6 & 7). Most of these faults step down to the east. Some faults step down to the west. Grabens can be recognized in some of those areas. The

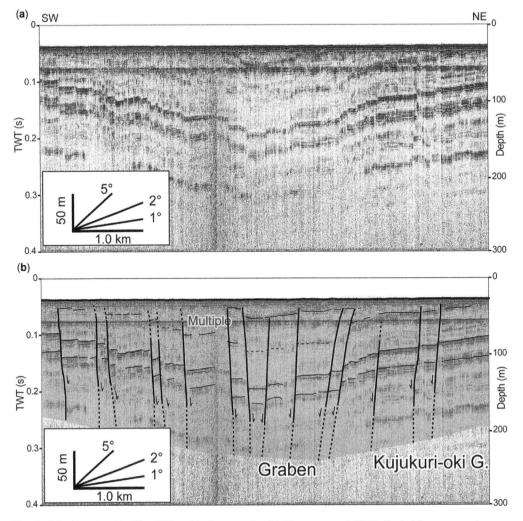

Fig. 7. Seismic section (**a**) of line 103b and its interpretation (**b**) (Furuyama *et al.* 2019). Dashed lines represent inferred parts. The seismic reflections are time profiles and the depth of reflections is calculated using 1500 m s^{-1}. The location of this profile is represented in Figure 3.

largest graben is 10 km off Hasunuma and is 6 km wide and 100 ms (75 m) deep (Figs 6 & 7). Normal faults in the Kujukuri-oki anticline step down to the east in its western limb and down to the west in its eastern limb, thus forming horsts and grabens around the axial region of the anticline. The north- to NNE-striking normal faults are consistent with the strike of the anticline.

Discussion

The recognition of geological structures in the offshore region east of the Boso Peninsula contributes to understanding of the subsidence of the uplift in

the eastern part of the Kanto Basin from the Early to Middle Pleistocene.

The Kazusa and Inubo Groups on land and the Kujukuri-oki Group (Kujukuri-oki Group and equivalent strata) were deposited in the forearc basin during the Early to Middle Pleistocene (Ogawa *et al.* 1985). The development of the Kujukuri-oki anticline can be estimated using the geological structures of those strata. The dips of the Kujukuri-oki Group in the western limbs of the Kujukuri-oki anticline (Figs 4 & 5) are consistent with the inclination of the Kazusa and Inubo Groups. This consistency indicates that the Kujukuri-oki Group and equivalent strata compose a homocline towards the NW. This observation indicates that the development of the

Kujukuri-oki anticline deformed these strata. Aiba and Tsuburaya (1981) recognized in seismic profiles that strata equivalent to the Kazusa Group abut against the topographic high of the lower strata near the Taito submarine canyon and show a homocline towards the NW on the west of that topographic high. They inferred that the Kazusa Group was originally deposited with an onlap pattern towards the basement. Later the Kazusa Group tilted with the deformation of the basement. This previous study also indicated that the development of the Kujukuri-oki anticline began after the deposition of the Kujukuri-oki Group.

Sedimentary facies on land allow us to estimate the amount and initiation of development of the Kujukuri-oki anticline. The Otadai Formation of the Kazusa Group and the Yokone Formation of the Inubo Group on land record a shallowing of the depositional environment. The Kazusa and Inubo groups deposited on shelf slopes to deep sea terraces estimated to be 1 km to 1.5 km in depth (Hirayama and Nakajima 1977; Sakai 1990) before sedimentation of the Otadai Formation and the Yokone Formation. The Otadai Formation shallowed at around 1.1 Ma and the Yokone Formation shallowed at around 0.9 Ma (Fujioka and Kameo 2004; Kazaoka et al. 2015). Therefore, the uplift forming the Kujukuri-oki anticline became enhanced after the end of the Early Pleistocene and the amount of uplift of the Kujukuri-oki anticline could be up to 1.5 km. This interval includes the time when the subduction of the Philippine Sea plate changed direction (Takahashi 2006). The change might have affected the development of the Kujukuri-oki anticline; however, further study with deeper seismic profiles is needed to gain an understanding of the relationship between the uplift below the Kujukuri-oki anticline and the plate motion.

The characteristics of normal faults of the Kujukuri-oki Group that strike north to NNE and almost step down to the east, are consistent with those in the Kazusa Group (Yamaji 2000) and Inubo Group (Sakai 1990). These north- to NNE-striking faults of the Kujukuri-oki Group are consistent with the strike of the anticline. This consistency indicates that the normal faults of the Kujukuri-oki Group are bending-moment faults (Yeats 1986; Ozaki et al. 2019) and developed with extensional stress orthogonal to the axis of the Kujukuri-oki anticline.

The development of normal faults in the Kazusa Group was estimated by analysis of map-scale normal faults on land (Yamaji 2000; Yamanaka et al. 2015; Otsubo et al. 2017). It has been divided into five stages (Yamanaka et al. 2015) including: (1) parallel extensional stress before the erosion that led to the Kurotaki unconformity; (2) parallel NE–SW-trending compressional stress between the

development of the Kurotaki unconformity and deposition of the lower Katsuura Formation of the Kazusa Group; (3) NW–SE-trending extensional stress between deposition of the upper Katsuura Formation and the Umegase Formation of the Kazusa Group; (4) north–south-trending extensional stress during deposition of the Kokumoto Formation; and (5) vertical compressional stress (Tsukahara and Kobayashi 1991). The consistency of the strikes of the normal faults in the survey area with that of the Kujukuri-oki anticline indicates that the normal faults could have developed at stage three (c. 2 Ma to c. 0.8 Ma) as defined above (Ozaki et al. 2019). This duration is conformable to the initiation of the Kujukuri-oki anticline inferred from sedimentary environmental changes of the Otadai and Yokone formations. The lack of normal faults in the Shimosa Group (Ishiwada et al. 1971) suggests that the tectonic activity forming the Kujukuri-oki anticline had attenuated during the Middle Pleistocene.

The interpretation of the Kujukuri-oki anticline and its tectonic evolution provide new insights into the Kashima–Boso uplift zone (Fig. 1b; Kaizuka 1987). In the Kanto region, the Kashima–Boso uplift zone has elevated the eastern part of the Kanto Basin since 1 Ma (Kaizuka 1987). This timing is consistent with the initiation of the development of the Kujukuri-oki anticline estimated in this study. The uplift forming the Kujukuri-oki anticline constructed the eastern edge of the Kashima–Boso uplift zone and tilted the Kujukuri-oki Group and equivalent strata towards the NW. Although sedimentary facies analyses of the Shimosa Group have also revealed the activity in the eastern part of the Kashima–Boso uplift zone after the Late Pleistocene (Nishikawa et al. 2000, 2001; Tamura et al. 2010), this uplift is attributed to another force that is different from the uplift of the Kujukuri-oki anticline.

Conclusion

First, we provided high-resolution seismic profiles and new tectonic information from the area east of the Boso Peninsula. We recognized two dominant geological structures in the survey area. The Kujukuri-oki anticline, more than 47 km long, has an approximately north–south-striking major anticline and dips towards the Kujukuri-oki Group. Normal faults with displacements of less than tens of metres are widely developed in the Kujukuri-oki Group. Their strikes are north–south or NNE–SSW, consistent with that of the Kujukuri-oki anticline. They compose the graben and horst features around the axis of the Kujukuri-oki anticline.

The development of the Kujukuri-oki anticline began after the deposition of the Kujukuri-oki Group. Shallowing of the Otadai and Yokone formations on land indicates that the initiation of that development could have started around the end of the Early Pleistocene. The tectonic activity forming the Kujukuri-oki anticline attenuated during the Middle Pleistocene. The axis of the Kashima–Boso uplift zone on the Boso Peninsula corresponded to the Kujukuri-oki anticline from the Early to Middle Pleistocene and the Boso Peninsula is at western limb of the anticline.

We will study the relationship of the geological structures of the shallow parts under the seafloor with those of the deep parts, such as plate motion, in the future and will resolve the tectonics of the eastern part of the Boso Peninsula during the Quaternary in detail.

Acknowledgements We thank the Daiwa Explosion & Consulting Co. Ltd and the crew of Verny 3 for their professional efforts in completing the seismic reflection surveys. We also thank Yukinobu Okamura, Takahiko Inoue and Junko Komatsubara of the Geological Survey of Japan for valuable comments and discussions. The early manuscript received the benefit of many constructive comments by Makoto Otsubo, three anonymous reviewers and the Editor Hiroshi Kitazato.

Author contributions SF: conceptualization (lead), investigation (lead), writing – original draft (lead); **TS**: conceptualization (supporting), investigation (equal), writing – review & editing (lead); **KA**: conceptualization (supporting), writing – review & editing (equal); **MO**: conceptualization (supporting), writing – review & editing (equal).

Funding This research received no specific grant from any funding agency in the public, commercial, or not-for-profit sectors.

Data availability The datasets generated during and/or analysed during the current study are available in the Geological Survey of Japan repository, https://www.gsj.jp/researches/project/coastal-geology/results/s-6.html.

References

Aiba, J. and Tsuburaya, H. 1981. Post-Tertiary unconformities off Sanriku–Joban–Chiba. *Chikyu Monthly*, **13**, 168–174 [in Japanese].

Fujioka, M. and Kameo, K. 2004. Correlation between the Obama Formation of the Inubou Group in the Choshi district and the Kiwada, Otadai and Umegase Formations of the Kazusa Group in the Boso Peninsula, central Japan, based on key tephra layers. *Journal of Geological Society of Japan*, **110**, 480–496 [in Japanese with English abstract], https://doi.org/10.5575/geosoc.110.480

Furuyama, S., Sato, T. and Arai, K. 2019. *Explanatory Notes of 1:200,000 Marine Geological Map along the Coastal Zone of Eastern Part of the Boso Peninsula*. Geological Survey of Japan, Seamless Geoinformation of Coastal Zone **S-6** [in Japanese with English abstract].

Hirayama, J. and Nakajima, T. 1977. Analytical study of turbidites, Otadai Formation Boso Peninsula, Japan. *Sedimentology*, **24**, 747–779, https://doi.org/10.1111/j.1365-3091.1977.tb01914.x

Ishiwada, Y., Mitsunashi, S., Shinada, Y. and Makino, T. 1971. *Geological Maps of Oil and Gas Field of Japan, 10, 'Mobara'*. Geological Survey of Japan.

Kaizuka, S. 1987. Quaternary crustal movements in Kanto, Japan. *Journal of Geography*, **96**, 223–240, https://doi.org/10.5026/jgeography.96.4_223

Kawai, K. 1961. Economic geological study on the southern Kanto gas-producing region. *Journal of the Japanese Association for Petroleum Technology*, **26**, 212–266 [in Japanese with English abstract], https://doi.org/10.3720/japt.26.212

Kazaoka, O., Suganuma, Y. *et al.* 2015. Stratigraphy of the Kazusa Group, Boso Peninsula: an expanded and highly-resolved marine sedimentary record from the Lower and Middle Pleistocene of central Japan. *Quaternary International*, **383**, 116–135, https://doi.org/10.1016/j.quaint.2015.02.065

Komatsubara, J. 2019. *Sedimentary Environments and Basal Topography of Postglacial Deposits in the Kujukuri Coastal Plain, Boso Peninsula, Central Japan*. Geological Survey of Japan, Seamless Geoinformation of Coastal Zone **S-6** [in Japanese with English abstract].

Maritime Safety Agency. 1986. *Basic Map of the Sea in Coastal Waters (1:50,000), Taito-saki, Tokyo, Japan*.

Maritime Safety Agency. 2000. *Basic Map of the Sea in Coastal Waters (1:50,000), Kujukuri Hama, Tokyo, Japan*.

Moriwaki, H. 1979. The landform evolution of the Kujukuri coastal plain, central Japan. *The Quaternary Research*, **18**, 1–16 [in Japanese with English abstract], https://doi.org/10.4116/jaqua.18.1

Nakazato, M. and Sato, H. 2001. Chronology of the Shimosa group and movement of the 'Kashima' uplift zone, central Japan. *The Quaternary Research*, **40**, 251–257 [in Japanese with English abstract], https://doi.org/10.4116/jaqua.40.251

Naruse, Y. 1959. On the formation of the Paleo-Tokyo Bay, geologic history of the Late Cenozoic Deposits in the Southern Kanto Region (1). *The Quaternary Research*, **1**, 143–155 [in Japanese with English abstract], https://doi.org/10.4116/jaqua.1.143

Nishida, S. 1984. *Calcareous Nanoplankton Biostratigraphy off the Boso Peninsula*. Cruise Report: Geological Investigation of the Junction Area of the Tohoku and Ogasawara Arcs, April–June 1980 (GH80-2 and 3 Cruises). Geological Survey of Japan, **19**, 67–73.

Nishida, N., Ajioka, T., Ikehara, K., Nakashima, R. and Utsunomiya, M. 2016. *Preliminary Report on a Coring Survey off the Kujukuri Strandplain, the Pacific Ocean, Japan*. Annual Report of Investigations on Geology and Active Faults in the Coastal Zone of Japan (FY2015) GSJ Interim Report, **71**, 13–23 [in Japanese with English abstract].

Nishida, N., Ajioka, T., Ikehara, K., Nakashima, R. and Utsunomiya, M. 2019. *Spatial Variation and Stratigraphy of the Marine Sediments off the East of the Boso Peninsula, Pacific Ocean, Japan*. Geological Survey of Japan, Seamless Geoinformation of Coastal Zone **S-6** [in Japanese with English abstract].

Nishikawa, T., Ito, M. and O'Hara, T. 2000. Spatial variation in the timing of a maximum flooding surface: an example from the Middle Pleistocene Yabu Formation in the Boso Peninsula, Japan. *Journal of Geological Society of Japan*, **106**, 15–30 [in Japanese with English abstract], https://doi.org/10.5575/geosoc.106.15

Nishikawa, T., Sugimoto, H. and Ito, M. 2001. Sea-level changes and tectonics documented in depositional systems in Paleo-Tokyo bay, Japan. *Quaternary Research*, **40**, 275–282 [in Japanese with English abstract], https://doi.org/10.4116/jaqua.40.275

Ogawa, Y., Horiuchi, K., Taniguchi, H. and Naka, J. 1985. Collision of the IZU arc with Honshu and the effects of oblique subduction in the Miura–Boso peninsulas. *Tectonophysics*, **119**, 349–379, https://doi.org/10.1016/0040-1951(85)90046-0

Okuda, Y. and Miyazaki, T. 1986. *Geological Map of Kashimanada 1:200,000*. Geological Survey of Japan, Marine Geology Map Series **27**.

Otsubo, M., Utsunomiya, M. and Miyakawa, M. 2017. Reactivation of map-scale faults in response to changes in crustal stress: Examples from Boso Peninsula, Japan. *Quaternary International*, **456**, 117–124, https://doi.org/10.1016/j.quaint.2017.05.057

Ozaki, M., Furuyama, S., Sato, T. and Arai, K. 2019. *1:200,000 Marine and Land Geological Map of the Eastern Coastal Zone of the Boso Peninsula and Its Explanation, Especially with Quaternary Crustal Deformation*. Geological Survey of Japan, Seamless Geoinformation of Coastal Zone **S-6** [in Japanese with English abstract].

Sakai, T. 1990. Upper Cenozoic in the Choshi district, Chiba Prefecture, Japan – Litho-, magneto- and radiolarian biostratigraphy. *Bulletin of the Faculty of General Education, Utsunomiya University*, **23**, 1–34 [in Japanese].

Seno, T. and Takano, T. 1989. Seismotectonics at the trench–trench–trench triple junction off central Honshu. *Pure and Applied Geophysics*, **129**, 27–40, https://doi.org/10.1007/BF00874623

Sugai, T., Matsushima, H. and Mizuno, K. 2013. Last 400ka landform evolution of the Kanto Plain: under the influence of concurrent glacio-eustatic sea level changes and tectonic activity. *Journal of Geography*, **122**, 921–948, https://doi.org/10.5026/jgeography.122.921

Suzuki, Y., Kodama, K. *et al*. 1995. *Explanatory Text of Geological Map of Tokyo Bay and Adjacent Area*. Geological Survey of Japan, Miscellaneous Geological Map Series **20** (2nd ed.) [in Japanese with English abstract].

Suzuki, T., Obara, M. *et al*. 2011. Identification of Lower Pleistocene tephras under Tokyo and reconstruction of Quaternary crustal movements, Kanto Tectonic Basin, central Japan. *Quaternary International*, **246**, 247–259, https://doi.org/10.1016/j.quaint.2011.06.043

Takahashi, M. 2006. Tectonic development of the Japanese islands controlled by Philippine Sea Motion. *Journal of Geography*, **115**, 116–123, https://doi.org/10.5026/jgeography.115.116

Takahashi, M., Suto, I., Ohki, J. and Yanagisawa, Y. 2003. Chronostratigraphy of the Miocene Series in the Choshi area, Chiba Prefecture, central Japan. *Journal of Geological Society of Japan*, **109**, 34–360 [in Japanese with English abstract], https://doi.org/10.5575/geosoc.109.345

Tamura, T., Murakami, F. and Watanabe, K. 2010. Holocene beach deposits for assessing coastal uplift of the northeastern Boso Peninsula, Pacific coast of Japan. *Quaternary Research*, **74**, 227–234, https://doi.org/10.1016/j.yqres.2010.07.009

Tanahashi, M. and Honza, E. 1983. *Geological Map of the East of Boso Peninsula 1:200,000*. Geological Survey of Japan, Marine Geology Map Series **24** [in Japanese with English abstract].

Tsukahara, H. and Kobayashi, Y. 1991. Crustal stress in the central and western parts of Honshu, Japan. *Journal of the Seismological Society of Japan*, **44**, 221–231 [in Japanese with English abstract], https://doi.org/10.4294/zisin1948.44.3_221

Yamaji, A. 2000. The multiple inverse method applied to meso-scale faults in mid-Quaternary fore-arc sediments near the triple trench junction off central Japan. *Journal of Structural Geology*, **22**, 429–440, https://doi.org/10.1016/S0191-8141(99)00162-5

Yamamoto, Y., Chiyonobu, S., Kamiya, N., Hamada, Y. and Saito, S. 2017. Structural characteristics of shallow portion of plate subduction zone: a forearc system in the southern Boso Peninsula, central Japan. *Journal of Geological Society of Japan*, **123**, 41–55 [in Japanese with English abstract], https://doi.org/10.5575/geosoc.2016.0057

Yamanaka, K., Sato, H. and Yamaji, A. 2015. Revisiting paleostresses in the east coast of Boso Peninsula, central Japan (Abstract). *The 122nd Annual Meeting of the Geological Society of Japan, 11–13 September, Nagano, Japan*, **285**.

Yeats, R.S. 1986. Faults related to folding with examples from New Zealand. *Royal Society of New Zealand Bulletin*, **24**, 273–292.

Bent incised valley formed in uplifting shelf facing subduction margin: case study off the eastern coast of the Boso Peninsula, central Japan

Tomoyuki Sato[1]*, Seishiro Furuyama[2], Junko Komatsubara[1], Masanori Ozaki[1] and Kazuo Yamaguchi[1]

[1]Geological Survey of Japan, National Institute of Advanced Industrial Science and Technology, 1-1-1 Higashi, Tsukuba, Ibaraki 305-8567, Japan

[2]Department of Marine Resources and Energy, Tokyo University of Marine Science and Technology, 4-5-7 Konan Minato-ku, Tokyo 108-8477, Japan

TS, 0000-0002-9310-2107; SF, 0000-0002-5723-2027; JK, 0000-0003-0099-6744; MO, 0000-0001-6457-6402

*Correspondence: tomoyuki-sato@aist.go.jp

Abstract: The Boso Peninsula is located in central Japan near the junction of the subduction boundary of three tectonic plates. A forearc basin has been developing there since 3 Ma and has been uplifting since 1 Ma. The basal surface of the Holocene deposits in the offshore area was investigated based on a seismic survey and is very similar to the adjacent land areas (the Iioka Plateau, the Kujukuri Plain and the Kazusa Hills). The basal surface in the Kujukuri Plain and its corresponding offshore area contains many incised valleys. Most of them extend southeastward, parallel to the direction from the hinterland to the ocean, but one incised valley (Kujukuri-oki Buried Valley) lies perpendicular to the others. A buried terrace is located SE of the valley and along the area where mudstone (of the Kiwada Formation) is distributed. The present observations indicate that differential erosion formed the terrace, after which the valley bent to follow the terrace. The rivers tend to be perpendicular to the strike of the sediment in the forearc basin owing to tectonic movement. Thus, the valley must have been incised into the underlying strata with a perpendicular strike and may have become bent in uplifting forearc basins.

Most large cities in the world are located on coastal plains, and the underground structure in these plains constitutes the foundation for city buildings. Thus, knowledge of the underground structures is useful in the assessment of ground subsidence owing to groundwater use and the estimation of earthquake damage (Olsen et al. 1995; Koketsu and Kikuchi 2000). For these purposes, the underground structure of coastal plains has been well studied (Matsuda 1974; Kaizuka et al. 1977). Because the coastal plains face shelves, the underground structure of the shelves can also be useful for these purposes. Additionally, the shelves themselves are also important because they support the bases of offshore structures, such as recently developed offshore wind power plants.

In shelves incised valleys are formed during geological stages with low sea-levels (Zaitlin et al. 1994). Sediments inside the valleys are younger and may have different grain sizes compared with those in sediments outside the valleys. Thus, the distribution and size of the valleys have a strong influence on the physical properties of the seafloor in shelves. Many studies on the formation processes and characteristics of incised valleys have been conducted, and formation models have been proposed (Ashley and Sheridan 1994; Zaitlin et al. 1994). Although these models can well explain incised valleys, they show wide variation owing to the variations of geological, tectonic and hydrological settings, and sediment supply. In particular, accounting for the effects of bedrock is complicated because it plays a key role in determining the lithology of the valley-fill deposits in shelves with thin sediment cover (Chaumillon et al. 2008); specifically, exposed bedrock strongly controls characteristics such as wave patterns and sediment supply. Futher, the shape of the valleys and their orientation are also controlled by the bedrock. For example, in the Bay of Biscay, the valleys are bent and parallel to the distribution direction of the Late Jurassic carbonate strata (Chaumillon et al. 2008). Tectonic movement also controls the orientation in some cases. The incised valleys in the East China Sea bend to the north along the Zhedong–Xihu depression because the river could not incise into Taiwan–Sinzi uplift belt (Li et al. 2005).

The aim of this study is to describe the shape of a valley incised into an uplifting shelf and discuss the formation processes and the relationship among the valley shape, geology and tectonic movements in the area. For this purpose, the formation process of

From: Asch, K., Kitazato, H. and Vallius, H. (eds) 2022. *From Continental Shelf to Slope: Mapping the Oceanic Realm.* Geological Society, London, Special Publications, **505**, 229–240.
First published online December 11, 2020, https://doi.org/10.1144/SP505-2019-117

the Holocene deposits and incised valleys during the Last Glacial Maximum (LGM) east off the Boso Peninsula, Japan, is discussed as a case study. Both the land and marine geology of the Boso Peninsula have been well studied. As a result, bent incised valleys have been identified (Maritime Safety Agency 1984*a*, *b*, 1986, 2000). We discuss the formation process of the valley in the context of the geological history with a newly conducted seismic reflection survey and control factors for the shape of incised valley.

Geological background

The study area is the eastern shelf of the northern part of the Boso Peninsula (Fig. 1), which is located in the eastern part of Japan and faces the Pacific Ocean (Fig. 1a). The Kujukuri Plain is located along the eastern coast of the Boso Peninsula, and the Iioka Plateau, Shimosa Plateau and Kazusa Hills surround the plain (Fig. 1b). The Iioka Plateau is located in the northeastern part of the Kujukuri Plain and extends NW–SE; its eastern end is on the Choshi Peninsula. Its elevation ranges from 30 to 60 m. Although it contains some valleys, most of the plateau is flat. The Shimosa Plateau is located behind the Kujukuri Plain and has an elevation of 20–130 m. It contains many valleys extending from NW to SE. The Kujukuri Plain is *c.* 56 km in length and extends in an arc shape in a NE–SW direction. The distance from the coast to the Shimosa Plateau ranges from 7 to 11 km. It is a strand plain where beach ridges have formed with wetlands between them (Moriwaki 1979). Many rivers flow along the plain and connect to the valleys in the Shimosa Plateau in the NW–SE direction. The Kazusa Hills are 50–370 m above sea-level and are distributed south of the Kujukuri Plain. The hills are the highest and steepest in this area. A NE–SW ridge line is present, reflecting the geological structure described later.

The marine area off the eastern coast of the Boso Peninsula is characterized by a broad shelf. This shelf is flat and dips to the SE, and the depth of its edge reaches *c.* 200 m. The width of the shelf is *c.* 35 km off the Iioka Plateau and *c.* 46 km off the Kujukuri Plain but narrows to *c.* 16 km off the Kazusa Hills. Two canyons have also developed in this area: the Katakai Submarine Canyon off the Kujukuri Plain and the Onjuku Submarine Canyon off the Kazusa Hills. Both cut the shelf slope and connect to the Boso Submarine Canyon (Maritime Safety Agency 1994). In the offshore region east of the peninsula, a triple junction of three plate subduction boundaries has developed, and the Philippine Sea and Pacific tectonic plates subduct west-northwestwards under the Eurasian

and North American plates, respectively. Intensive investigations have revealed the tectonic histories around this area (e.g. Kaizuka 1987; Shishikura 2001; Tamura *et al.* 2007, 2010) because many earthquakes are related to the subduction (Headquarters for Earthquake Research Promotion 2014), although Tokyo, the capital of Japan, is located in an adjacent area. Owing to the subduction of the Philippine Sea plate, a forearc basin that extends WNW to ESE and parallel to the subduction boundary has been developing since 3 Ma. In this basin, the Lower to Middle Pleistocene Kazusa Group was deposited (Figs 1 & 2; Kaizuka 1987; Suzuki 2002). The southern boundary of the basin is uplifting northward and is called the Mineoka belt (Takahashi *et al.* 2003). The eastern part of the basin, which is called the Kashima–Boso uplift zone, has also been uplifting westward for 1 Ma and is parallel to the subduction boundary of the Pacific plate (Kaizuka 1987). The Kujukuri-oki anticline developed along the uplift zone east of the Boso Peninsula during the Early Pleistocene (Furuyama *et al.* 2020). The forearc basin has been undergoing uplift since the initiation of the Kashima–Boso uplift zone. The average rate of uplift since Marine Isotope Stage (MIS) 5e is 0.69–0.79 mm a^{-1} in the southern part of the peninsula and 0.45–0.53 mm a^{-1} in the northern part based on the altitude of the marine terrace formed during MIS 5e (Okuno *et al.* 2014). The easternmost part of the forearc basin is located east of the Boso Peninsula, where these two uplifting areas meet owing to the plate subduction at that location. The depth of the basement of the basin was revealed near there on land based on the distribution of the Kazusa Group (Suzuki 2002) and in marine formations based on a seismic reflection survey (Tanahashi and Murakami 1984; Fig. 2). In addition, the basal surface of the Holocene deposits was also revealed based on a seismic reflection survey (Maritime Safety Agency 1984*a*, *b*, 1986, 2000). Many incised valleys are observed on the basal surface.

The Kazusa, Inubou and Shimosa groups and the Holocene deposits are mainly distributed in the northern part of the Boso Peninsula (Fig. 3; Suzuki *et al.* 1995). The Kazusa Group was formed in the Early to Middle Pleistocene, is located in the Shimosa Plateau and Kazusa Hills, and is divided into the Kurotaki, Katsuura, Namihana, Ohara, Kiwada, Otadai, Umegase, Kokumoto, Kakinokidai, Chonan, Mandano, Kasamori and Kongochi formations in ascending order (Tokuhashi and Endo 1984) (Figs 3 & 4). These sediments are mainly composed of mudstone and alternations of sandstone and mudstone. In the lower part of the Kazusa Group (from the Katsuura to the Kiwada formations), the sedimentary environment is abyssal plain, submarine

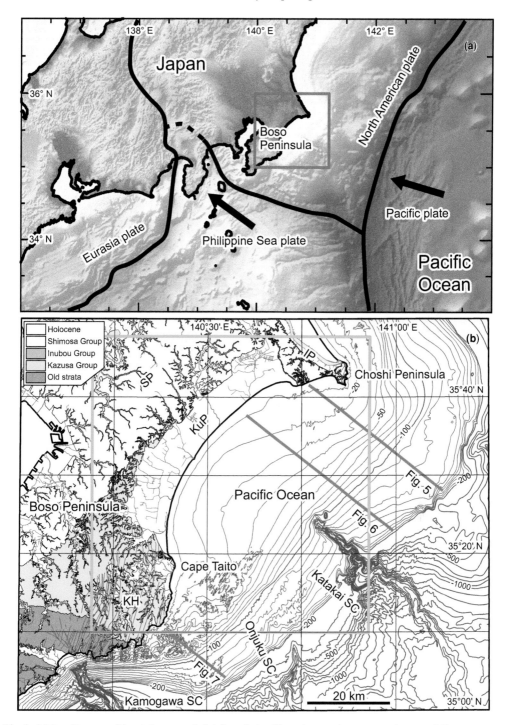

Fig. 1. (a) Locality map of the study area and plate boundaries. The red rectangle represents the area of (b). (b) Geological map of the eastern coastal zone of the Boso Peninsula (after Ozaki *et al.* 2019). Lines in the marine area represents the localities of the seismic section shown in Figures 5, 6 and 7. KuP, Kujukuri Plain; IP, Iioka Plateau; SP, Shimosa Plateau; KH, Kazusa Hills; SC, Submarine Canyon.

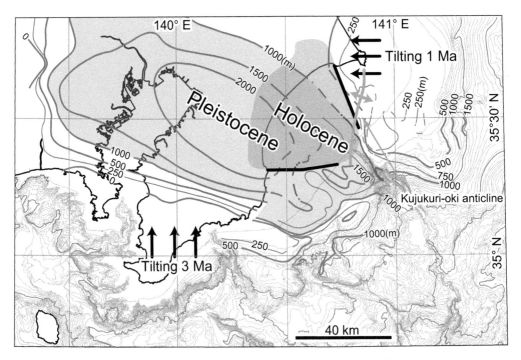

Fig. 2. Locality map of the sedimentary basins during the Pleistocene and the Holocene. The red area represents the basin during the Early to Middle Pleistocene when the Kazusa Group was deposited (Tanahashi and Murakami 1984; Suzuki 2002). The red contours represent the thickness of the Lower to Middle Pleistocene Kazusa Group. The blue area represents the area where the Holocene is thick.

fan and slope base, and the sedimentary environment of the middle and upper parts is shallower than these formations and the shelf (Katsura 1984; Ito 1992; Ito and Katsura 1992).

The Inubou Group was formed in the Late Pliocene to Middle Pleistocene and is located in the Iioka Plateau and divided into the Naarai, Kasuga, Obama, Yokone, Kurahashi and Toyosato formations in ascending order (Fig. 3; Sakai 1990). This group mainly consists of mudstone and alternations of sandstone and mudstone. In addition to these groups, the Jurassic Atago Unit, the Cretaceous Choshi Group and the Lower Miocene Senninzuka and Metogahana formations are distributed in the Choshi Peninsula (Takahashi *et al.* 2003). According to tephra chronology, the Inubou Group in the Iioka Plateau is correlated to the Kazusa Group (Ozaki *et al.* 2019).

The Holocene deposits distributed in the Kujukuri Plain are well studied based on borehole cores and ^{14}C dating (Masuda *et al.* 2001; Tamura *et al.* 2003, 2008, 2010; Komatsubara 2019). The plain is a strand plain that prograded since the mid-Holocene (Moriwaki 1979; Tamura *et al.* 2003). Sediment is supplied to the plain from the rivers and Byobugaura sea cliffs and Cape Taito, which are located NE and

SW of the plain and are being eroded (Sunamura and Horikawa 1977).

The Holocene basal surface in the northern part of the Kujukuri Plain has been revealed in detail by borehole cores (Komatsubara 2019). The altitude of the surface is −10 m in the southern part of the plain and −25 m in the central and northern parts. Additionally, many valleys incise the basal surface. These valleys connect to valleys in the Shimosa Plateau and extend southeastward. The altitudes of the bottom of the valleys range from −40 to −55 m around the coastline (Yamaguchi *et al.* 2019).

The geology off the eastern coast of the Boso Peninsula is divided into the Kujukuri-oki Group, the Katakai-oki Group, the lower Holocene and the upper Holocene in ascending order based on a seismic survey by Furuyama *et al.* (2019) (Fig. 3). (The suffix 'oki' means 'off' in Japanese, so 'Kujukuri-oki' means 'off Kujukuri'.) The Kujukuri-oki Group is widely distributed in this area, and continuous stratifications were developed and cut by small-scale normal faults in this group. It has been shown that this group was formed during the Early Pleistocene (CN14a Subzone; Okada and Bukry 1980) and is correlated to the Kazusa Group on land based on calcareous nonnoplankton (Nishida

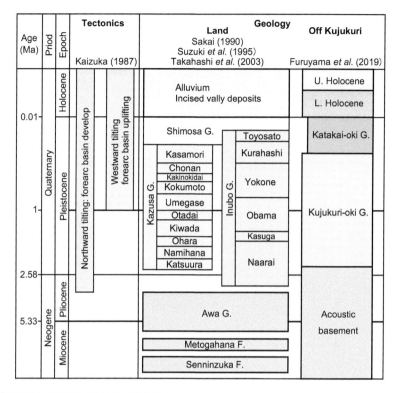

Age (Ma)	Priod	Epoch	Tectonics Kaizuka (1987)		Geology			Off Kujukuri Furuyama *et al.* (2019)
					Land Sakai (1990) Suzuki *et al.* (1995) Takahashi *et al.* (2003)			

Fig. 3. Stratigraphic correlation of sea and land areas.

et al. 2019). Additionally, this group was deformed by the Kujukuri-oki anticline, causing the strata in the western part of the anticline to incline towards the NW. The anticline extends from the Choshi Peninsula to the shelf edge around the Katakai Submarine Canyon, and strikes north–south (Furuyama *et al.* 2020; Figs 2 & 4). The upper boundary is eroded and covered by the Katakai-oki Group, as well as the lower and upper Holocene in places. The Katakai-oki Group is distributed around the shelf edge and unconformably covers the Kujukuri-oki Group. The group has continuous stratification and shows downlapping and toplapping patterns. The Holocene deposits show continuous horizontal stratification and are divided into upper and lower sections. The Holocene deposits can be correlated to the stratigraphy in the plain according to altitude, and the ages are based on the results of ^{14}C dating of borehole cores. According to the age, the lower Holocene deposits are older than 6 ka and the upper Holocene is younger than 6 ka, when progradation of the plain started (Moriwaki 1979). The lower Holocene is distributed only in the incised valleys, as will be described later. The upper Holocene widely covers this area except for the central area of the shelf and the region off the Kazusa Hills (Furuyama *et al.* 2019). The basal surface of

the Holocene deposits was well revealed by the Maritime Safety Agency (1984a, b, 1986, 2000) with a seismic reflection survey having 900 m horizontal intervals, as discussed later.

The forearc basin and the outer-arc highlands have been developed in the study area owing to the subduction of the Philippine Sea plate (Kaizuka 1987). The tectonic movement forming the basin continues until the present. The amount of uplift was estimated by the altitude of the marine terrace formed during MIS 5e (Koike and Machida 2001), and Okuno *et al.* (2014) calculated the average rate of uplift considering the effect of glacio-hydro isostacy using the results. The rates are 0.69–0.79 mm a^{-1} in the southmost part of the peninsula and 0.45–0.53 mm a^{-1} in the NW. The average uplift rate after the mid-Holocene was also estimated by the altitude of the marine terrace formed then (Shishikura 2001). As a result, the most southern part uplifted 30 m and Kazusa Hill uplifted 10 m. These results mean that the northward tilting that has formed the forearc basin since 3 Ma (Kaizuka 1987) continues until the present.

The relative sea-level during the LGM east of the Kujukuri Plain can be estimated to *c.* −90 m because the eustatic sea-level was −120 m (Waelbroeck *et al.* 2002) and the amount of uplift is *c.* 30 m. The latter value has been calculated on the assumption that the uplift rate after the mid-Holocene (10 m since 6 ka) determined by Shishikura (2001) did not change after the LGM.

Dataset and methods

The Maritime Safety Agency (1984*a*, *b*, 1986, 2000) has published a bathymetric chart of the study area with a scale of 1:50 000. In addition to the chart, contour maps of the Holocene basal surface have been published (Maritime Safety Agency 1984*a*, *b*, 1986, 2000; Fig. 4). These maps were based on seismic reflection surveys using a sonoprobe. The vertical resolution is better than 1 m. The directions of the survey lines were north to south and east to west and the interval was 900 m. The contour intervals were 10 m. We used these data as a relief map of the

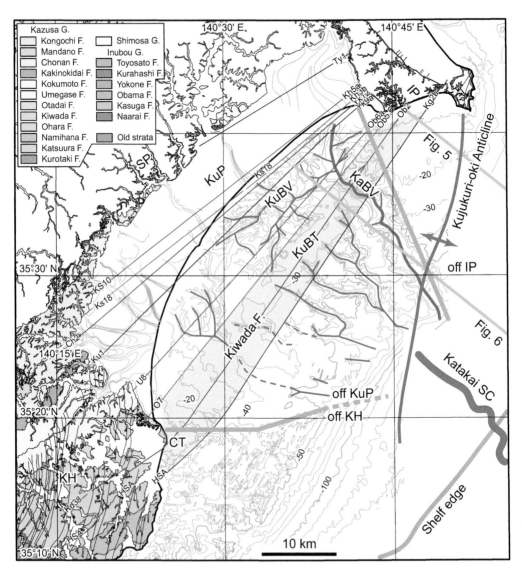

Fig. 4. Contour map of the basal surface of the Holocene in the eastern coastal zone of Boso Peninsula (based on Maritime Safety Agency 1984*a*, *b*, 1986, 2000; Nanayama *et al.* 2016; Komatsubara 2019). Fine lines represent tephra correlation (Ozaki *et al.* 2019). Blue lines in the marine area represents incised valleys. Thick grey lines represent the boundaries of the areas based on the characteristics of the basal surface. KuP, Kujukuri Plain; IP, Iioka Plateau; SP, Shimosa Plateau; KH, Kazusa Hills; SC, Submarine Canyon; CT, Cape Taito; KuBT, Kujukuri-oki Buried Terrace; KuBV, Kujukuri-oki Buried Valley; KaBV, Katakai Buried Valley.

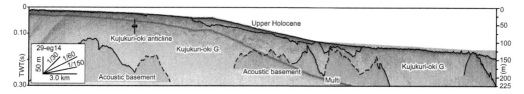

Fig. 5. Seismic section off Iioka Plateau. Locality is shown in Figures 1 and 4.

Holocene basal surface. Although the vertical and horizontal precisions are excellent, internal structures below the Holocene deposits are difficult to interpret in this dataset. Therefore, 1100 km of seismic profiles were interpreted to discuss the formation process of the surface based on the internal structure below the Holocene deposits from the viewpoint of geological history. The profiles were collected from 27 August to 4 October 2014 and from 30 June to 27 July 2015 using a boomer and a 24-channel streamer cable with a spacing of 3.125 m. The profiles have a horizontal resolution of 3.125 m and a vertical resolution of a few metres. Seismic data were processed by bandpass filtering (100–2000 Hz), deconvolution, gain normalization, normal moveout and stacking. This dataset is also discussed in Furuyama *et al.* (2019) for stratigraphy and Furuyama *et al.* (2020) for tectonic history.

Basal surface of the Holocene deposits

The topography of the Holocene basal surface on the eastern coast of the Boso Peninsula is described below. The region is divided into three areas based on their topographical characteristics (Fig. 4).

First, in the south-southeastern region of the Iioka Plateau in the northeastern part of the study area, the surface is flat and gently slopes south-southeastward with a slope of several thousandths. The altitude of the plateau ranges from 0 m at the coast to −50 m in the area 20 km off the Iioka Plateau, and is generally higher than the area off the Kujukuri Plain described next. The Holocene deposits form a thin

layer, and the Kujukuri-oki Group (Lower to Middle Pleistocene) is exposed on the seabed near the coastline (Fig. 5) (Furuyama *et al.* 2019).

Next, off the Kujukuri Plain in the central part of the study area, the basal surface is deeper than in the other two areas, and many incised valleys are present (Fig. 4). The altitudes of the valley bottoms range from −40 to −50 m near the coastline and reach −80 m near the shelf edge. The lower Holocene with horizontal stratification fills these valleys (Fig. 6).

Finally, off the Kazusa Hills in the southern part of the study area, there are severe undulations in the Holocene basal surface (Fig. 4). The undulations have NE–SW ridge lines similar to those for the Kazusa Hills. The Kujukuri-oki Group is exposed in the ridge lines, whereas the Holocene deposits exist only in the valleys between the ridge lines (Fig. 7).

The Holocene basal surfaces in these three areas extend from the land topography to the marine area. Because the relative sea-level during the last glacial period, when the Holocene basal surface was formed, was *c.* 90 m lower than present-day levels, the basal surface was on land at that time. The high flat surface off the Iioka Plateau was simply an extension of the plateau. Similarly, the low surface carved with many valleys off the Kujukuri Plain and the high undulating surface off the Kazusa Hills were extensions of these respective formations. The surfaces that formed the plateau, plain and hills during the glacial stage were submerged and covered by Holocene deposits as a result of the rising sea-level during deglaciation. The Shimosa Group covers the Iioka Plateau, while the extended marine area is covered by the Kujukuri-oki Group, which is older. The

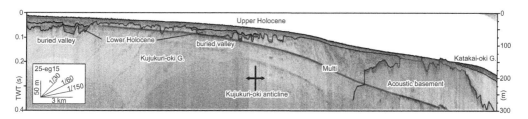

Fig. 6. Seismic section off Kujukuri Plain. Locality is shown in Figures 1 and 4.

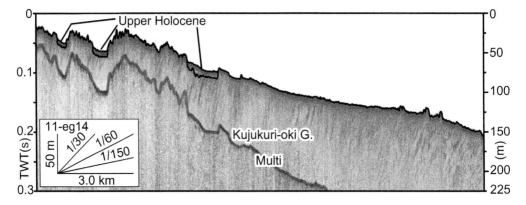

Fig. 7. Seismic section off Kazusa Hills. Locality is shown in Figure 1.

strata that can be correlated to the Shimosa Group seem to be eroded because the surface of the shelf in this area was eroded after the LGM (Sunamura 1978). The Kujukuri Plain in the LGM extended to a fan shape, with the Katakai Submarine Canyon at the top (Figs 2 & 4). This shape is similar to the shape in the easternmost part of the basin during the Pleistocene (Tanahashi and Murakami 1984; Suzuki 2002) and located *c.* 20 km north from it (Fig. 2). The shift northward is concordant with the northward tilting (Kaizuka 1987; Shishikura 2001).

Incised valleys in the Holocene basal surface

The topography of the Holocene basal surface off the Kujukuri Plain, which contains well-developed incised valleys, is described in detail below. The area is characterized by the development of a terrace and incised valleys. There are four incised valleys extending southeastward and one incised valley extending northeasternward. First, the four valleys extending SE are described. The most western one is connected to the incised valley beneath the land. This valley is broader and gentler than others, especially in the lower part. The second and third valleys from the west are not connected to the valleys beneath the land. The heads are located at the southeastern edge of the Kujukuri-oki Buried Terrace described later. The depth of these valleys is 10–20 m. The width is less than 3 m. The easternmost valley is the longest valley in this area. It extends from the coastline to the shelf edge, is connected to the Kakatai Submarine Canyon and is 40 km long. Its depth is more than 30 m and its width is 4 km. We named this valley the Katakai Buried Valley (KaBV). We use the term 'buried' instead of 'incised' to separate the geological interpretation from topographical nomencleture. Next, an incised valley extends northeastward and is located along

the coast line. This valley is connected to the KaBV, its length is 20 km, and its depth is 30 m. We named this valley the Kujukuri-oki Buried Valley (KuBV). The buried terrace is located between KuBV and the heads of two valleys extending SE. The flat surface of the terrace is 30 m below sealevel, and is 5 km wide and 15 km long. We named this terrace the Kujukuri-oki Buried Terrace (KuBT).

The altitude of the lower end of these incised valleys is *c.* −90 m, which indicates that the relative sea-level during the LGM was −90 m (Chiba Prefectural Historical Research Foundation 2004). The sea-level is concordant with the estimated level using the average uplift rate after the mid-Holocene (Shishikura 2001).

Is the valley bent by tectonic or lithological control?

The distribution and direction of the incised valley off the Kujukuri Plain is discussed here from the perspective of lithology and tectonics. The observations made in this study may help shed some light on whether tectonic or lithological influences played a larger role in defining the directionality of the channel.

The hinterland of the study area is the Kazusa Hills and Shimosa Plateau located on its northwestern side and rivers flowing to the SE, and the present coastline and shelf edge extend in the NE–SW direction, perpendicular to the rivers. This indicates that the rivers that formed the incised valleys probably also extended from NW to SE. Most of the incised valleys in the Holocene basal surface also extend from NW to SE; only the KuBV is perpendicular to this general direction. This unique direction seems to have been created by some constraints or forces acting on the formation.

The first possible cause of this unique directionality that was considered was tectonic control. An anticline developed in the Kujukuri-oki Group, which is the basement of the Holocene deposits (Furuyama *et al.* 2019). However, the anticline strikes NNE to SSW and intersects the incised valley at an angle of *c.* 30°; additionally, there is also an incised valley near the axis of the anticline (Fig. 6). These factors indicate that the anticline was not a major factor influencing the formation of the incised valley, and that the unique direction of the KuBV cannot be explained by tectonic deformation.

Next, the lithology of the area was considered. The Kujukuri-oki Group strikes NE–SW (Furuyama *et al.* 2019) and is almost parallel to the extension of the incised valley and the KuBT. The Kazusa Group, which is correlated to the Kujukuri-oki Group and is distributed in the Kazusa Hills, also strikes NE–SW (Suzuki *et al.* 1995). Furthermore, the ridge lines of the Kazusa Hills and the basal surface of the Holocene deposits off the Kazusa Hills are parallel to the strike direction, indicating that the incised valley and the terrace could have been controlled by the strike, as in these areas.

Differential erosion is one possible formation process that could have defined the formation of the topography. The Kiwada Formation, which consists of mudstone, is probably distributed in the terrace based on the strike and tephra correlation found in the Kazusa Hills and the Choshi Peninsula (Ozaki *et al.* 2019; Fig. 4). However, other formations consist of alternations of sandstone and mudstone. Therefore, the terrace could have formed as a result of the mudstone it contains, as mudstone erodes less readily than alternating formations. In addition, rivers could have bent to follow the terrace. As a similar example of how bedrock controls valley formation, the valley in the Bay of Biscay is incised into soft marly sediment and is oriented parallel to the Late Jurassic carbonate strata (Chaumillon *et al.* 2008).

Direction of rivers and strata in forearc basin

The incised valley bent because a river incised the basement in semi-consolidated sediment, with perpendicular strike. The conditions for this incision event were considered.

First, the occurrence condition of the intersection of rivers and the strike of sedimentary deposits was considered. The basement of the area is deposited in the forearc basin owing to plate subduction. The forearc basin is aligned parallel to the subduction boundary because the formation process for the basin is controlled by subduction. The strike is mainly controlled by tectonic deformation and also

tends to be parallel because the deformation axis is also controlled by subduction stress. In contrast, the direction from the hinterland to the basin tends to be perpendicular to the sediment strike because the hinterland is generally formed by uplift owing to the subduction, and rivers are also perpendicular to the strike because they generally flow according to the gradient of the terrain. Therefore, rivers near the subduction boundary tend to be perpendicular to the strike of the sediment.

However, the rivers do not need to incise into older and consolidated sediment while the forearc basin is developing. Rivers even at low sea-level stages develop within new strata because sediments aggrade during basin development. Therefore, uplift after a basin is filled is essential for the direction of the basement to affect a river.

From this evaluation, a forearc basin near the subduction boundary that uplifted after basin filling is essential for the occurrence of a bending event. The forearc basin in the study area was formed owing to the subduction of the Philippine Sea plate after 3 Ma and uplifted since 1 Ma owing to the subduction of the Pacific plate (Kaizuka 1987). As a result, the Lower to Middle Pleistocene strata deposited between 3 and 1 Ma with a strike parallel to the boundary cropped out and bent the river as it incised into the deposits during the LGM.

The same bending event was observed in the back arc basin of the East China Sea facing the Okinawa trough, which is also a subduction boundary of Philippine Sea plate. The depression and uplift belts have developed parallel to the trough (Emery and Niino 1967; Wageman *et al.* 1970) and perpendicular to the Changjiang River located in adjacent land. The incised valley extends eastward, parallel to the Changjiang River, which connects the valley and bends to north along the Zhedong–Xihu depression because the river could not incise into the Taiwan–Sinzi uplift belt (Li *et al.* 2005). This incised valley is in a back arc basin, but the orientation of the basin and river is controlled by plate subduction in the same way as in the forearc basin off the Boso Peninsula, as described in this paper. Thus, bending events in incised valleys are possibly common in basins near a plate subduction boundary with active uplifting movement.

Conclusion

This paper described the basal surface of the Holocene deposits off the eastern coast of the Boso Peninsula based on seismic sections as a case study of a forearc basin. This surface is considered to represent the topography formed during the LGM, when the relative sea-level fell to −90 m. The surface was divided into three areas with characteristics similar to their corresponding adjacent land areas: the Iioka Plateau, the Kujukuri Plain and the Kazusa Hills.

Off the Kujukuri Plain, many incised valleys extending in a NW–SE direction cut the basal surface. However, the KuBV is perpendicular to this general direction because of the KuBT extending in a SW–NE direction. This terrace lies parallel to the strike of the basement and consists of mudstone of the Kiwada Formation. The observations made in this study indicate that the terrace formed by differential erosion and that the valley bent to follow the terrace.

The conditions for bending were that the strike of the basement was perpendicular to the direction from the hinterland to the sea, and the area was uplifted to crop out the basement. A bending event like that in this study area may occur in a forearc basin when it begins to be uplifted after being filled.

Acknowledgements We are grateful to Daiwa Exploration & Consulting Co. Ltd, and the crew of Verny 3 for their professional efforts in conducting seismic reflection surveys. We acknowledge anonymous reviewers and the editor for their constructive comments. We also thank Yo Iwabuchi of the Japan Coast Guard and Naohisa Nishida of Tokyo Gakugei University for their valuable comments and discussions.

Author contributions TS: conceptualization (lead), data curation (supporting), investigation (equal), methodology (equal), writing – original draft (lead), writing – review & editing (lead); **SF**: data curation (equal), investigation (equal), methodology (equal), writing – original draft (supporting), writing – review & editing (supporting); **JK**: conceptualization (equal), data curation (equal), investigation (supporting), writing – original draft (supporting); **MO**: data curation (equal), investigation (equal), writing – original draft (equal); **KY**: conceptualization (supporting), data curation (equal), investigation (supporting).

Funding This research is a part of the Geological Mapping Project in Coastal Area by the Geological Survey of Japan and received no specific grant from any funding agency in the public, commercial, or not-for-profit sectors.

Data availability The datasets generated during and/or analysed during the current study are available in the Seamless Geoinformation of Coastal Zone 'Eastern Coastal Zone of Boso Peninsula', https://www.gsj.jp/researches/project/coastal-geology/results/s-6.html.

References

Ashley, G.M. and Sheridan, R.E. 1994. Depositional model for valley fills on a passive continental margin. *In*: Dalrymple, R.W., Boyd, R.J. and Zaitlin, B.A. (eds) *Incised-valley Systems: Origin and Sedimentary Sequences*. SEPM, Special Publications, **51**, 285–301.

Chaumillon, E., Proust, J.N., Menier, D. and Weber, N. 2008. Incised-valley morphologies and sedimentary-fills within the inner shelf of the Bay of Biscay (France): a synthesis. *Journal of Marine Systems*, **72**, 383–396, https://doi.org/10.1016/j.jmarsys.2007.05.014

Chiba Prefectural Historical Research Foundation 2004. Natural history 8: changing nature in Chiba Prefecture. *History of Chiba Prefecture*, **47**, 830 [in Japanese with English abstract].

Emery, K.O. and Niino, H. 1967. Stratigraphy and petroleumprospects of Korea strait and the East China Sea. *Report of Geophysical Exploration*, **1**, Korea Geological Survey, 249–263.

Furuyama, S., Sato, T. and Arai, K. 2019. *1:200,000 Marine Geological Map along the Coastal Zone of Eastern Part of Boso Peninsula. Seamless Geoinformation of Coastal Zone, S-6*. Geological Survey of Japan, Tsukuba [in Japanese with English abstract].

Furuyama, S., Sato, T., Arai, K. and Ozaki, M. 2020. The tectonic evolution of the Early to Middle Pleistocene in the East Off Boso Peninsula, Japan. *Geological Society, London, Special Publications*, **505**, https://doi.org/10.1144/SP505-2019-116

Headquarters for Earthquake Research Promotion 2014. *Long-Time Evaluation of the Activities of the Earthquake along the Sagami Trough*, 2nd edn. https://www.jishin.go.jp/main/chousa/kaikou_pdf/sagami_2.pdf [last accessed 10 February 2020, in Japanese].

Ito, M. 1992. High-frequency depositional sequences of the upper part of the Kazusa Group, a middle Pleistocene forearc basin fill in Boso Peninsula, Japan. *Sedimentary Geology*, **76**, 155–175, https://doi.org/10.1016/0037-0738(92)90081-2

Ito, M. and Katsura, Y. 1992. Inferred glacio-eustatic control for high-frequency depositional sequences of the Plio-Pleistocene Kazusa Group, a forearc basin fill in Boso Peninsula, Japan. *Sedimentary Geology*, **80**, 67–75, https://doi.org/10.1016/0037-0738(92)90032-M

Kaizuka, S. 1987. Quaternary crustal movements in Kanto, Japan. *Journal of Geography*, **96**, 223–240 [in Japanese with English abstract]. https://doi.org/10.5026/jgeography.96.4_223

Kaizuka, S., Naruse, Y. and Matsuda, I. 1977. Recent formations and their basal topography in and around Tokyo Bay, Central Japan. *Quaternary Research*, **8**, 32–50, https://doi.org/10.1016/0033-5894(77)90055-2

Katsura, Y. 1984. Depositional environments of the Plio-Pleistocene Kazusa Group, Boso Peninsula, Japan. *Science Reports of the Institute of Geoscience, University of Tsukuba, Section B: Geological Sciences*, **5**, 69–104.

Koike, K. and Machida, H. 2001. *Atlas of Quaternary Marine Terraces in the Japanese Islands*. University of Tokyo Press [in Japanese].

Koketsu, K. and Kikuchi, M. 2000. Propagation of seismic ground motion in the Kanto basin, Japan, *Science*, **288**, 1237–1239, https://doi.org/10.1126/science.288.5469.1237

Komatsubara, J. 2019. Sedimentary environments and basal topography of postglacial deposits in the Kujukuri

Coastal Plain, Boso Peninsula, central Japan. *Seamless Geoinformation of Coastal Zone, S-6.* Geological Survey of Japan, Tsukuba [in Japanese with English abstract].

Li, G.X., Liu, Y., Yang, Z.G., Yue, S.H., Yang, W.D. and Han, X.B. 2005. Ancient Changjiang channel system in the East China Sea continental shelf during the last glaciation. *Science in China (Series D: Earth Sciences),* **48**, 1972–1978, https://doi.org/10.1360/04yd0053

Maritime Safety Agency 1984a. *Basic Map of the Sea in Coastal Waters (1:50,000), Kamogawa-wan, Tokyo, Japan* [in Japanese with English abstract].

Maritime Safety Agency 1984b. *Basic Map of the Sea in Coastal Waters (1:50,000), Inubo-saki, Tokyo, Japan* [in Japanese with English abstract].

Maritime Safety Agency 1986. *Basic Map of the Sea in Coastal Waters (1:50,000), Taito-saki, Tokyo, Japan* [in Japanese with English abstract].

Maritime Safety Agency 1994. *Basic Geological Map of Shelf (1:500,000), Boso-Izu, Tokyo, Japan* [in Japanese with English abstract].

Maritime Safety Agency 2000. *Basic Map of the Sea in Coastal Waters (1:50,000), Kujukuri-hama, Tokyo, Japan* [in Japanese with English abstract].

Masuda, F., Fujiwara, O., Sakai, T. and Araya, T. 2001. Relative sea-level changes and coseismic uplifts over six millennia, preserved in beach deposits of the Kujukuri strand plain, Pacific coast of the Boso Peninsula, Japan. *Chigaku Zasshii,* **110**, 650–664, https://doi.org/10.5026/jgeography.110.5_650 [in Japanese with English abstract].

Matsuda, I. 1974. Distribution of the Recent Deposits and Buried Landforms in the Kanto Lowland, Central Japan. *Geographical Report of Tokyo Metropolitan University,* **9**, 1–36.

Moriwaki, H. 1979. The landform evolution of the Kujukuri coastal plain, central Japan. *The Quaternary Research,* **18**, 1–16 [in Japanese with English abstract]. https://doi.org/10.4116/jaqua.18.1

Nanayama, F., Nakazato, H., Ooi, S. and Nakashima, R. 2016. *Geology of the Mobara District 1:50,000.* Geological Survey of Japan, **101** [in Japanese with English abstract].

Nishida, N., Ajioka, T., Ikehara, K., Nakashima, R. and Utsunomiya, M. 2019. Spatial variation and stratigraphy of the marine sediments off the east of the Boso Peninsula, Pacific Ocean, Japan. *Seamless Geoinformation of Coastal Zone, S-6.* Geological Survey of Japan, Tsukuba [in Japanese with English abstract].

Okada, H. and Bukry, D. 1980. Supplementary modification and introduction of code numbers to the low-latitude coccolith biostratigraphic zonation (Bukry, 1973, 1975). *Marine Micropaleontology,* **5**, 321–325, https://doi.org/10.1016/0377-8398(80)90016-X

Okuno, J., Nakada, M., Ishii, M. and Miura, H. 2014. Vertical tectonic crustal movements along the Japanese coastlines inferred from late Quaternary and recent relative sea-level changes. *Quaternary Science Reviews,* **91**, 42–61, https://doi.org/10.1016/j.quascirev.2014.03.010

Olsen, K., Archuleta, R. and Matarese, J. 1995. Three-dimensional simulation of a magnitude 7.75 earthquake on the San Andreas fault. *Science,* **270**, 1628–1632, https://doi.org/10.1126/science.270.5242.1628

Ozaki, M., Furuyama, S., Sato, T. and Arai, K. 2019. 1:200,000 Marine and land geological map of the eastern coastal zone of the Boso Peninsula and its explanation, especially with Quaternary crustal deformation. *Seamless Geoinformation of Coastal Zone, S-6.* Geological Survey of Japan, Tsukuba [in Japanese with English abstract].

Sakai, T. 1990. Upper Cenozoic in the Choshi district, Chiba Prefecture, Japan. Litho-, magneto- and radiolarian biostratigraphy. *Bulletin of the Faculty of General Education, Utsunomiya University,* **23**, 1–34 [in Japanese with English abstract].

Shishikura, M. 2001. Crustal movements in the Boso Peninsula from the analysis of height distribution of the highest Holocene paleo-shoreline. *Annual Report on Active Fault and Paleoearthquake Researches,* **1**, Active Fault Research Center, GSJ, 273–285, http://dl.ndl.go.jp/info:ndljp/pid/10960118 [in Japanese with English abstract].

Sunamura, T. 1978. A model of the development of continental shelves having erosional origin. *Geological Society of America Bulletin,* **89**, 504–510, https://doi.org/10.1130/0016-7606(1978)89<504:AMOTDO>2.0.CO;2

Sunamura, T. and Horikawa, K. 1977. Sediment budget in Kujukuri coastal area, Japan. *Coastal Sediments '77.* American Society of Civil Engineers, 475–487.

Suzuki, H. 2002. *Underground Geological Structures Beneath Kanto Plain, Japan.* Research Report of National Research Institute for Earth Science and Disaster Prevention, Japan, **63**, 1–19.

Suzuki, Y., Kodama, K. *et al.* 1995. Explanatory text of geological map of Tokyo Bay and adjacent area. *Miscellaneous Geological Map, Series 20,* 2nd edn. Geological Survey of Japan, Tsukuba [in Japanese with English abstract].

Takahashi, M., Suto, I., Ohki, J. and Yanagisawa, Y. 2003. Chronostratigraphy of the Miocene Series in the Choshi area, Chiba Prefecture, central Japan. *Journal of Geological Society of Japan,* **109**, 345–360, https://doi.org/10.5575/geosoc.109.345

Tamura, T., Masuda, F., Sakai, T. and Fujiwara, O. 2003. Temporal development of prograding beach–shoreface deposits: the Holocene of Kujukuri coastal plain, eastern Japan. *Marine Geology,* **198**, 191–207, https://doi.org/10.1016/S0025-3227(03)00123-3

Tamura, T., Nanayama, F., Saito, Y., Murakami, F., Nakashima, R. and Watanabe, K. 2007. Intrashoreface erosion in response to rapid sea-level fall: depositional record of a tectonically uplifted strand plain, Pacific coast of Japan. *Sedimentology,* **54**, 1149–1162, https://doi.org/10.1111/j.1365-3091.2007.00876.x

Tamura, T., Murakami, F., Nanayama, F., Watanabe, K. and Saito, Y. 2008. Ground-penetrating radar profiles of Holocene raised-beach deposits in the Kujukuri strand plain, Pacific coast of eastern Japan. *Marine Geology,* **248**, 11–27, https://doi.org/10.1016/j.margeo.2007.10.002

Tamura, T., Murakami, F. and Watanabe, K. 2010. Holocene beach deposits for assessing coastal uplift of the northeastern Boso Peninsula, Pacific coast of Japan. *Quaternary Research,* **74**, 227–234, https://doi.org/10.1016/j.yqres.2010.07.009

Tanahashi, M. and Murakami, F. 1984. Continuous seismic reflection profiling survey on the southeastern offshore of the Boso Peninsula. *Cruise Report: Geological Investigation of the Junction Area of the Tohoku and Ogasawara Arcs, April-June 1980 (GH80-2 and 3 Cruise)*. Geological Survey of Japan, **19**.

Tokuhashi, S. and Endo, H. 1984. *Geology of the Anesaki District 1:50,000*. Geological Survey of Japan, Tsukuba [in Japanese with English abstract].

Waelbroeck, C., Labeyrie, L. *et al.* 2002. Sea-level and deep water temperature changes derived from benthic foraminifera isotopic records. *Quaternary Science Reviews*, **21**, 295–305, https://doi.org/10.1016/S0277-3791(01)00101-9

Wageman, J.M., Hilde, T.W.C. and Emery, K.O. 1970. Stracturalframe works of East China Sea and Yellow Sea. *AAPG Bulletin*, **54**, 1611–1643.

Yamaguchi, K., Ito, S. and Kinoshita, S. 2019. Shallow subsurface structure in the Kujukuri coastal plain by seismic reflection surveys. *Seamless Geoinformation of Coastal Zone, S-6*. Geological Survey of Japan, Tsukuba [in Japanese with English abstract].

Zaitlin, B.A., Dalrymple, R.W. and Boyd, R. 1994. The stratigraphic organisation of incised valley systems associated with relative sea- level change. *In*: Dalrymple, R.W., Boyd, R.J. and Zaitlin, B.A. (eds) *Incised Valley Systems: Origin and Sedimentary Sequences*. SEPM Special Publication, **51**. SEPM, 45–60.

Application of spatial distribution patterns of multi-elements in geochemical maps for provenance and transfer process of marine sediments in Kyushu, western Japan

Atsuyuki Ohta*, Noboru Imai, Yoshiko Tachibana and Ken Ikehara

Geological Survey of Japan, National Institute of Advanced Industrial Science and Technology, Central 7, Tsukuba, Ibaraki 305-8568, Japan

AO, 0000-0002-0770-3273; KI, 0000-0003-3906-4303

*Correspondence: a.ohta@aist.go.jp

Abstract: We investigated the origin of marine sediments and their transfer by water currents using the spatial distribution patterns of multi-elements in terrestrial and marine areas of Kyushu, western Japan. Quaternary volcanic material covers Cretaceous granitic rock and Jurassic–Paleogene sedimentary rock of an accretionary complex in this region. Cluster analysis based on chemical compositions identified the origin of marine sediments from stream sediments originating from the above lithologies. Pyroclastic-flow deposits associated with caldera formation, particularly that of the Kikai Caldera (7.3 ka), were characterized by low Cr/Ti and La/Yb ratios. In contrast, the La/Yb ratio was very high in sediments derived from granitic rock and sedimentary rock of the accretionary complex. The spatial distributions of low Cr/Ti and La/Yb ratios suggest that marine sediments containing pyroclastic materials, which are found within an 80 km radius of the Kikai Caldera, were distributed on the shelf and transported northeastwards by a branch of the Kuroshio Current. The continuous distribution of the medium Cr/Ti and high La/Yb ratios from the land to the coast, slope and deep basin on the Pacific Ocean side suggests that sediments supplied from the terrestrial area were transferred by gravitational transport from the shelf to the deep basin.

The Geological Survey of Japan, National Institute of Advanced Industrial Science and Technology (AIST), has created geochemical maps showing the spatial distribution patterns of the concentrations of multiple elements throughout the country based on stream and marine sediments (Imai *et al.* 2004; Imai *et al.* 2010). These maps provide fundamental information on the behaviour of the elements in nature, and are widely used for environmental assessment and mineral exploration (e.g. Webb *et al.* 1978; Darnley *et al.* 1995). Several subcontinental and cross-border geochemical mapping projects have been conducted recently to further nationwide geochemical mapping (Xie and Chen 2001; Salminen *et al.* 2005; De Vos *et al.* 2006; De Caritat and Cooper 2011; Reimann *et al.* 2014a, b; Smith *et al.* 2014).

Japanese geochemical mapping was initially designed to provide natural background concentrations of elements to quantitatively elucidate anthropogenic contamination in urban regions and coastal sea zones (Imai *et al.* 2004; Imai *et al.* 2010). Nonetheless, the spatial distribution patterns of elements across land and marine areas may be further applicable to the study of the particle transfer process from land to the coastal sea, as well as in a marine environment. Although Japan is surrounded by oceans, we still have little information about the characteristics of the surface sediment on shelves, slopes and basins

in the island arc region in terms of the erosion, transport and deposition of the sediments. Thus, we have applied geochemical maps to the provenance and transfer analyses of marine sediments (Ohta and Imai 2011) to reveal objectively the complex factors affecting the spatial distributions of elements in marine sediments using analysis of variance (ANOVA), cluster analysis and Mahalanobis' generalized distances (Ohta *et al.* 2007, 2010, 2013, 2017b).

We successfully identified the long-range transport process of modern silty sediments on the shelf and slope by oceanic and tidal currents using Cr and Ni geochemical maps (Ohta *et al.* 2007, 2010, 2017a). Silty sediments originating from ultramafic rocks can be easily isolated from those with a different origin because the rocks greatly elevate the Cr and Ni concentrations of stream and marine sediments. In contrast, granule and sandy sediments on the shelf are mainly composed of sediments that formed during the Quaternary regression and transgression cycles (e.g. Park and Yoo 1988; Ikehara 1993); most of them formed under hydrographical conditions that are different to the modern environment. It is therefore difficult to directly identify their origin, evaluate the corresponding relationship between marine and terrestrial materials, and to examine the transportation process of sandy sediments (Ohta *et al.* 2007, 2010).

From: Asch, K., Kitazato, H. and Vallius, H. (eds) 2022. *From Continental Shelf to Slope: Mapping the Oceanic Realm.* Geological Society, London, Special Publications, **505**, 241–270.
First published online December 22, 2020, https://doi.org/10.1144/SP505-2019-87

Thus, the Kyushu region, southwestern Japan, was selected as a study area in which sandy sediments containing pyroclastic deposits supplied by large-scale eruptions from the Kikai Caldera (7.3 ka) (Maeno and Taniguchi 2007; Fujihara and Suzuki-Kamata 2013) are widely distributed in the southern marine area of Kyushu Island (Ikehara 2014) (Fig. 1). In this region, there are various bedforms such as subaqueous sand dunes, sand ribbons and ripples (Ikehara 1988, 1993; Ikehara and Kinoshita 1994). Shelf sandbodies with these bedforms are presumed to have formed during the transgression

after the last glacial maximum and are also presently active owing to the branch currents of the Kuroshio Current (Fig. 1). Therefore, the Kyushu region is a suitable place to apply our geochemical datasets in order to find the range over which the coarse pyroclastic deposit moved, or the boundary separating pyroclastic deposits from the sandy sediments derived from terrestrial areas without volcanic materials using the geochemical data of sediments. In this study, we elucidate in particular the following three items using marine and terrestrial geochemical maps: (1) the relationship between marine sediments

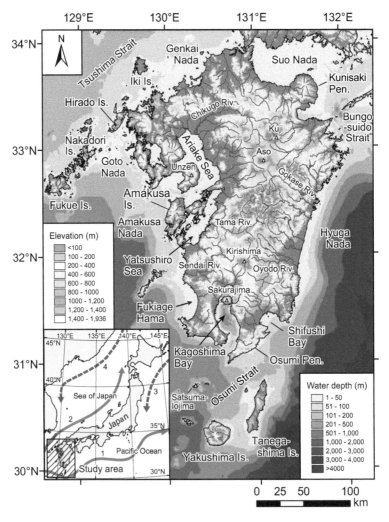

Fig. 1. Geographical names for terrestrial and marine areas. Open triangular symbols indicate active volcanoes that have erupted over the past century. The red lines numbered 1 and 2 represent the Kuroshio Current and the Tsushima Current, respectively. The blue dotted lines numbered 3 and 4 represent the Oyashio Current and the Liman Current, respectively. The gradation map of elevation was created using the digital elevation model provided by the Geospatial Information Authority of Japan. The bathymetric depth map was delineated using a dataset provided by the Japan Oceanographic Data Centre.

and adjacent terrestrial materials; (2) the dispersion processes of volcanic materials erupted by very large eruptions; and (3) and the transportation of granules and sandy sediments by water currents.

Study area

Marine topography and hydrographical condition

Figure 1 shows the geographical names of the features within the study area, as well as a geological map of the area. The water depth in the northern part of the study area is generally less than 200 m. The Goto Nada, Tsushima Strait and Genkai Nada in the northwestern part are continental shelf areas, where 'Nada' means an open sea where a strong ocean current, fast tidal current or severe wave makes marine navigation challenging. The Suo Nada is in the western part of the Seto Inland Sea and is generally less than 60 m deep. It passes through the Bungo–Suido Strait to the Pacific side of Kyushu Island (Hyuga Nada). The Hyuga Nada is characterized by a narrow shelf with a depth that increases steeply towards the SE. Submarine landslide surfaces are found along the slope of the Hyuga Nada (Ikehara 2000). The Amakusa and Yatsushiro seas, the largest inner bays of Kyushu mainland, have extreme tidal ranges and connect to the Amakusa Nada. The water depth of the Amakusa Nada gently increases towards the SW from the SW area of Kyushu mainland and connects to the East China Sea. Kagoshima Bay, with a maximum water depth of 80 m, is part of the Aira Caldera and is associated with thermal activity. The Osumi Strait is a marine area between the Osumi Peninsula and the islands of Yakushima and Tanegashima, where the water depth is less than 500 m. The SE area off Tanegashima is a steep slope associated with a submarine canyon. The strong oceanic currents, which are branches of the Kuroshio Current, prevail on the coast of Kyushu mainland. The Kuroshio Current flows from SW to NE in the the SW areas of the Yakushima and Tanegashima islands. A branch of the Kuroshio Current also flows from SW to NE through the Tsushima and Osumi straits.

Marine sedimentology

Most of the marine sediments in the study area are mainly from the Holocene and Pleistocene. Coarse sand and gravel from the Goto Nada, Tsushima Strait, Genkai Nada, Bungo–Suido Channel and Osumi Strait are composed of sediments that formed during Quaternary regression and transgression cycles (Ohshima et al. 1975, 1982; Ikehara 2000, 2013, 2014). Bedforms such as ripples and current

lineations have been found in the Tsushima Strait, Genkai Nada, Osumi Strait and NE of Tanegashima Island; these can be explained by changes in sea level after the last glacial maximum and a strong bottom current related to the oceanic current that moves surface sediments (Ikehara 1988, 1992, 1993; Nishida and Ikehara 2013). Submarine topography and bedforms in the Bungo–Suido Channel are influenced by post-glacial sea-level changes and related changes with strong tidal energies (Ikehara 1998). Silt and clay found in the depths of the Hyuga Nada and in the slope and basin off Tanegashima Island are hemipelagic deposits. Modern silty sediments derived from the adjacent terrestrial area are found in the shallows of inner bays such as the Ariake and Yatsushiro seas, the Suo Nada, and nearshore regions. The shelf and upper slope of the Hyuga Nada are covered by sandy deposits, with some turbidites intercalated in the slope sediments (Ikehara 2000).

Terrestrial area and riverine system

Mainland Kyushu is mountainous and comprises many active volcanoes, such as Kujusan, Asosan, Unzendake, Kirishimayama, Sakurajima and Satsuma–Iojima (see the small triangles in Fig. 1). It is also associated with many isolated islands such as Iki, Hirado, Fukue, Nakadori, Amakusa, Tanegashima and Yakushima, and its major rivers include the Chikugo, Kuma, Gokase, Oyodo and Sendai. The sediment yield, drainage basin area and river water discharge rate of the major rivers in the study area are summarized in Appendix Table A1. The sediment yield of each river system is calculated using the drainage basin area and the average rate of erosion in each river basin (after Akimoto et al. 2009). In the current environment, rivers on Kyushu Island supply 49% of the total sediment yield to the western region, including the Ariake Sea, Yatsushiro Sea and Amakusa Nada (Appendix Table A1). The Hyuga Nada and Suo Nada receive sediment discharges of 29 and 17% from Kyushu Island, respectively (Appendix Table A1). Little sediment is discharged from the rivers on Kyushu Island to the northern regions of the Genkai Nada, Goto Nada and Tsushima Strait, or to the southern region including the Osumi Strait and Kagoshima Bay (Appendix Table A1).

Terrestrial geology and metalliferous deposits

Figure 2 shows a geological map of SW Japan that has been simplified from the Geological Map of Japan 1:1 000 000 scale (Geological Survey of Japan, AIST 1992). The details of the geology of the Kyushu region have been provided by Miyazaki et al. (2016). A Permian accretionary complex is located in the NE; and Jurassic, Cretaceous and

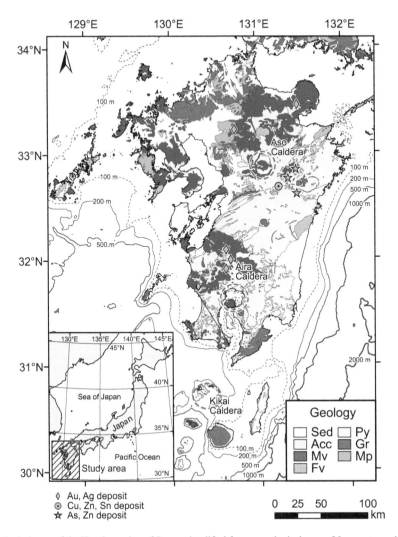

Fig. 2. Geological map of the Kyushu region of Japan, simplified from a geological map of Japan at a scale of 1:1 000 000 (Geological Survey of Japan 1992). Sed, sediment and sedimentary rock; Acc, sedimentary rocks of accretionary complexes associated with metagabbro, metabasalt and ultramafic rock; Mv, mafic volcanic rocks; Fv, felsic volcanic rocks; Py, pyroclastic-flow deposit and debris; Gr, granitic rocks; Mp, metamorphic rocks (mostly high-pressure type). Circles with a dot and star symbol indicate a skarn-type deposit bearing Cu, Zn, As and Sn; diamond symbols indicate a hydrothermal-type deposit bearing As and Au. The bathymetric depth contours were delineated using a dataset provided by the Japan Oceanographic Data Centre.

Paleogene accretionary complexes are located in the central and SE parts of the study area. The Paleogene unit is also found in the Tanegashima and Yakushima islands. These accretionary complexes are exposed across 23% of the study area. The Permian unit mainly consists of mudstone and sandstone associated with blocks of limestone, basaltic rock and chert. The Jurassic unit is associated with exotic rocks such as serpentinite mélange, metamorphic rock and ultramafic rock. The Cretaceous and Paleogene units consist mainly of accreted turbidites (mudstone to sandstone) associated with exotic blocks of basaltic rock. Carboniferous–Permian and Triassic–Jurassic high-pressure-type metamorphic rock is distributed in the north. These metamorphic rocks are dominated by pelitic, mafic and psammitic schists; and their exposed area is less than 4% of the total. Cretaceous high-temperature-type metamorphic rock is found in the central region, although the area of distribution is only 0.5% of the area. Cretaceous granitic rock is intruded in the northern region and comprises only 5% of the area.

In contrast, Neogene granitic rock is intruded in the Osumi Peninsula and the island of Yakushima, and crops out sporadically in accretionary complexes in 3% of the area.

Neogene–Quaternary basaltic, andesitic and dacitic volcanic rocks are distributed in both the northern and southern regions of Kyushu; they crop out across 22% of the area. They are associated with huge pyroclastic-flow deposits, debris and tephra that were erupted by large-scale caldera-forming processes; they comprise only 17% of the area. In particular, Aso and Ito pyroclastic rocks erupted from the Aso and Aira calderas mainly at about 90 and 29 ka, respectively; they are distributed widely in the central and the southern parts, respectively (Nakada et al. 2016). The Kikai Caldera to the south of the Kyushu mainland was formed by large-scale violent eruptions at about 7.3 ka. Koya pyroclastic flows from the Kikai Caldera were extended to 40–80 km (Maeno and Taniguchi 2007; Fujihara and Suzuki-Kamata 2013). Asosan, Sakurajima and Satsuma–Iojima are volcanoes associated with the Aso, Aira and Kikai calderas, respectively (see the triangular symbols in Fig. 1).

Pre-Cretaceous non-accretionary sedimentary rocks are found in the central part with an exposure of less than 3% across the total area. Paleogene fore-arc sedimentary rocks with coal beds are distributed in the northern and northwestern regions of the Kyushu mainland and in the Amakusa islands (5% of the area). Neogene and Quaternary unconsolidated sediments are distributed widely in the downstream basin of the Chikugo and Oyodo rivers but are rather restricted to a small area elsewhere; they comprise 17% of the study area.

Figure 2 also shows the distribution of major and economically mined deposits in Kyushu (Sudo et al. 2003). Many Au and Ag mineral deposits are hydrothermal or disseminated, and are closely related to Neogene and Quaternary volcanic activity. Some skarn deposits bearing Cu, Zn, As and Sn were formed in Paleozoic limestone bedrocks of the Jurassic accretionary complex where the Neogene granitic rocks intruded.

Materials and methods

Samples, sampling methods and processing

Figure 3 shows the sampling location of the Kyushu region. For a geochemical mapping project carried out during 1999–2004 by the Geological Survey of Japan, AIST, 366 stream sediment samples were collected from locations on Kyushu Island (Imai et al. 2004) (Fig. 3). In addition, 191 stream sediment samples were collected from 23 isolated small islands around the Kyushu mainland from 2013 to 2014 for a high-density geochemical map (Ohta 2018); of these, 25 samples with wide watershed areas were used for this study (see Appendix B). The collected stream sediments were dried in air and sieved through a 180 µm screen. Magnetic minerals contained in the samples were removed using a magnet to minimize the effect of magnetic mineral accumulation (Imai et al. 2004; Ohta 2018).

A total of 504 marine sediment samples were collected from the marine areas around the Kyushu mainland during cruises GH83-1, GH84-1 and GH85-2 in 1983, 1984 and 1985, respectively (Ikehara 2000, 2001, 2013, 2014). In addition, 102 marine sediment samples were collected from the Tsushima Strait, Seto Inland Sea, Ariake and Yatsushiro seas, and Amakusa Nada in 2005, and from Kagoshima Bay in 2007 (Imai et al. 2010). These 606 samples were collected using a Kinoshita-type grab sampler (RIGO Inc. Co.), and the uppermost 3 cm layer of each sediment sample was separated, air dried, ground with an agate mortar and pestle, and retained for chemical analysis.

The particle sizes of 334 marine sediments were determined on the basis of the median particle diameter of the surface sediments (Ikehara 2013, 2014). The median particle diameter was not measured for the remaining 272 samples; their classification is based on a visual inspection of texture. The elemental concentrations in the marine sediments change with variations in particle sizes. As a general rule, CaO, MnO, total $Fe_2O_3(Fe_2O_3^T)$, Sr and Co are high in coarse and medium sand because these soils contain calcareous materials and coatings of Fe–Mn oxides; the trace elements are high in silt and clay because of the less effective dilution effect in the region (Ohta et al. 2010; Ohta and Imai 2011). Thus, marine sediments were classified roughly as coarse sediment comprising pebbles, granules and medium–coarse–very coarse sands; fine sands comprising very fine and fine sand; and silt and clay consisting of coarse silt, fine silt and clay (Fig. 3).

Geochemical analysis

We compiled the geochemical data of 53 elements in 391 stream sediments and 606 marine sediments around the Kyushu region that had been reported in Imai et al. (2004, 2010) and Ohta (2018). The details of the digestion of the samples, the measurement of elemental concentrations and the quality control of measurement values are described in Imai et al. (2004, 2010) and Ohta (2018). Each sample was digested using HF, HNO_3 and $HClO_4$. The concentrations of Na_2O, MgO, Al_2O_3, P_2O_5, K_2O, CaO, TiO_2, MnO, $Fe_2O_3^T$, V, Sr and Ba were determined using inductively coupled plasma atomic emission spectrometry (ICP-AES), while those of Li, Be, Sc, Cr, Co, Ni, Cu, Zn, Ga, Rb, Y, Nb, Zr, Mo, Cd, Sn, Sb, Cs, the lanthanides (Ln: La–Lu),

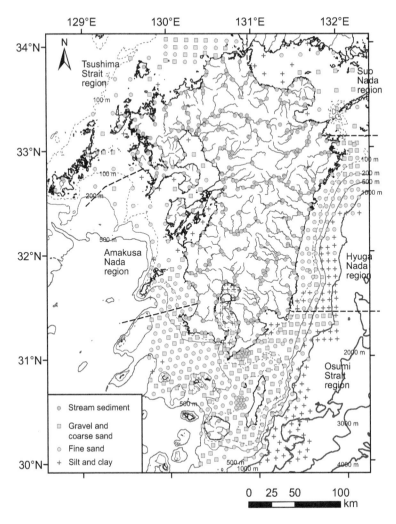

Fig. 3. Sampling locations for the stream and marine sediments. The regions of the Tsushima and Osumi straits and the Suo Amakusa and Hyuga nadas are delineated by zonal classifications represented by broken lines.

Ta, Hf, Tl, Pb, Bi, Th and U were determined by using ICP mass spectrometry (ICP-MS) (Imai *et al.* 2004, 2010; Ohta 2018). The As and Hg analyses of stream sediment samples from the Kyushu mainland were subcontracted to ALS Chemex in Vancouver, BC (Imai *et al.* 2004). The As concentrations in the stream sediment samples collected from isolated islands were determined using ICP-MS after the digestion of samples using HF, HNO_3, $HClO_4$ and $KMnO_4$ (Ohta 2018). The Hg concentrations in stream sediments on isolated islands and marine sediments were determined by using an atomic absorption spectrometer equipped with the direct thermal decomposition of a sample (Imai *et al.* 2010; Ohta 2018). A summary of the geochemical characteristics of the marine and stream sediments in the

Kyushu region is given in Table 1. However, the Zr and Hf concentrations were not used in this study because the heavy mineral fraction, especially that of zircon, was not digested by the $HF–HNO_3–HClO_4$ solution. Additionally, the Na_2O concentration of marine sediments was also used as a guide because such sediments were not desalinated in this study.

Geochemical map preparation

Geochemical maps of the Kyushu mainland have been recreated as mesh maps associated with the data of the surrounding isolated islands. They were created using geographical information system software (ArcGIS 10.5: Environmental Systems

Table 1. *Summary for the geochemistry of marine and stream sediments in the Kyushu region, Japan*

	Unit	Marine sediment (n = 606)					Stream sediment (n = 391)				
		Min.	Median	Mean	Max.	MAD	Min.	Median	Mean	Max.	MAD
Na_2O	wt%	0.545	2.61	2.62	5.48	0.45	0.548	2.18	2.18	5.44	0.42
MgO	wt%	0.413	2.68	3.03	9.69	0.49	0.554	2.34	2.63	7.61	0.758
Al_2O_3	wt%	0.631	10.27	9.84	17.96	1.26	3.32	12.41	12.37	27.99	1.77
P_2O_5	wt%	0.0227	0.114	0.122	0.504	0.026	0.045	0.150	0.170	0.767	0.038
K_2O	wt%	0.193	1.53	1.46	2.68	0.34	0.519	1.71	1.70	4.71	0.352
CaO	wt%	0.405	7.48	8.84	39.4	2.40	0.173	2.33	2.53	23.3	1.18
TiO_2	wt%	0.0242	0.503	0.627	2.92	0.118	0.234	0.79	0.925	4.81	0.222
MnO	wt%	0.011	0.079	0.109	0.973	0.031	0.013	0.126	0.139	0.524	0.0413
$Fe_2O_3^T$	wt%	0.304	4.54	5.51	35.4	1.03	1.50	5.81	6.36	16.0	1.40
Li	mg kg⁻¹	4.70	32.5	35.5	122	12.1	13.2	36.7	38.4	112	8.55
Be	mg kg⁻¹	0.105	1.11	1.09	3.05	0.26	0.715	1.49	1.56	5.96	0.233
Sc	mg kg⁻¹	0.955	11.1	13.9	71.9	2.57	2.18	13.6	15.2	61.6	4.04
V	mg kg⁻¹	3.93	88.2	106	1030	26.9	27.6	114	135	474	37.0
Cr	mg kg⁻¹	8.11	28.7	31.9	77.1	9.98	7.46	52.9	68.8	550	21.9
Co	mg kg⁻¹	1.29	9.26	10.8	54.0	2.41	3.66	14.6	16.3	68.4	4.25
Ni	mg kg⁻¹	3.82	12.3	15.3	62.7	3.71	3.86	20.4	26.4	301	8.99
Cu	mg kg⁻¹	1.22	10.5	13.4	303	4.21	5.19	28.0	35.1	737	8.98
Zn	mg kg⁻¹	7.57	73.9	77.8	330	13.1	39.3	114	128	990	23.2
Ga	mg kg⁻¹	0.702	13.2	12.4	20.9	1.50	7.52	17.4	17.4	31.4	1.51
As	mg kg⁻¹	0.2	5.6	6.0	31	1.9	0.7	8.0	18	2010	4.0
Rb	mg kg⁻¹	4.66	52.3	51.7	115	16.8	12.3	72.0	71.5	170	17.0
Sr	mg kg⁻¹	26.7	351	450	3000	105	46.2	155	175	528	47.9
Y	mg kg⁻¹	1.98	14.0	14.5	35.9	3.23	5.56	18.0	18.5	46.8	4.69
Nb	mg kg⁻¹	0.456	4.95	5.00	24.9	1.19	3.51	9.21	10.4	55.0	1.82
Mo	mg kg⁻¹	0.133	0.798	0.958	8.77	0.304	0.0853	1.20	1.47	16.5	0.35
Cd	mg kg⁻¹	0.013	0.0695	0.0794	0.454	0.0205	0.027	0.127	0.175	3.74	0.037
Sn	mg kg⁻¹	0.098	1.48	1.54	7.53	0.474	0.863	2.55	4.53	194	0.546
Sb	mg kg⁻¹	0.074	0.423	0.465	2.34	0.121	0.137	0.596	0.963	33.3	0.196
Cs	mg kg⁻¹	0.133	2.64	2.79	6.74	1.17	0.41	4.49	4.61	14.9	1.24
Ba	mg kg⁻¹	31.3	246	256	696	67.2	111	385	379	980	54.9
La	mg kg⁻¹	2.52	12.9	13.0	27.0	2.91	5.90	19.1	21.6	194	2.72
Ce	mg kg⁻¹	4.31	27.4	27.3	58.6	6.05	12.5	35.5	41.3	564	5.89
Pr	mg kg⁻¹	0.537	3.23	3.16	6.71	0.582	1.55	4.47	5.04	38.0	0.597
Nd	mg kg⁻¹	2.11	13.4	12.9	27.2	1.99	6.64	17.9	19.9	137	2.41

(Continued)

Table 1. *Continued.*

	Unit	Marine sediment (n = 606)					Stream sediment (n = 391)				
		Min.	Median	Mean	Max.	MAD	Min.	Median	Mean	Max.	MAD
Sm	mg kg^{-1}	0.375	2.93	2.78	5.65	0.425	1.47	3.72	4.05	23.5	0.570
Eu	mg kg^{-1}	0.142	0.722	0.704	1.08	0.087	0.334	0.928	0.941	1.94	0.150
Gd	mg kg^{-1}	0.366	2.75	2.66	4.92	0.447	1.21	3.46	3.68	16.3	0.580
Tb	mg kg^{-1}	0.062	0.463	0.458	1.02	0.088	0.177	0.589	0.611	1.78	0.120
Dy	mg kg^{-1}	0.325	2.34	2.37	5.69	0.485	1.07	3.02	3.11	7.99	0.679
Ho	mg kg^{-1}	0.061	0.455	0.469	1.10	0.099	0.195	0.584	0.600	1.39	0.147
Er	mg kg^{-1}	0.195	1.35	1.39	3.25	0.305	0.496	1.70	1.75	4.22	0.463
Tm	mg kg^{-1}	0.029	0.215	0.221	0.543	0.051	0.070	0.269	0.275	0.655	0.076
Yb	mg kg^{-1}	0.197	1.34	1.37	3.39	0.320	0.493	1.65	1.71	4.11	0.478
Lu	mg kg^{-1}	0.030	0.195	0.203	0.473	0.050	0.064	0.244	0.253	0.619	0.0768
Ta	mg kg^{-1}	0.002	0.471	0.456	1.61	0.123	0.129	0.746	0.828	3.55	0.158
Hg	mg kg^{-1}	0.007	0.043	0.061	0.393	0.029	0.010	0.060	0.142	13.7	0.030
Tl	mg kg^{-1}	0.028	0.341	0.343	0.838	0.115	0.075	0.484	0.485	1.81	0.100
Pb	mg kg^{-1}	3.55	16.3	16.2	68.7	3.17	8.77	22.6	32.8	1660	4.81
Bi	mg kg^{-1}	0.039	0.194	0.226	1.60	0.084	0.036	0.268	0.387	15.0	0.091
Th	mg kg^{-1}	0.435	4.46	4.51	26.6	1.49	1.52	6.22	7.51	67.6	1.35
U	mg kg^{-1}	0.232	1.09	1.08	4.51	0.264	0.503	1.48	1.65	7.76	0.311

Min., minimum; Max., maximum; MAD, median absolute deviation.

Research Institute (ESRI), Japan Corporation, Tokyo, Japan) after Ohta *et al.* (2004). The stream sediment is considered a composite sample of the products of weathering and soil and rock erosion in the watershed area upstream of the sampling site (Howarth and Thornton 1983). A sampling site is presumed to express the average chemical concentrations in a drainage basin. The watershed area for each sample was calculated in ArcGIS based on a digital elevation model (50 m mesh data) obtained from the Geospatial Information Authority of Japan. The marine geochemical maps were created by interpolating data points using radial basis functions (Aguilar *et al.* 2005). The resultant marine geochemical maps were combined with the existing land geochemical maps as shown in Figure 4a–l. The elemental concentrations were classified separately for the stream and marine sediments because their chemical and mineralogical compositions differ significantly, as detailed by Ohta *et al.* (2004, 2010). The following percentile ranges were used for classification of the elemental concentration intervals in the colour image maps: $0 \leq x \leq 5$, $5 < x \leq 10$, $10 < x \leq 25$, $25 < x \leq 50$, $50 < x \leq 75$, $75 < x \leq 90$, $90 < x \leq 95$ and $95 < x \leq 100\%$, where x represents the elemental concentration according to Reimann (2005). This class selection is advantageous in that the same range of percentiles (e.g. 90–95%) implies the same statistical weight even at different numerical scales (Reimann 2005).

Results

Spatial distribution patterns of concentrations of multi-elements in the terrestrial area

Geochemical maps of the 12 elements K_2O, CaO, MnO, TiO_2, Cr, Cs, La, Yb, Cu, Cd, Sb and Pb are shown in Figure 4a–l. They are representative and characteristic distribution patterns of elements that were obtained by considering the similarity of the spatial distribution patterns of 49 elements across land and sea. The Li, K_2O, Rb, Cs and Tl contents are high in the northern, central and southeastern parts of Kyushu mainland and in Yakushima Island, where accretionary complex, granitic rock and felsic volcanic rock crop out. However, the concentrations are low in the central part of Kyushu, where mafic volcanic rock, pyroclastic rock and metamorphic rock are distributed (K_2O and Cs: Fig. 4). The spatial distributions of MgO, CaO, TiO_2, MnO, $Fe_2O_3^T$, Sc, V, Co and Sr are opposite to those of Li, K_2O, Rb, Cs and Tl (CaO and $Fe_2O_3^T$: Fig. 4). Their high concentration areas correspond mainly to areas covered by the mafic volcanic rocks and pyroclastic rocks, particularly near Aso volcano (Fig. 2). The Cr and Ni concentrations are high in the northern-central and

central parts of Kyushu mainland. High-pressure-type metamorphic rock and mafic rocks of the accretionary complex crop out sporadically in these areas (Cr: Fig. 4). High enrichment of Be, Na_2O, rare earth elements (REEs), Nb, Ta, Th and U was found in the northern and southern parts of the islands of Kyushu and Yakushima, which are underlain by granitic rocks (La and Yb: Fig. 4). In addition, Y and heavy REEs (HREEs), such as Gd, Tb, Dy, Ho, Er, Tm, Yb and Lu, were also enriched in the pyroclastic rock outcrops of central and southern Kyushu Island (Yb: Fig. 4). The metalliferous skarn-type deposits bearing Cu, Zn, As and Sn (circle with a dot and star symbols in Fig. 4) caused extremely high concentrations of Cu, Zn, As, Mo. Cd. Sn, Sb, Hg, Pb and Bi (Cu, Cd, Sb and Pb: Fig. 4). However, the Au–Ag deposits (diamond symbols in Fig. 4) did not appear to contribute to the enrichment of these elements.

Spatial distribution patterns of concentrations of multi-elements in the marine environment

Silt and clay sediments in the Suo and Hyuga nadas, southeastern Osumi Strait, and Yatsushiro Bay are commonly enriched in Li, Be, Cr, Ni, Cu, Nb, Mo, Cd, Sn, Sb, Cs, Ta, Hg, Tl, Pb, Bi and U. Silt and clay sediments in the basin SE of Tanegashima Island have high concentrations of Be, K_2O, Cr, MnO, Ni, Cu, As, Rb, Mo, Cd, Sn, Sb, Cs, Ba, light REEs (LREEs), Hg, Tl, Pb, Bi and Th. The enrichment of MnO, Ni, Cu, Mo, Cd, Sn, Sb, Pb and Bi in the deep seas was caused by early diagenetic processes under a static depositional environment; their concentrations have a definite correlation with the MnO concentration (e.g. Klinkhammer 1980; Shaw *et al.* 1990).

The coarse sediments and fine sands in the Tsushima Strait, Genkai Nada and Goto Nada are rich in CaO and Sr, and poor in other elements, as a result of calcareous deposits such as shell fragments and foraminifera tests. The coarse and fine sands in the Amakusa Nada are somewhat enriched in MgO, Cr and Co, although their chemical compositions are similar to those in the Tsushima Strait, Genkai Nada and Goto Nada. The sandy sediments in the southern part of the Suo Nada near the Kunisaki Peninsula are enriched in MgO, CaO, TiO_2, $Fe_2O_3^T$, Sc, V, Co and Sr, and are poor in K_2O, Rb and Cs; the sediments in the northern part of the Suo Nada are in contrast to those near the Kunisaki Peninsula.

The fine sands in the Hyuga Nada are rich in Li, Be, K_2O, Cr, Ni, Rb, Ba, LREEs and Nb, and poor in P_2O_5, CaO, MnO, Sr, HREEs and Cd. High concentrations of As, Sn, Hg, Pb and Bi, in particular, were found in samples collected within a 16 km radius of the mouth of the Gokasegawa River, where upper

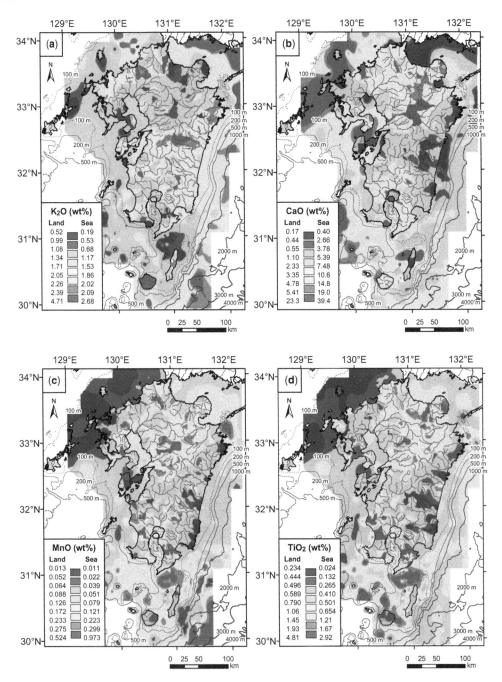

Fig. 4. Spatial distributions of elemental concentrations in the Kyushu region: (**a**) K_2O, (**b**) CaO, (**c**) MnO, (**d**) $Fe_2O_3^T$.

river basin skarn-type deposits bearing Cu, Zn, As and Sn occur. Granules and coarse sands in the Osumi Strait are highly enriched in MgO, P_2O_5, TiO_2, MnO, $Fe_2O_3^T$, Sc, V, Co, Zn, Sr, Mo, Cd and

HREEs. Marine sediments occurring between the Osumi Peninsula and Tanegashima Island are, in particular, enriched in MgO, P_2O_5, TiO_2, MnO, $Fe_2O_3^T$, Sc, V and Co; however, these elements are

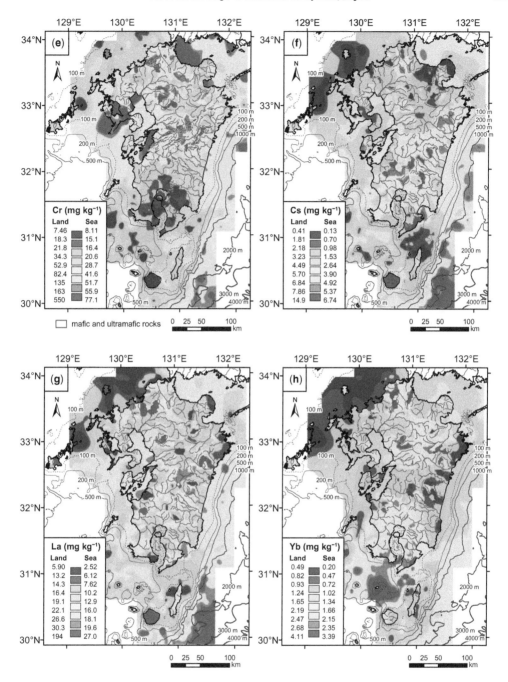

Fig. 4. *Continued.* (**e**) Cr, (**f**) Cs, (**g**) La, (**h**) Yb.

less abundant in stream sediments derived from the adjacent terrestrial area in which unconsolidated sediment and sedimentary rock, granitic rocks, and sedimentary rocks of accretionary complexes are the dominant lithologies. In contrast, the fine sands of the Osumi Strait are more enriched in CaO and Sr, and are poorer in MgO, TiO$_2$, V, Fe$_2$O$_3^T$, Co and , Co and Zn than the coarse sands owing to dilution by biogenic carbonate materials. The sediments in Kagoshima Bay have chemical compositions similar

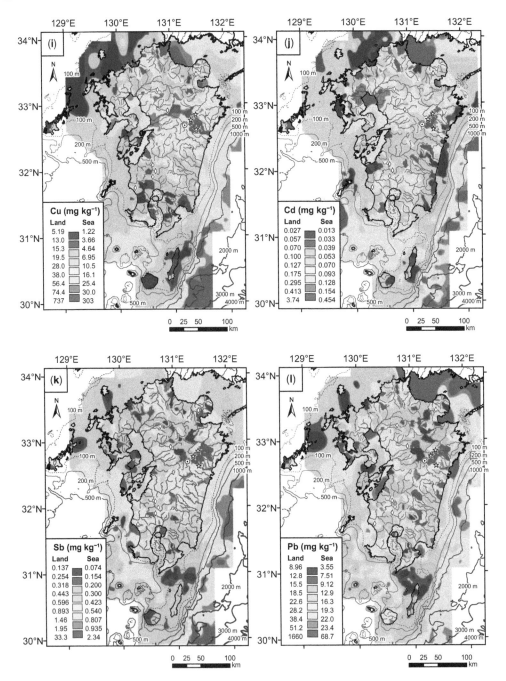

Fig. 4. *Continued.* (**i**) Cu, (**j**) Cd, (**k**) Sb and (**l**) Pb. Star, circle and diamond symbols indicate the major metalliferous deposits shown in Figure 1.

to those of the coarse sediments in the Osumi Strait irrespective of particle size; furthermore, they are highly enriched in As, Cd, Mo, Sb, Hg and Bi owing to hydrothermal activity (Sakamoto 1985). Almost all of the elements are less abundant in the fine sands of the slope to the SE of Tanegashima Island.

Discussion

Differences in elemental concentrations of stream sediments grouped by dominant lithology distributed in the watershed of stream sediment samples

The geochemistry of stream sediment is determined predominantly by the dominant lithology distributed in the catchment area, as we have previously suggested (e.g. Ohta *et al.* 2004). As a first approximation, we assume that when a specific rock type underlies more than half of the area of the river basin, it is representative of the geology of the watershed and mainly controls the chemical composition of the stream sediment (Ohta *et al.* 2004). Unconsolidated sediment and sedimentary rock (Sed); an accretionary complex comprising mainly sandstone, mudstone and a mélange matrix (Acc); granitic rock (Gr); felsic volcanic rock (Fv); mafic volcanic rock (Mv); a pyroclastic-flow deposit (Py); and high-pressure-type metamorphic rock (Mp) were considered as typical lithologies in the study area. When no specific rock type extensively covered a catchment basin, the sample was classified as other (Oth). Table 2 shows the median of the elemental concentration of stream sediment classified to the parent lithology.

An analysis of variance (ANOVA) and Bonferroni's multiple comparison test were applied to the geochemical datasets to objectively examine the influence of the lithology on the chemical compositions of stream sediments (Hochberg and Tamhane 1987; Nagata and Yoshida 1997; Miller and Miller 2010). These statistical tests assume that the data follow a normal distribution. However, the data distribution for each elemental concentration followed neither a normal nor a log-normal distribution when the Shapiro–Wilk test was applied to our dataset (Shapiro and Wilk 1965). In such cases, approaching data symmetry is the second-best alternative. Accordingly, data distribution with a skewness closer to zero, which indicated a symmetrical distribution, was used for the statistical tests (Ohta *et al.* 2005). The concentrations of the 42 elements were transformed into common logarithms; those of Al_2O_3, Ga, Rb, Er, Tm and Yb were unchanged for the statistical tests (Table 3).

The null hypothesis of the one-way ANOVA is that no significant difference exists in elemental concentrations among the seven categories: Sed, Acc, Gr, Fv, Mv, Py and Mp. Table 3 shows the significance probability (P) and effect size (η^2) of the one-way ANOVA analysis. The effect size (η^2) was used to form a plausible estimation of the P value irrespective of sample number because a statistical test with a large amount of input data is highly sensitive to very small differences (Richardson 2011; Fritz *et al.*

2012). In this study, we concluded that the null hypothesis was rejected for the elemental concentration data with $P < 0.01$ and $\eta^2 > 0.14$. Table 4 shows that the null hypothesis was rejected for all elements except Cu, Zn, As, Cd, Sn, Hg and Pb, which indicates that most elements are significantly influenced by the parent lithology. The concentrations of these elements were influenced by metalliferous deposits rather than by a specific lithology, as described above.

Next, we examined whether the parent lithology significantly affects the concentrations of elements, except Cu, Zn, As, Cd, Sn, Hg and Pb, in the stream sediment using Bonferroni's test at the 0.01 confidence level (Hochberg and Tamhane 1987). Table 4 presents the results of direct pairwise comparisons. The elements are grouped as felsic element enrichment in rhyolitic–dacitic and granitic rocks, mafic element enrichment in andesite–basaltic and gabbroic rocks, and sulfurous elements concentrated in a sulfide deposit for descriptive purposes. The stream sediments derived from Acc are highly rich in K_2O, Li, Rb, Cs, Ba and Tl compared with those from Mv, Py and Mp. Moreover, they are poor in Na_2O, P_2O_5, CaO, Sc, V, TiO_2, MnO, $Fe_2O_3^T$, Sr, Nb, Eu and HREEs (Gd–Lu) compared with those from Fv, Mv, Py and Gr. The sediments derived from Mv and Py are significantly richer in MgO, CaO, Sc, TiO_2, V, $Fe_2O_3^T$, Co and Sr than those from Sed, Acc and Gr, and are significantly poorer in alkali metal elements Nb, Ba, Tl, Th and U than those from Sed, Acc and Gr. Although the stream sediments derived from Py have chemical compositions similar to those from Mv, they are significantly poorer in Cr and Ni. The stream sediments derived from Gr show greater enrichment in alkali metal elements Be, Ga, REEs, Ta, Th and U than the other samples, although they are poor in Cr and Ni. The sediments derived from Mp are highly enriched in Cr and Ni, and are poorer than the other samples in K_2O, Nb and Ta. The samples derived from Fv have few systematic geochemical features compared with those of other samples. These results are consistent with the visual interpretations of the spatial distribution patterns of elemental concentrations (Fig. 4a–l).

In terms of the sulfurous elements, the samples derived from Fv are significantly more enriched in Sb than those from Mv, Py and Gr; the sediments from Gr are richer in Bi than those from Sed and Mv; and the sediments from Mv and Py are richer in Mo and Bi than those from Sed and Gr. The hydrothermal activity associated with Mv, Fv and Gr may have enhanced the concentrations of the sulfurous elements. However, the spatial distributions of the major hydrothermal-type deposits do not appear to be comparable to areas in which sulfurous elements are highly concentrated (Cd and Sb: Fig. 4).

Table 2. *Median elemental concentration of stream sediment for each parent lithology*

	Unit	Sed n = 41	Acc n = 83	Fv n = 9	Mv n = 48	Py n = 56	Gr n = 29	Mp n = 13	Oth n = 111
Na_2O	wt%	1.76	1.76	1.86	2.23	2.45	3.03	2.68	2.30
MgO	wt%	1.72	1.65	1.53	3.43	3.17	2.06	2.63	2.64
Al_2O_3	wt%	9.88	12.30	12.00	12.70	13.90	14.80	9.31	12.60
P_2O_5	wt%	0.131	0.114	0.157	0.159	0.164	0.180	0.167	0.160
K_2O	wt%	1.78	2.10	1.73	1.18	1.27	2.17	1.44	1.68
CaO	wt%	0.87	0.72	2.05	3.47	3.35	2.56	2.11	2.55
TiO_2	wt%	0.67	0.57	1.08	1.10	1.12	0.64	0.64	0.84
MnO	wt%	0.076	0.096	0.123	0.152	0.159	0.116	0.101	0.140
$Fe_2O_3^T$	wt%	4.56	4.67	6.10	7.78	7.70	5.25	4.98	6.31
Li	mg kg^{-1}	41	43	40	26	28	49	30	36
Be	mg kg^{-1}	1.5	1.7	1.4	1.2	1.3	2.0	1.2	1.5
Sc	mg kg^{-1}	9.0	10	14	17	18	10	11	15
V	mg kg^{-1}	77	88	140	170	170	85	88	130
Cr	mg kg^{-1}	61	53	47	78	30	31	110	50
Co	mg kg^{-1}	13	12	12	20	19	12	15	17
Ni	mg kg^{-1}	24	22	13	26	10	12	50	20
Cu	mg kg^{-1}	24	31	28	27	26	22	41	30
Zn	mg kg^{-1}	110	97	130	120	130	110	130	120
Ga	mg kg^{-1}	14	17	17	19	18	22	14	17
As	mg kg^{-1}	5.4	8.7	8.0	6.0	5.9	10	7.0	9.0
Rb	mg kg^{-1}	75	90	75	41	52	100	65	70
Sr	mg kg^{-1}	100	110	190	230	220	160	130	170
Y	mg kg^{-1}	13	12	17	17	22	26	15	21
Nb	mg kg^{-1}	9.1	8.2	8.4	11	10	12	7.2	9.2
Mo	mg kg^{-1}	0.81	1.1	1.5	1.3	1.4	0.75	0.88	1.3
Cd	mg kg^{-1}	0.13	0.11	0.15	0.13	0.14	0.12	0.12	0.13
Sn	mg kg^{-1}	2.3	2.6	2.9	2.4	2.3	4.0	2.5	2.6
Sb	mg kg^{-1}	0.55	0.81	1.2	0.46	0.47	0.38	0.61	0.63
Cs	mg kg^{-1}	4.0	5.5	5.5	2.8	3.5	5.0	4.0	4.9
Ba	mg kg^{-1}	380	430	370	340	360	380	280	380
La	mg kg^{-1}	20	19	18	18	18	24	16	19
Ce	mg kg^{-1}	35	38	36	30	36	44	29	37
Pr	mg kg^{-1}	4.4	4.5	4.1	4.0	4.3	5.5	3.7	4.6
Nd	mg kg^{-1}	17	18	16	17	18	22	15	19
Sm	mg kg^{-1}	3.3	3.5	3.4	3.5	3.8	5.0	3.2	3.9
Eu	mg kg^{-1}	0.73	0.78	0.90	0.98	0.99	1.1	0.84	0.99
Gd	mg kg^{-1}	2.9	3.0	3.2	3.2	3.7	4.7	3.0	3.8
Tb	mg kg^{-1}	0.43	0.48	0.58	0.55	0.68	0.82	0.52	0.66
Dy	mg kg^{-1}	2.2	2.2	3.0	2.8	3.7	4.2	2.6	3.4
Ho	mg kg^{-1}	0.42	0.41	0.57	0.55	0.74	0.83	0.51	0.67
Er	mg kg^{-1}	1.2	1.2	1.7	1.6	2.2	2.3	1.5	2.0
Tm	mg kg^{-1}	0.18	0.18	0.25	0.26	0.35	0.30	0.23	0.32
Yb	mg kg^{-1}	1.2	1.1	1.5	1.6	2.3	1.7	1.4	2.0
Lu	mg kg^{-1}	0.18	0.16	0.22	0.25	0.33	0.24	0.20	0.29
Ta	mg kg^{-1}	0.64	0.73	1.0	0.69	0.82	1.1	0.35	0.76
Hg	mg kg^{-1}	0.061	0.070	0.070	0.065	0.040	0.030	0.070	0.060
Tl	mg kg^{-1}	0.45	0.57	0.59	0.34	0.42	0.54	0.36	0.50
Pb	mg kg^{-1}	21	23	24	20	21	28	21	23
Bi	mg kg^{-1}	0.20	0.31	0.34	0.16	0.27	0.37	0.20	0.29
Th	mg kg^{-1}	5.9	6.5	6.1	4.7	6.1	8.3	5.2	6.7
U	mg kg^{-1}	1.4	1.4	1.7	1.0	1.6	2.5	1.3	1.7

Table 3. *Skewness of the unchanged and the log-transformed data, data transformation, and significance probability (P) and size effect (η^2) of one-way ANOVA*

	Skewness		Data transformation	One-way ANOVA	
	Unchanged	Log-transformed		P	η^2
Na_2O	0.5	−1.0	Unchanged	<0.01	0.29
MgO	1.2	0.0	Log-transformed	<0.01	0.34
Al_2O_3	0.5	−1.1	Unchanged	<0.01	0.23
P_2O_5	2.9	0.6	Log-transformed	<0.01	0.15
K_2O	0.7	−0.6	Log-transformed	<0.01	0.45
CaO	3.7	−0.6	Log-transformed	<0.01	0.52
TiO_2	2.9	0.8	Log-transformed	<0.01	0.37
MnO	1.6	−0.4	Log-transformed	<0.01	0.22
Fe_2O_3	1.1	0.0	Log-transformed	<0.01	0.35
Li	1.3	0.0	Log-transformed	<0.01	0.31
Be	3.3	0.9	Log-transformed	<0.01	0.41
Sc	1.8	−0.1	Log-transformed	<0.01	0.36
V	1.5	0.2	Log-transformed	<0.01	0.39
Cr	3.3	0.2	Log-transformed	<0.01	0.27
Co	1.9	0.1	Log-transformed	<0.01	0.24
Ni	5.6	0.3	Log-transformed	<0.01	0.25
Cu	12	0.9	Log-transformed	<0.01	0.06
Zn	5.8	1.0	Log-transformed	0.013	0.06
Ga	0.5	−0.7	Unchanged	<0.01	0.37
As	17	1.1	Log-transformed	<0.01	0.09
Rb	0.3	−0.9	Unchanged	<0.01	0.48
Sr	1.2	0.1	Log-transformed	<0.01	0.39
Y	0.6	−0.4	Log-transformed	<0.01	0.31
Nb	3.7	1.1	Log-transformed	<0.01	0.18
Mo	5.5	0.4	Log-transformed	<0.01	0.15
Cd	9.8	1.1	Log-transformation	0.39	0.02
Sn	12	2.7	Log-transformed	0.02	0.05
Sb	12	1.4	Log-transformed	<0.01	0.26
Cs	1.0	−0.9	Log-transformed	<0.01	0.24
Ba	0.7	−0.7	Unchanged	<0.01	0.16
La	7.6	2.6	Log-transformed	<0.01	0.24
Ce	9.2	2.7	Log-transformed	<0.01	0.22
Pr	7.2	2.5	Log-transformed	<0.01	0.24
Nd	7.0	2.2	Log-transformed	<0.01	0.23
Sm	6.1	1.6	Log-transformed	<0.01	0.24
Eu	0.6	−0.5	Log-transformed	<0.01	0.23
Gd	4.3	0.7	Log-transformed	<0.01	0.23
Tb	1.7	−0.1	Log-transformed	<0.01	0.25
Dy	1.0	−0.2	Log-transformed	<0.01	0.28
Ho	0.5	−0.4	Log-transformed	<0.01	0.29
Er	0.4	−0.5	Unchanged	<0.01	0.28
Tm	0.4	−0.5	Unchanged	<0.01	0.29
Yb	0.4	−0.4	Unchanged	<0.01	0.30
Lu	0.4	−0.4	Unchanged	<0.01	0.31
Ta	2.6	0.0	Log-transformed	<0.01	0.20
Hg	17	0.8	Log-transformed	<0.01	0.08
Tl	1.6	−1.0	Log-transformed	<0.01	0.28
Pb	15	3.0	Log-transformed	<0.01	0.10
Bi	13	0.9	Log-transformed	<0.01	0.21
Th	6.3	1.6	Log-transformed	<0.01	0.25
U	2.9	0.5	Log-transformed	<0.01	0.26
Cr/Ti	4.1	−0.2	Log-transformed	<0.01	0.40
La/Yb	6.5	1.4	Log-transformed	<0.01	0.36

Bold fonts in column *P* indicate that the null hypothesis is rejected for $P = 0.01$. Bold fonts in the η^2 column indicate that a factor with $\eta^2 > 0.14$ has a significant effect on the elemental concentrations of the sediments.

Table 4. *Results of the Bonferroni multiple comparison procedure at the 0.01 confidence interval*

A*	B*	Felsic elements A > B	Felsic elements A < B	Mafic elements A > B	Mafic elements A < B	Sulfophile elements A > B	Sulfophile elements A < B
Sed	Acc		Cs		Al		As, Bi
Sed	Fv				Ca, Sc		
Sed	Mv	K, Li, Rb, Tl, Th, U	Eu		Mg, Al, Ca, Ti, Mn, Fe, Sc, V, Co, Ga, Sr		Mo
Sed	Py	K, Li, Rb	Na, Eu, Ta	Cr, Ni	Mg, Al, Ca, Ti, Mn, Fe, Sc, V, Ga, Sr, Y, Tb–Lu		Mo
Sed	Gr		Na, K, Rb, Nb, La–Eu, Ta, Th, U	Cr, Ni	Al, Ca, Ga, Sr, Y, Gd–Lu	Sb	Bi
Sed	Mp		Na		Mg, Cr		
Acc	Sed	Cs		Al		As, Bi	
Acc	Fv				Ca, Ti		
Acc	Mv	K, Li, Rb, Cs, Ba, Tl, Th	Na, Nb, Eu		Mg, P, Ca, Ti, Mn, Fe, Sc, V, Co, Ga, Sr, Y, Ho–Lu,	Sb, Pb, Bi	
Acc	Py	K, Li, Rb, Cs, Ba, Tl	Na, Nb, Eu	Cr, Ni	Mg, P, Ca, Ti, Mn, Fe, Sc, V, Co, Sr, Y, Gd–Lu	As, Sb, Pb	
Acc	Gr		Na, Nb, La–Eu, Ta, Th, U	Cr, Ni	Al, P, Ca, Ga, Sr, Y, Gd–Lu	Mo, Sb	
Acc	Mp	K, Rb, Ba, Ta, Tl	Na		Ca, Cr, Ni		
Mv	Sed	Eu	K, Li, Rb, Tl, Th, U	Mg, Al, Ca, Ti, Mn, Fe, Sc, V, Co, Ga, Sr		Mo	
Mv	Acc	Na, Nb, Eu	K, Li, Rb, Cs, Ba, Tl, Th	Mg, P, Ca, Ti, Mn, Fe, Sc, V, Co, Ga, Sr, Y, Ho–Lu			Sb, Pb, Bi
Mv	Fv		K, Rb, Cs, Tl				Sb
Mv	Py		U	Cr, Ni	Ho–Lu		Bi
Mv	Gr		Na, K, Li, Rb, Cs, La–Sm, Ta, Tl, Th, U	Mg, Ti, Fe, Sc, V, Cr, Co, Ni	Ga, Y, Gd–Ho	Mo	Bi
Mv	Mp	Nb, Ta		Ti, Fe, V, Ga, Sr	Ni		
Fv	Sed			Ca, Sc			
Fv	Acc			Ca, Ti			
Fv	Mv	K, Rb, Cs, Tl				Sb	
Fv	Py					Sb	
Fv	Gr		Na, La–Sm, Th, U		Ga, Gd	Sb	
Fv	Mp	Ta			Cr, Ni		
Py	Sed	Na, Eu, Ta	K, Li, Rb	Mg, Al, Ca, Ti, Mn, Fe, Sc, V, Ga, Sr, Y, Tb–Lu	Cr, Ni	Mo	
Py	Acc	Na, Nb, Eu	K, Li, Rb, Cs, Ba, Tl	Mg, P, Ca, Ti, Mn, Fe, Sc, V, Co, Sr, Y, Gd–Lu	Cr, Ni		As, Sb, Pb
Py	Fv						Sb

(Continued)

Table 4. *Continued.*

		Felsic elements		Mafic elements		Sulfophile elements	
A*	B*	A > B	A < B	A > B	A < B	A > B	A < B
Py	Mv	U		Ho–Lu	Cr, Ni	Bi	
Py	Gr		Na, K, Li, Rb, Cs, La–Sm, Tl, Th, U	Mg, Ti, Fe, Sc, V, Co	Ga, Gd	Mo	
Py	Mp	Nb, Eu, Ta		Al, Ti, Fe, Sc, V, Ga, Sr, Tm–Lu	Cr, Ni		
Gr	Sed	Na, K, Rb, Nb, La–Eu, Ta, Th, U		Al, Ca, Ga, Sr, Y, Gd–Lu	Cr, Ni	Bi	Sb
Gr	Acc	Na, Nb, La–Eu, Ta, Th, U		Al, P, Ca, Ga, Sr, Y, Gd–Lu	Cr, Ni		Mo, Sb
Gr	Fv	Na, La–Sm, Th, U		Ga, Gd			Sb
Gr	Mv	Na, K, Li, Rb, Cs, La–Sm, Ta, Tl, Th, U		Ga, Y, Gd–Ho	Mg, Ti, Fe, Sc, V, Cr, Co, Ni	Bi	Mo
Gr	Py	Na, K, Li, Rb, Cs, La–Sm, Tl, Th, U		Ga, Gd	Mg, Ti, Fe, Sc, V, Co		Mo
Gr	Mp	K, Rb, Nb, Ba, La–Eu, Ta, Th, U		Al, Ga, Gd, Tb, Dy	Cr, Ni		
Mp	Sed	Na		Mg, Cr			
Mp	Acc	Na	K, Rb, Ba, Ta, Tl	Ca, Cr, Ni			
Mp	Fv		Ta	Cr, Ni			
Mp	Mv		Nb, Ta	Ni	Ti, Fe, V, Ga, Sr		
Mp	Py		Nb, Eu, Ta	Cr, Ni	Al, Ti, Fe, Sc, V, Ga, Sr, Tm–Lu		
Mp	Gr		K, Rb, Nb, Ba, La–Eu, Ta, Th, U	Cr, Ni	Al, Ga, Gd, Tb, Dy		

*Abbreviations of Sed, Acc, Fv, Mv, Py, Gr, and Mp are the same as in Table 2. A > B (or A < B) indicates that elemental concentrations of sediments originated from A are significantly higher (or lower) than those of sediments originated from B.

Differences in elemental concentrations in coastal sea sediments by region and particle sizes

We examined the supplementary process of clastic deposits from terrestrial to marine environments based on the chemical compositions of stream and marine sediments. Those of the marine sediments were influenced by various factors: (1) particle transport from the land to the sea; (2) the dilution effect of quartz and biogenic calcareous materials; (3) increases in the concentrations of alkali metal ions in silty and clayey sediments; (4) the transportation of sediments by gravity flow or oceanic currents; (5) early diagenetic processes in the deep sea; (6) denudation or resedimentation of basement rocks; and (7) contamination from human activity (Ohta et al. 2010; Ohta et al. 2017a). Thus, we used a step-by-step approach.

The chemical compositions of marine sediments differ between regions because they originate from adjacent terrestrial materials. Some marine sediments are formed by the denudation or resedimentation of basement rocks, which may belong to any of the lithologies in the adjacent terrestrial area. This factor is referred to herein as a regional difference. Additionally, the elemental concentrations in the marine sediments vary with particle size, caused by the dilution of large amounts of biogenic calcareous materials in the coarse grains and by an increase in clay minerals in the fine grains. This factor is referred to herein as the particle size effect. These two effects are common in all marine sediments.

The marine sediments were grouped as those from the Tsushima Strait and Goto Nada, those from the Suo Nada, those from the Amakusa Nada including the Ariake and Yatsuhiro seas, those from the Hyuga Nada, and those from the Osumi Strait; the zonal classification is given in Figure 3. They were further classified into coarse sediments, fine sands and silt–clay. The median concentration of elements in the marine sediments grouped by each region and particle size were also calculated (Table 5). The silty sediments of Kagoshima Bay were separated from the silts and clays of the Osumi Strait because there are clear differences in their chemical compositions.

A two-way ANOVA test was applied to identify the factors that significantly change the chemical compositions of marine sediments. Table 6 presents the variance ratios (F), probabilities (P) and effect size (η^2) owing to the regional effect (factor A), particle size (factor B) and the interaction effect (factor A × B). The interaction effect (A × B) refers to the effect that one factor has on another (Miller and Miller 2010). When the estimated probability (P) was lower than 0.01 and the effect size (η^2) was larger than 0.14, we concluded that the factor made a significant difference in the chemical composition (Table 4). The ANOVA results suggest that the chemical compositions of 29 elements in marine sediments change significantly between different regions. The regional difference effect was not significant for K_2O, Ni, Rb, Sb, Cs, Ba, Hg, Bi and U, whose concentrations were determined simply by particle size (factor B). The results suggest that these elements cannot be used to understand the influence of terrestrial materials on coastal sea sediments. The concentrations of 17 elements in the marine sediments are strongly controlled by the particle effect; however, 15 elements, such as Cr and La, are also significantly influenced by the regional effect. Therefore, it is more effective to subdivide marine sediments by region and particle size when evaluating their geochemical similarities and differences.

Identification of probable source materials for marine sediments using cluster analysis

We identified the probable source materials for the marine sediments by using cluster analysis with stream sediments grouped by the dominant lithology. The compositional data of the marine and stream sediments need to be standardized or transformed appropriately to examine the similarity and differences in their chemical compositions when using cluster analysis (Templ *et al.* 2008). Although various transformation procedures for compositional

data have been proposed by Aitchison (1982), Aitchison *et al.* (2000) and Pawlowsky-Glahn and Egozcue (2006), Ohta *et al.* (2017a) proposed a log transformation of the enrichment factor (EF) for geochemical datasets using following equation:

$$\log EF = alr(C)_{sample} - alr(C)_{UCC} \quad (1)$$

where EF is the enrichment factor for the upper continental crust (UCC) components (Taylor and McLennan 1995), and alr(C) represents the additive log-ratio transformation log $([C]/[Al_2O_3])$. The logarithmic EF is a normalized value for all elements with respective different ranges in concentration and is also effective in removing the dilution effect (Ohta *et al.* 2013).

Figure 5 shows dendrograms expressing the distances between datasets determined using the Ward method (Ward 1963). The dataset excludes sea salt (Na_2O), biogenic carbonate materials (CaO and Sr), and elements of K_2O, Ni, Rb, Cs, Ba and U that are not significantly impacted by the regional effect (Table 6). Furthermore, heavy metals related to mining and anthropogenic activities – including Cu, Zn, As, Mo, Cd, Sn, Sb, Hg, Pb and Bi – were also excluded from the analysis because their concentrations do not reflect the geochemistry of the parent lithology (Table 4). Figure 5a uses a dataset of the above 31 elements. Figure 5b uses a dataset of immobile elements, including Sc, TiO_2, Cr, Nb, Y, Ln, Ta and Th, which are not strongly influenced by the weathering process. Coarse sediments and fine sands in the Osumi Strait and silt in Kagoshima Bay, which are commonly observed features in both Figure 4a and b, were plotted in the same group as stream sediments derived from Py. The fine sands in the Hyuga and Amakusa nadas were clustered with stream sediments derived from Acc. The silt–clay sediments in the Amakusa and Hyuga nadas and the Osumi Strait were plotted in the same group and were related, remotely, to stream sediments. The marine sediments in the Tsushima Strait and Goto and Suo nadas were also remotely related to stream sediments. Unfortunately, not all of the source materials of the marine sediments could be identified using cluster analysis.

Discrimination study of stream and marine sediments using Cr/Ti and La/Yb ratios

Finally, we visualized the particle transfer process in marine environments using the chemical compositions of the sediments. The concentration ratios of immobile elements, such as Ti/Zr, La/Sc, Th/Sc and Cr/Th, were used as indices of the source rock composition of sandstone and mudstone (e.g. Roser 2000). Simultaneously, the Ti/Nb, Cr/Th

Table 5. *Median elemental concentrations of marine sediments classified by region and particle size*

	Tsushima Strait		Suo Nada			Amakusa Nada			Hyuga Nada			Osumi Strait			
	C. sed.*	F. sand*	C. sed.*	F. sand*	Silt	C. sed.*	F. sand*	Silt	C. sed.*	F. sand*	Silt	C. sed.*	F. sand*	Silt	Silt in Bay*
n	21	25	9	7	8	11	37	11	13	74	35	137	137	64	17
wt%															
Na_2O	1.43	1.79	2.94	2.82	3.91	2.14	2.64	3.07	2.41	2.47	3.05	2.32	2.66	3.43	3.68
MgO	1.46	2.23	3.04	2.22	3.10	3.23	2.41	3.08	3.03	2.61	2.32	4.00	2.50	2.54	3.52
Al_2O_3	4.39	5.34	9.09	7.44	8.16	7.39	10.38	9.63	9.64	10.44	10.78	9.94	10.46	11.44	11.34
P_2O_5	0.054	0.078	0.099	0.085	0.139	0.108	0.080	0.126	0.079	0.091	0.109	0.141	0.118	0.123	0.177
K_2O	1.51	1.35	1.95	2.12	1.63	0.796	1.39	1.56	1.58	1.82	1.97	0.955	1.44	1.99	1.47
CaO	12.5	13.5	4.86	1.58	1.72	14.6	8.47	8.24	4.84	4.64	5.16	8.16	9.48	7.58	5.94
TiO_2	0.093	0.174	0.437	0.384	0.538	0.387	0.376	0.406	0.589	0.500	0.483	0.870	0.514	0.478	0.672
MnO	0.019	0.022	0.119	0.066	0.088	0.092	0.056	0.052	0.102	0.066	0.047	0.144	0.089	0.064	0.122
Fe_2O_3	1.29	1.84	5.09	3.54	4.27	4.10	3.84	3.58	5.21	5.03	4.16	7.61	4.50	4.38	5.24
$mg\ kg^{-1}$															
Li	16	20	26	61	110	28	41	77	30	50	52	21	26	44	27
Be	0.42	0.66	1.2	1.5	1.6	0.82	1.2	1.2	1.1	1.4	1.5	0.84	1.0	1.4	1.1
Sc	2.2	3.7	9.1	6.4	8.9	9.1	9.3	9.8	13	11	11	19	12	11	15
V	19	24	55	46	74	55	56	60	94	89	90	140	85	92	120
Cr	17	27	38	40	62	29	32	60	36	42	47	24	22	47	16
Co	2.4	3.7	14	9.7	11	9.3	6.9	8.2	11	11	8.0	14	8.3	8.9	11
Ni	8.4	11	15	17	24	12	13	24	14	19	26	10	11	27	7.3
Cu	3.2	5	9.5	13	28	7.1	7.1	16	6.9	14	22	8.2	9.5	28	19
Zn	17	27	77	83	130	56	58	78	70	79	78	87	66	78	98
Ga	4.6	6.1	14	14	16	9.8	13	13	13	14	15	13	12	14	15
As	6.5	3.7	6.8	5.8	9.2	7.1	5.8	4.8	7.7	7.2	6.1	4.2	4.9	6.2	6.9
Rb	53	50	41	87	43	28	51	53	57	65	69	30	51	84	22
Sr	650	680	380	130	130	740	420	400	320	260	250	360	450	330	230
Y	4.8	7.1	13	13	13	11	12	13	11	11	14	16	16	16	18
Nb	1.8	2.7	6.2	6.1	8.8	5.1	5.4	6.0	4.8	6.0	6.8	4.0	4.2	6.3	5.3
Mo	0.26	0.35	0.62	0.85	1.5	0.61	0.62	0.98	0.44	0.66	0.87	0.65	0.90	1.1	1.6
Cd	0.037	0.046	0.071	0.084	0.24	0.057	0.055	0.084	0.046	0.052	0.11	0.072	0.075	0.10	0.11
Sn	0.35	0.56	1.5	2.4	3.5	1.0	1.2	1.8	1.6	1.9	2.5	1.1	1.3	2.1	2.0
Sb	0.31	0.33	0.34	0.41	0.48	0.32	0.50	0.45	0.38	0.49	0.85	0.26	0.40	0.74	0.46
Cs	1.0	1.3	2.1	4.0	4.4	1.5	2.9	3.6	2.0	3.6	4.8	1.3	2.3	5.3	2.4
Ba	290	240	300	260	210	170	260	190	300	310	340	160	230	420	210

(*Continued*)

Table 5. *Continued.*

	Tsushima Strait		Suo Nada			Amakusa Nada			Hyuga Nada			Osumi Strait			
	C. sed.*	F. sand*	C. sed.*	F. sand*	Silt	C. sed.*	F. sand*	Silt	C. sed.*	F. sand*	Silt	C. sed.*	F. sand*	Silt	Silt in Bay*
La	5.1	8.9	13	13	16	11	14	17	12	16	16	10	12	19	11
Ce	12	20	28	28	27	22	29	35	25	33	34	22	26	38	22
Pr	1.2	2.0	3.1	3.1	3.8	2.4	3.2	3.8	3.0	3.7	3.9	2.7	3.1	4.3	2.8
Nd	4.6	8.0	13	13	15	10	12	15	13	15	15	12	13	17	12
Sm	0.84	1.5	2.7	2.7	3.1	2.1	2.6	3.1	2.8	3.0	3.2	2.9	3.0	3.5	2.8
Eu	0.30	0.43	0.76	0.62	0.64	0.73	0.71	0.73	0.77	0.69	0.69	0.78	0.75	0.76	0.76
Gd	0.85	1.3	2.6	2.4	2.8	2.1	2.4	2.9	2.5	2.6	2.9	2.9	2.9	3.1	3.0
Tb	0.14	0.22	0.45	0.41	0.49	0.35	0.41	0.49	0.39	0.43	0.47	0.50	0.50	0.51	0.56
Dy	0.70	1.1	2.3	2.1	2.5	1.8	2.1	2.4	2.0	2.1	2.3	2.7	2.6	2.6	3.0
Ho	0.13	0.21	0.44	0.41	0.46	0.35	0.39	0.44	0.38	0.39	0.45	0.53	0.52	0.50	0.63
Er	0.41	0.60	1.3	1.3	1.4	1.1	1.2	1.3	1.1	1.1	1.3	1.6	1.6	1.5	1.9
Tm	0.061	0.092	0.21	0.21	0.22	0.17	0.19	0.21	0.17	0.17	0.21	0.25	0.25	0.24	0.32
Yb	0.39	0.56	1.3	1.3	1.3	1.1	1.2	1.2	1.1	1.1	1.3	1.5	1.5	1.5	2.0
Lu	0.054	0.080	0.19	0.19	0.20	0.16	0.16	0.18	0.16	0.15	0.19	0.23	0.23	0.22	0.29
Ta	0.13	0.24	0.58	0.60	0.85	0.46	0.49	0.61	0.36	0.54	0.61	0.35	0.40	0.60	0.56
Hg	0.017	0.033	0.043	0.080	0.14	0.025	0.052	0.12	0.018	0.073	0.14	0.015	0.032	0.12	0.069
Tl	0.28	0.29	0.48	0.60	0.59	0.17	0.35	0.47	0.38	0.44	0.51	0.18	0.29	0.49	0.42
Pb	11	13	19	23	31	13	16	20	16	18	21	12	16	20	19
Bi	0.08	0.11	0.22	0.3	0.69	0.13	0.15	0.31	0.18	0.24	0.42	0.13	0.19	0.33	0.44
Th	1.4	2.5	3.8	5.9	6.2	2.4	4.7	5.7	3.4	5.5	6.4	2.6	4.2	6.9	4.1
U	0.53	0.94	1.0	1.6	1.6	0.68	1.2	1.6	0.89	1.1	1.7	0.72	1.1	1.3	1.2
Cr/Ti	0.034	0.027	0.013	0.018	0.019	0.013	0.014	0.022	0.009	0.014	0.015	0.004	0.006	0.016	0.004
La/Yb	13.2	14.0	10.7	10.8	11.4	11.2	11.7	13.6	11.6	14.8	12.8	6.6	8.2	12.6	5.6

*C. sed, F. sand and silt in Bay indicate coarse sediments, fine sand and silt collected from Kagoshima Bay, respectively.

Table 6. Skewness of the unchanged and the log-transformed data, data transformation, and significance probability (P) and size effect (η^2) of two-way ANOVA

	Skewness		Data transformation	P			η^2			Major factor
	Unchanged	Log-transformed		Region	Particle size	Interaction effect	Region	Particle size	Interaction effect	
MgO	1.9	0.2	Log-transformed	**<0.01**	**<0.01**	**<0.01**	**0.15**	0.12	0.12	Region
Al$_2$O$_3$	−1.0	−2.9	Unchanged	**<0.01**	**<0.01**	0.06	**0.37**	0.02	0.02	Region
P$_2$O$_5$	2.4	0.1	Log-transformed	**<0.01**	**<0.01**	**<0.01**	**0.22**	0.03	0.07	Region
K$_2$O	−0.5	−1.6	Unchanged	**<0.01**	**<0.01**	**<0.01**	0.12	**0.26**	0.08	Particle size
CaO	2.2	−0.3	Log-transformed	**<0.01**	0.02	**<0.01**	**0.28**	0.01	0.05	Region
TiO$_2$	2.1	−0.5	Log-transformed	**<0.01**	**<0.01**	**<0.01**	**0.40**	0.04	0.06	Region
MnO	3.5	0.2	Log-transformed	**<0.01**	**<0.01**	**<0.01**	**0.36**	0.05	0.03	Region
Fe$_2$O$_3$	2.9	−0.5	Log-transformed	**<0.01**	**<0.01**	**<0.01**	**0.36**	0.06	0.07	Region
Li	1.4	−0.3	Log-transformed	**<0.01**	**<0.01**	**<0.01**	**0.28**	**0.26**	0.03	Both
Be	−0.1	−1.6	Unchanged	**<0.01**	**<0.01**	**<0.01**	**0.27**	**0.20**	0.03	Region
Sc	2.4	−0.4	Log-transformed	**<0.01**	**<0.01**	**<0.01**	**0.46**	0.05	0.06	Region
V	4.4	−0.3	Log-transformed	**<0.01**	**<0.01**	**<0.01**	**0.43**	0.05	0.03	Region
Cr	0.6	−0.1	Log-transformed	**<0.01**	**<0.01**	**<0.01**	**0.21**	**0.22**	0.09	Both
Co	2.0	−0.4	Log-transformed	**<0.01**	**<0.01**	**<0.01**	**0.36**	0.05	0.07	Region
Ni	1.9	0.5	Log-transformed	**<0.01**	**<0.01**	**<0.01**	0.11	**0.41**	0.07	Particle size
Cu	12.5	0.3	Log-transformed	**<0.01**	**<0.01**	**<0.01**	**0.14**	**0.35**	0.04	Particle size
Zn	1.6	−1.1	Log-transformed	**<0.01**	**<0.01**	**<0.01**	**0.43**	0.01	0.08	Region
Ga	−1.3	−2.9	Unchanged	**<0.01**	**<0.01**	**<0.01**	**0.42**	0.03	0.02	Region
As	1.7	−1.0	Log-transformed	**<0.01**	**<0.01**	**<0.01**	0.08	0.03	0.04	Region
Rb	0.0	−1.3	Unchanged	**<0.01**	**<0.01**	**<0.01**	0.04	**0.30**	0.13	Particle size
Sr	3.4	0.6	Log-transformed	**<0.01**	**<0.01**	**<0.01**	**0.18**	0.03	0.04	Region
Y	0.5	−0.8	Unchanged	**<0.01**	0.93	0.05	**0.33**	0.00	0.02	Region
Nb	1.8	−1.6	Log-transformed	**<0.01**	**<0.01**	**<0.01**	**0.21**	0.10	0.03	Region
Mo	4.9	0.5	Unchanged	**<0.01**	**<0.01**	0.50	**0.17**	**0.15**	0.01	Both
Cd	3.1	0.2	Log-transformed	**<0.01**	**<0.01**	**<0.01**	**0.14**	**0.18**	0.05	Particle size
Sn	1.5	−1.0	Log-transformed	**<0.01**	**<0.01**	**<0.01**	**0.35**	**0.18**	0.02	Region
Sb	1.7	−0.3	Log-transformed	**<0.01**	**<0.01**	**<0.01**	0.05	**0.31**	0.05	Particle size
Cs	0.3	−0.9	Unchanged	**<0.01**	**<0.01**	**<0.01**	0.09	**0.49**	0.06	Particle size
Ba	0.6	−1.1	Unchanged	**<0.01**	**<0.01**	**<0.01**	0.06	**0.29**	**0.18**	Particle size
La	0.0	−1.2	Unchanged	**<0.01**	**<0.01**	**<0.01**	**0.14**	**0.32**	0.07	Particle size
Ce	−0.1	−1.2	Unchanged	**<0.01**	**<0.01**	**<0.01**	**0.14**	**0.27**	0.07	Particle size
Pr	−0.3	−1.6	Unchanged	**<0.01**	**<0.01**	**<0.01**	**0.20**	**0.25**	0.05	Particle size
Nd	−0.5	−1.8	Unchanged	**<0.01**	**<0.01**	**<0.01**	**0.24**	**0.18**	0.04	Region
Sm	−0.7	−1.9	Unchanged	**<0.01**	**<0.01**	0.01	**0.30**	0.08	0.02	Region

(Continued)

Table 6. *Continued.*

	Skewness		Data transformation	P			η^2			Major factor
	Unchanged	Log-transformed		Region	Particle size	Interaction effect	Region	Particle size	Interaction effect	
Eu	−0.9	−2.1	Unchanged	**<0.01**	0.94	**<0.01**	**0.35**	0.00	0.03	Region
Gd	−0.4	−1.7	Unchanged	**<0.01**	**<0.01**	0.05	**0.32**	0.02	0.02	Region
Tb	−0.1	−1.4	Unchanged	**<0.01**	0.09	0.04	**0.32**	0.01	0.02	Region
Dy	0.2	−1.2	Unchanged	**<0.01**	0.56	0.02	**0.32**	0.00	0.02	Region
Ho	0.4	−1.1	Unchanged	**<0.01**	0.71	0.02	**0.32**	0.00	0.02	Region
Er	0.4	−1.0	Unchanged	**<0.01**	0.85	0.03	**0.33**	0.00	0.02	Region
Tm	0.4	−1.0	Unchanged	**<0.01**	0.96	0.03	**0.32**	0.00	0.02	Region
Yb	0.5	−0.9	Unchanged	**<0.01**	0.97	0.07	**0.32**	0.00	0.02	Region
Lu	0.5	−0.9	Unchanged	**<0.01**	0.98	0.03	**0.32**	0.00	0.02	Region
Ta	0.3	−3.1	Unchanged	**<0.01**	**<0.01**	**<0.01**	**0.18**	**0.17**	0.03	Both
Hg	1.5	−0.5	Log-transformed	**<0.01**	**<0.01**	**<0.01**	0.06	**0.38**	0.02	Particle size
Tl	0.1	−1.2	Unchanged	**<0.01**	**<0.01**	**<0.01**	**0.16**	**0.34**	0.05	Particle size
Pb	1.7	−0.8	Log-transformed	**<0.01**	**<0.01**	0.03	**0.14**	**0.24**	0.02	Particle size
Bi	3.2	−0.1	Log-transformed	**<0.01**	**<0.01**	0.15	0.13	**0.30**	0.01	Particle size
Th	1.9	−0.8	Log-transformed	**<0.01**	**<0.01**	**<0.01**	**0.14**	**0.31**	0.03	Particle size
U	1.3	−0.5	Log-transformed	**<0.01**	**<0.01**	0.92	0.07	**0.32**	0.00	Particle size
Cr/Ti	2.8	−0.2	Log-transformed	**<0.01**	**<0.01**	**<0.01**	**0.41**	**0.18**	0.05	Region
La/Yb	0.3	−0.7	Unchanged	**<0.01**	**<0.01**	**<0.01**	**0.26**	0.09	0.07	Region

Bold fonts in column *p* indicate that the null hypothesis is rejected for *p* = 0.01. Bold fonts in the $\eta 2$ column indicate that a factor with $\eta 2 > 0.14$ has a significant effect on the elemental concentrations of the sediments.

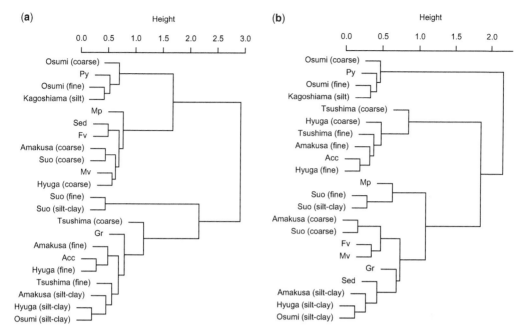

Fig. 5. (a) Cluster dendrogram obtained using a dataset of 32 elements; only those in marine sediments have been affected by terrestrial materials. (b) Cluster dendrogram obtained using a dataset of 21 immobile elements. Sed, sediment and sedimentary rock; Acc, sedimentary rocks of accretionary complexes associated with metagabbro, metabasalt and ultramafic rock; Mv, mafic volcanic rocks; Fv, felsic volcanic rocks; Py, pyroclastic-flow deposit and debris; Gr, granitic rocks; Mp, metamorphic rocks (mostly high-pressure type). Amakusa, Hyuga, Kagoshima, Osumi, Suo and Tsushima indicate the marine regions of the Amakusa Nada, Hyuga Nada, Kagoshima Bay, Osumi Strait, Suo Nada, and Tsushima Nada including Goto Nada and Genkai Nada, respectively. Coarse, fine and silt indicate coarse sediment, fine sand and silt including clay, respectively.

and La/Yb ratios were obtained for the provenance indicators of silt- and clay-sized marine sediments in the Yellow Sea (e.g. Yang *et al.* 2003). Although many combinations are possible for the provenance analysis, combinations of elemental concentration ratios effective in identifying contribution of source components are used (e.g. Roser 2000).

The cluster analysis suggested that pyroclastic rocks and accretionary complexes are the parent lithologies significantly affecting the surrounding marine sediments (Fig. 4a, b). As discussed in an earlier subsection in this Discussion, ANOVA and multiple comparison tests have suggested that stream sediments derived from Py have similar chemical compositions to those from Mv, which have significantly high concentrations of mafic elements such as TiO_2 and $Fe_2O_3^T$; nevertheless, Cr and Ni are significantly enriched in sediments from Mp but are extremely poor in those from Py (Tables 2 and 4). Therefore, the concentration ratio of immobile elements, such as Cr and Ti, is likely to be useful in distinguishing sediments that originate from Py, Mv and Mp. The alkaline metal elements Ba and Tl are significantly enriched in stream sediments derived

from Acc; however, their concentrations in marine sediments are predominantly determined by the particle size effect rather than the regional effect (Table 6). HREEs are significantly poor in stream sediments derived from Acc, and are highly enriched in those from Py and Mv. Therefore, the La/Yb ratio can be effective in highlighting the influence of sediments derived from Acc.

Figure 5a and b present scatter diagrams that plot the Cr/Ti and La/Yb ratios of stream sediments classified by the dominant lithology. The Cr/Ti ratios of stream sediments derived from Mp varied significantly, as did the La/Yb ratios of stream sediments derived from Sed and Gr (Fig. 6a). In general, the Cr/Ti ratio was low for stream sediments derived from Py and Fv (<0.9), and high for sediments from Mp (>1.4) (Fig. 6b). The La/Yb ratio was high for stream sediments derived from Acc (>13) and low for sediments from Py (<10) (Fig. 6b). The stream sediments derived from Mv are characterized by a low–middle La/Yb ratio (<15) and a wide range in the Cr/Ti ratio (0.5–3.0).

Figure 6c shows a scatter diagram plotting the regionally grouped Cr/Ti and La/Yb ratios of the

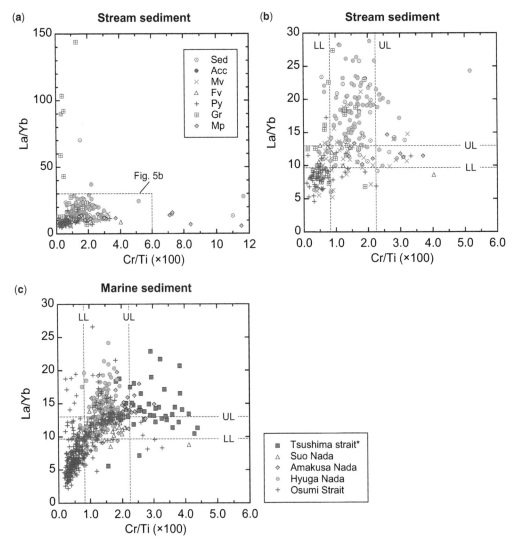

Fig. 6. Scatter diagram of Cr/Ti (×100) and La/Yb ratios in the Kyushu region. (**a**) Stream sediments derived from sediment and sedimentary rock; sedimentary rocks of accretionary complexes associated with metagabbro, metabasalt and ultramafic rock; mafic and felsic volcanic rocks; pyroclastic-flow deposits and debris; and granitic and metamorphic rocks (mostly high-pressure type). (**b**) Maximum values for the *x*-axis and *y*-axis were set to 6.0 and 30, respectively, for the stream sediment datasets. (**c**) Marine sediments classified by region. Sed, sediment and sedimentary rock; Acc, sedimentary rocks of accretionary complexes associated with metagabbro, metabasalt and ultramafic rock; Mv, mafic volcanic rocks; Fv, felsic volcanic rocks; Py, pyroclastic-flow deposit and debris; Gr, granitic rocks; Mp, metamorphic rocks (mostly high-pressure type). UL and LL indicate the upper and lower limits of the concentration ratios, respectively. Two samples from the Tsushima Strait with outliers of Cr/Ti (×100) ratio (>8.0) were excluded.

marine sediments. Sediments from the Hyuga Nada have a high La/Yb ratio (10–20) and an average Cr/Ti ratio (1.0–2.0), which correspond to stream sediments derived from Acc. In contrast, marine sediments with low ratios of both Cr/Ti (<1.0) and La/Yb (<9.0) were collected mostly from the Osumi Strait; these are comparable to Cr/Ti and La/Yb ratios of stream sediments derived from Py. Some samples collected near the Osumi Strait have an average Cr/Ti ratio (1.0–2.0) and an average–high La/Yb ratio (12–20), which are comparable to those of samples from the Hyuga Nada. Furthermore, some of the outliers among the samples from the Osumi Strait have a low Cr/Ti ratio (<1.0) and a high La/Yb ratio (>10). Marine sediments from the Tsushima Strait – including the Genkai and Goto nadas – and the Amakusa Nada are characterized as having high ratios of both Cr/Ti (2.0–4.0) and La/Yb (>10). Sediments collected from the Suo Nada have average ratios for both Cr/Ti (1.0–2.0) and La/Yb (10–13).

Particle transfer process in marine environment and provenance of marine sediments

Figure 7 shows the spatial distribution patterns of the Cr/Ti and La/Yb ratios of marine and stream sediments. To visualize the ratios that characterized the dominant lithology, their high and low ratios are highlighted by shaded regions. The ratios for the stream sediments varied widely (Fig. 6a, b) owing to the mineralogical and chemical heterogeneities of source rock, the variation in the mineralogical composition of the stream sediment in the riverbed, and the influence of coexistent lithologies except for the dominant one in the watershed. Thus, to identify the specific characteristics of stream sediment

classified by the dominant lithology, the threshold values were defined by using their calculated median values (Med) and median absolute deviation (MAD). For this purpose, the MAD/0.6745 is a useful robust estimate of standard deviation (σ') (Miller and Miller 2010). The lower threshold values of the Cr/Ti and La/Yb ratios were defined as 0.814 and 9.67, respectively, which were calculated as Med + σ' of the Cr/Ti and La/Yb ratios in stream sediments derived from Py. In contrast, the higher threshold values of the Cr/Ti and La/Yb ratios were defined as 2.24 and 13.0, respectively, which were calculated as Med + σ' for the Cr/Ti ratios and Med − σ' for the La/Yb ratios in sediments derived from Acc. For reference purposes, these threshold values are also shown in Figure 6b and c.

Figure 7 shows that samples with lower Cr/Ti and La/Yb ratios are distributed in the Osumi Strait, including Kagoshima Bay. The spatial distribution of these samples within a radius of 80 km of the Kikai Caldera corresponds to Koya pyroclastic deposits. Samples with lower Cr/Ti and La/Yb ratios in Fukiage Hama and Kagoshima Bay and off the Satsuma Peninsula may have been supplied by adjacent terrestrial materials, which originated from Ito pyroclastic rocks through a riverine system or by coastal erosion. The shaded region with lower

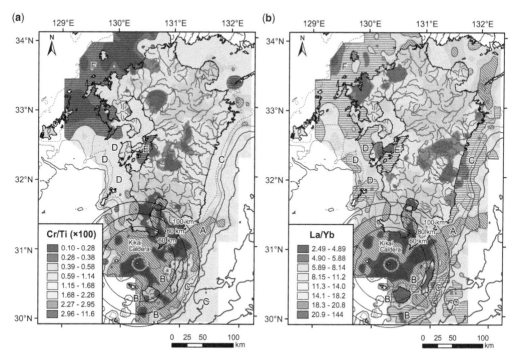

Fig. 7. Spatial distributions of (**a**) Cr/Ti (×100) and (**b**) La/Yb ratios in the Kyushu region. The shaded areas with diagonal lines indicate that the Cr/Ti and La/Yb ratios are less than 0.81 and 9.7, respectively; those with horizontal lines indicate ratios greater than 2.2 and 13.0, respectively.

Cr/Ti and La/Yb ratios continues from the area off of the Osumi Peninsula to the southern part of the Hyuga Nada, which is about 170 km from the Kikai Caldera in a straight line (area A in Fig. 7a, b). This distribution indicates that fine and coarse sands containing pumice have been conveyed by a strong oceanic current flow, which was reported by Ikehara (1988, 1993). However, part of the sediments from the Osumi Strait might have originated from older pyroclastic rocks supplied mainly from the Aira Caldera that were denudated and moved by sea-level fall during the last glacial age (Inouchi 1981).

In contrast, samples collected from the western and southern regions of the island of Yakushima and from the channel between the Yakushima and Tanegashima islands, which are described as outliers in the Osumi Strait samples (Fig. 6c), have a low Cr/Ti ratio but a high La/Yb ratio (area B in Fig. 7a, b). This feature suggests that sediments supplied from Gr in the island of Yakushima prevail in these regions because stream sediments from Gr have a low Cr/Ti ratio and a high La/Yb ratio (Fig. 6b). Furthermore, low Cr/Ti and La/Yb ratios were not found in the sediments from the southeastern region of the island of Tanegashima. Therefore, these islands appear to serve as high barriers for pyroclastic-flow deposits from the Kikai Caldera (Geshi 2009).

The fine sand and silt distributed in the Hyuga Nada and in the southeastern regions of the island of Tanegashima have middle–high La/Yb ratios and medium Cr/Ti ratios, which correspond to stream sediment derived from Acc (area C in Fig. 7a, b). Modern sandy sediments supplied through rivers were found near the coast according to the continuous spatial distribution patterns of high concentrations of heavy metals related to skarn-type deposits (Cu, Sb and Pb: Fig. 4). Furthermore, most samples from the Hyuga Nada were collected from water depths below the wave base. Therefore, their spatial distribution across the shelf, slope and basin can be explained by conveyance into deeper areas via gravitational transport including small-scale turbidity currents induced by storm waves and river floods. Indeed, small-scale turbidity currents occur with high frequency several times per year (e.g. Milliman and Kao 2005; Xu *et al.* 2010). In contrast, submarine landslides or sheet flow induced by subduction-zone earthquakes occur only once every hundred to thousand years. The spatial distributions of areas A and C in Figure 7 clearly show that sandy sediments originating from Py erupted from the Kikai Caldera at around 7.3 ka (Maeno and Taniguchi 2007; Fujihara and Suzuki-Kamata 2013). These sediments were conveyed by oceanic currents after the last glacial maximum (Ikehara 1992) before overlapping and mixing with surface sediments derived from Acc in the southeastern region of the study area.

The cluster analysis results suggest that silty sediments from the Amakusa Nada, including the Ariake and Yatsushiro seas, have no close relationship to any stream sediment. The silty sediments of the Yatsushiro Sea are modern sediments supplied mainly from the Kuma River, in which the watershed is dominantly covered by sedimentary rock from accretionary complexes associated with metabasalt and ultramafic rock. Therefore, their chemical compositions would be influenced by sediments originating from Acc with high La/Yb ratios, and those from metabasalt and ultramafic rock extremely enriched in MgO, Cr and Ni (Cr: Fig. 4).

Coarse sediments of the Tsushima Strait, Genkai Nada and Goto Nada have high Cr/Ti and La/Yb ratios. These regions receive a modern sediment discharge of only 4% from Kyushu Island (Appendix Table A1). Therefore, their chemical compositions are very loosely related to those of any stream sediment classified by the dominant lithology, as suggested by the cluster analysis (Fig. 5). However, some stream sediments from Amakusa Island were derived from Sed; those of the northern regions of Kyushu Island were derived from Acc; and those of the northwestern regions of Kyushu Island were derived from Mv, Sed and Mp with comparable Cr/Ti and La/Yb ratios (Figs 6a, b & 7). These sample sites are adjacent to the Amakusa Nada, Goto Nada, Tsushima Strait and Genkai Nada. Park and Yoo (1988) used seismic profile data to determine that most of the Tsushima Strait was exposed and eroded during the last glacial age. Nishida and Ikehara (2013) reported that the evolution of the depositional processes in the Genkai Nada was caused by sea-level change after the last glacial maximum and inflow of the Tsushima Warm Current, a branch of the Kuroshio Current, into the Sea of Japan through the Tsushima Strait. Accordingly, we simply conclude that the sedimentary layers in the Tsushima Strait originated as parent rock with high Cr/Ti and La/Yb ratios, as is the case with silty sediments in the Yatsushiro Sea, and were denudated, conveyed by the strong oceanic current and deposited at each stage during the transgression age after the last glacial maximum.

Summary

Geochemical data of 53 elements in 391 stream sediment samples and 606 marine sediment samples were used to study the provenance and transfer process of fine and coarse sands around the Kyushu region of western Japan. The chemical compositions of the stream sediments were determined by the lithology dominantly distributed in the watershed area. One-way ANOVA and a multiple comparison test revealed that: (1) stream sediments derived from sedimentary rocks of accretionary complexes are rich in K_2O, Li, Rb, Cs, Ba and Tl; (2) those from mafic volcanic rocks and pyroclastic rocks have a high abundance of MgO, P_2O_5, CaO, Sc, V, TiO_2, MnO, $Fe_2O_3^T$, Co, Sr and HREEs (Gd–Lu); (3) those from granitic rocks are enriched in alkali metal elements, Be, Ga, REEs, Ta, Th and U; and (4) skarn-type deposits have highly elevated concentrations of Cu, Zn, As, Mo, Cd, Sn, Sb, Hg, Pb and Bi in a small region.

The concentrations of K_2O, Ni, Rb, Sb, Cs, Ba, Hg, Bi and U in the marine sediments were determined simply by the particle size of the clastics rather than by their source materials. The other elements in the marine sediments were influenced by their origin. Cluster analysis using the chemical composition of the stream and marine sediments suggested that the coarse sediments and fine sands of the Osumi Strait originate from pyroclastic rocks that erupted from the Kikai Caldera, and that the fine sands from the Hyuga Nada and Amakusa Nada originate from sedimentary rocks of accretionary complexes distributed in the adjacent terrestrial area. Silt–clay sediments are highly enriched in alkaline metal elements, MnO, Ni and heavy metals, such as Cu and Mo, the compositions of which are weakly related to terrestrial materials. However, the silts of Kagoshima Bay have chemical compositions similar to those of stream sediments derived from pyroclastic-flow deposits, and are enriched in As, Mo, Cd, Sb, Hg and Bi owing to seawater hydrothermal activity.

The concentration ratios of immobile elements are effective at negating the dilution effect of quartz and calcareous materials in coarse marine sediment, and are used as an index of source rock composition in marine sediments. According to the geochemical feature of stream sediment classified to the parent lithology, Cr/Ti and La/Yb ratios are used as the most effective indicators of the dynamic transfer processes of marine sediments in the study area. The spatial distribution of the lower Cr/Ti and La/Yb ratios suggest that sandy sediments originating from pyroclastic-flow deposits are distributed within an 80 km radius of the Kikai Caldera, and that these have subsequently been conveyed to the Hyuga Nada via a strong oceanic current. The spatial distribution of fine sands with an average Cr/Ti ratio and a high La/Yb ratio suggest that sandy sediments on the shelf, which originate from accretionary complexes in the adjacent terrestrial area, have been conveyed to the deep sea by gravitational transport.

Acknowledgements The authors thank Takashi Okai, Masumi Mikoshiba, Mitsuo Manaka, Ran Kubota and Atsunori Nakamura for their useful suggestions, which have helped the authors to improve the manuscript. The authors would also like to thank the Japan Oceanographic Data Centre (JODC) for providing data files.

Author contributions AO: writing – original draft (lead); **NI**: project administration (lead); **YT**: data curation (supporting); **KI**: validation (equal).

Funding This research received no specific grant from any funding agency in the public, commercial, or not-for-profit sectors.

Data availability All data generated or analysed during this study are included in this published article (and its supplementary information files).

Appendix A

Table A1. *Sediment yield and river water discharge data for each river system of the study area*

River system	Average rate of sediment yield* $m^2/km^2/$ year	Drainage basin area† km^2	Sediment yield $m^3/year$ $(\times 10^3)$	Water discharge in 2000† $m^3/year$ $(\times 10^6)$	Discharged area	Relative rate to total sediment yield
Onga	122	695	85	581	Genkai Nada	3.2%
Matsu-ura	115	275	32	248	Genkai Nada	1.2%
Chikugo	144	2315	333	2470	Ariake Sea	12%
Yabe	139	460	64	440	Ariake Sea	2.4%
Rokkaku	122	95	12	93	Ariake Sea	0.4%
Kase	112	256	29	308	Ariake Sea	1.1%
Hon-myo	140‡	36	5	52	Ariake Sea	0.2%
Kikuchi	124	906	112	1248	Ariake Sea	4.2%
Shirakawa	333	477	159	685	Ariake Sea	5.9%

(Continued)

Table A1. *Continued.*

River system	Average rate of sediment yield* $m^2/km^2/$ year	Drainage basin area† km^2	Sediment yield m^3/year $(\times 10^3)$	Water discharge in 2000† m^3/year $(\times 10^6)$	Discharged area	Relative rate to total sediment yield
Midorikawa	140	681	95	940	Ariake Sea	3.6%
Kama	137	1856	254	3288	Yatsushiro Sea	9.5%
Yamakuni	183	483	88	418	Suo Nada	3.3%
Oita	230	494	114	498	Suo Nada	4.3%
Ono	164	1239	203	1616	Suo Nada	7.6%
Banjo	192	278	53	392	Suo Nada	2.0%
Gokase	320	1044	334	1922	Hyuga Nada	13%
Omaru	436	396	173	960	Hyuga Nada	6.5%
Oyodo	171	1564	267	3348	Hyuga Nada	10%
Sendai	192	1348	259	2529	Amakusa Nada	9.7%
Kimotsuki	185	450	83	1050	Kagoshima Bay	3.1%

*Akimoto *et al.* (2009)
†Ministry of Land, Infrastructure, Transport and Tourism (http://www.mlit.go.jp/river/toukei_chousa/, accessed in Nov. 24, 2017)
‡Average rate of sediment yield of Takase River is assumed to be the same as that of Midorikawa River.

Appendix B

Of the 191 stream sediments collected from remote islands near the Kyushu mainland, 25 samples include Am04, Am14, Am23, Am26, Am33, and Am43 from Amakusa Island; Fke02, Fke04, and Fke10 from Fukue Island; Hr04 and Hr09 from Hirado Island; Iki01 from Ikinoshima Island; Ng02 from Nagashima Island; Nk03 and Nk08 from Nakadori Island; Tn02, Tn04, Tn08, Tn21, and Tn27 from Tanegashima Island; and Yk01, Yk10, Yk14, Yk18, and Yk22 from Yakushima Island (Ohta 2018).

References

Aguilar, F.J., Agüera, F., Aguilar, M.A. and Carvajal, F. 2005. Effects of terrain morphology, sampling density, and interpolation methods on grid DEM accuracy. *Photogrammetric Engineering and Remote Sensing*, **71**, 805–816, https://doi.org/10.14358/pers.71.7.805

Aitchison, J. 1982. The statistical analysis of compositional data. *Journal of the Royal Statistical Society. Series B (Methodological)*, **44**, 139–177, https://doi.org/10.1111/j.2517-6161.1982.tb01195.x

Aitchison, J., Barceló-Vidal, C., Martin-Fernández, J.A. and Pawlowsky-Glahn, V. 2000. Logratio analysis and compositional distance. *Mathematical Geology*, **32**, 271–275, https://doi.org/10.1023/A:100752 9726302

Akimoto, T., Kawagoe, S. and Kazama, S. 2009. Estimation of sediment yield in Japan by using climate projection model. *Proceedings of Hydraulic Engineering*, **53**, 655–660 (in Japanese with English abstract).

Darnley, A.G., Björklund, A. *et al.* 1995. *A Global Geochemical database for Environmental and Resource Management: Recommendations for International Geochemical Mapping.* UNESCO, Paris.

De Caritat, P. and Cooper, M. 2011. *National Geochemical Survey of Australia: The Geochemical Atlas of Australia.* Geoscience Australia, Record, **2011/20**.

De Vos, W., Tarvainen, T. *et al.* 2006. *Geochemical Atlas of Europe. Part 2 – Interpretation of Geochemical Maps, Additional Tables, Figures, Maps, and Related Publications.* Geological Survey of Finland, Espoo, Finland.

Fritz, C.O., Morris, P.E. and Richler, J.J. 2012. Effect size estimates: current use, calculations, and interpretation. *Journal of Experimental Psychology: General*, **141**, 2–18, https://doi.org/10.1037/a0024338

Fujihara, M. and Suzuki-Kamata, K. 2013. Glass composition and emplacement mode of Koya pyroclastic flow deposit and its proximal equivalent. *Bulletin of the Volcanological Society of Japan*, **58**, 489–498, https://doi.org/10.18940/kazan.58.4_489 (in Japanese with English abstract).

Geological Survey of Japan, AIST. 1992. *Geological Map of Japan, 1:1 000 000*, 3rd, edn. Geological Survey of Japan, AIST, Tsukuba, Japan.

Geshi, N. 2009. Distribution and flow mechanisms of the 7.3 ka Koya pyroclastic flow deposits covering Yakushima Island, Kagoshima Prefecture. *Journal of Geography (Chigaku Zasshi)*, **118**, 1254–1260, https://doi.org/10.5026/jgeography.118.1254 (in Japanese with Englsih abstract).

Hochberg, Y. and Tamhane, A.C. 1987. *Multiple Comparison Procedures.* John Wiley & Sons, New York.

Howarth, R.J. and Thornton, I. 1983. Regional geochemical mapping and its application to environmental studies. *In:* Thornton, I. (ed.) *Applied Environmental Geochemistry.* Academic Press, London, 41–73.

Ikehara, K. 1988. Ocean current generated sedimentary facies in the Osumi Strait, south of Kyushu, Japan. *Progress in Oceanography*, **21**, 515–524, https://doi.org/10.1016/0079-6611(88)90025-0

Ikehara, K. 1992. Formation of duned sand bodies in the Osumi Strait, south of Kyushu, Japan. *Journal of the Sedimentological Society of Japan*, **36**, 37–45, https://doi.org/10.14860/jssj1972.36.37

Ikehara, K. 1993. Modern sedimentation in the shelf to basin areas around southwest Japan, with special reference to the relationship between sedimentation and oceanographic conditions. *Bulletin of the Geological Survey of Japan*, **44**, 283–349.

Ikehara, K. 1998. Sequence stratigraphy of tidal sand bodies in the Bungo Channel, southwest Japan. *Sedimentary Geology*, **122**, 233–244, https://doi.org/10.1016/s0037-0738(98)00108-0

Ikehara, K. 2000. *Sedimentological Map of Hyuga-Nada. 1:200 000*. Marine Geology Map Series, **54**. Geological Survey of Japan, AIST, Tsukuba, Japan.

Ikehara, K. 2001. *Sedimentological Map of Hibiki-Nada. 1:200 000*. Marine Geology Map Series, **56**. Geological Survey of Japan, AIST, Tsukuba, Japan.

Ikehara, K. 2013. *Sedimentological Map offshore of Cape Noma Misaki. 1:200 000*. Marine Geology Map Series, **79**. Geological Survey of Japan, AIST, Tsukuba, Japan.

Ikehara, K. 2014. *Sedimentological Map of the vicinity of Tanegashima. 1:200 000*. Marine Geology Map Series, **84**. Geological Survey of Japan, AIST, Tsukuba, Japan.

Ikehara, K. and Kinoshita, Y. 1994. Distribution and origin of subaqueous dunes on the shelf of Japan. *Marine Geology*, **120**, 75–87, https://doi.org/10.1016/0025-3227(94)90078-7

Imai, N., Terashima, S. *et al*. 2004. *Geochemical Map of Japan*. 1st edn. Geological Survey of Japan, AIST, Tsukuba, Japan, https://gbank.gsj.jp/geochemmap/

Imai, N., Terashima, S. *et al*. 2010. *Geochemical Map of Sea and Land of Japan*. Geological Survey of Japan, AIST, Tsukuba, Japan, https://gbank.gsj.jp/geochemmap/

Inouchi, Y. 1981. Sediments and Quaternary sedimentological history of the Osumi Strait and its vicinity, in relation to the evolution of the Osumi Strait. *Bulletin of the Geological Survey of Japan*, **32**, 693–716 (in Japanese with English abstract).

Klinkhammer, G.P. 1980. Early diagenesis in sediments from the eastern equatorial Pacific. II. Pore water metal results. *Earth and Planetary Science Letters*, **49**, 81–101, https://doi.org/10.1016/0012-821X(80)90151-X

Maeno, F. and Taniguchi, H. 2007. Spatiotemporal evolution of a marine caldera-forming eruption, generating a low-aspect ratio pyroclastic flow, 7.3 ka, Kikai caldera, Japan: implication from near-vent eruptive deposits. *Journal of Volcanology and Geothermal Research*, **167**, 212–238, https://doi.org/10.1016/j.jvolgeores.2007.05.003

Miller, J.C. and Miller, J.N. 2010. *Statistics and Chemometrics for Analytical Chemistry*. 6th edn. Pearson Education Canada, Toronto, Canada.

Milliman, J.D. and Kao, S.J. 2005. Hyperpycnal discharge of fluvial sediment to the ocean: impact of Super-Typhoon Herb (1996) on Taiwanese rivers. *Journal of Geology*, **113**, 503–516, https://doi.org/10.1086/431906

Miyazaki, K., Ozaki, M., Saito, M. and Toshimitsu, S. 2016. The Kyushu–Ryukyu Arc. *In*: Moreno, T.,

Wallis, S., Kojima, T. and Gibbons, W. (eds) *The Geology of Japan*. Geological Society, London, 139–174, https://doi.org/10.1144/GOJ.6

Nagata, Y. and Yoshida, M. 1997. *Basic Theory of Multiple Comparison Procedures*. Scientist Inc., Tokyo (in Japanese).

Nakada, S., Yamamoto, T. and Maeno, F. 2016. Miocene–Holocene volcanism. *In*: Moreno, T., Wallis, S., Kojima, T. and Gibbons, W. (eds) *The Geology of Japan*. Geological Society, London, 273–308, https://doi.org/10.1144/GOJ.11

Nishida, N. and Ikehara, K. 2013. Holocene evolution of depositional processes off southwest Japan: response to the Tsushima Warm Current and sea-level rise. *Sedimentary Geology*, **290**, 138–148, https://doi.org/10.1016/j.sedgeo.2013.03.012

Ohshima, K., Nakao, S., Mitsushio, H., Yuasa, M. and Kuroda, K. 1975. Sea bottom sediments. *In*: Inoue, E. (ed) *Goto-nada Sea and Tsushima Strait Investigations Northwestern Kyushu 1972–1973*, Geological Survey of Japan, AIST, 35–39.

Ohshima, K., Inoue, E., Onodera, K., Yuasa, M. and Kuroda, K. 1982. Sediments of the Tsushima Strait and Goto-nada Sea, northwestern Kyushu. *Bulletin of the Geological Survey of Japan*, **33**, 321–350 (in Japanese with English abstract).

Ohta, A. 2018. Geochemical mapping of remote islands around Kyushu, Japan. *Bulletin of the Geological Survey of Japan*, **69**, 233–263, https://doi.org/10.9795/bullgsj.69.233

Ohta, A. and Imai, N. 2011. Comprehensive survey of multi-elements in coastal sea and stream sediments in the Island Arc Region of Japan: mass transfer from terrestrial to marine environments. *In*: El-Amin, M. (ed.) *Advanced Topics in Mass Transfer*. InTech, London, 373–398, https://doi.org/10.5772/14251

Ohta, A., Imai, N., Terashima, S., Tachibana, Y., Ikehara, K. and Nakajima, T. 2004. Geochemical mapping in Hokuriku, Japan: influence of surface geology, mineral occurrences and mass movement from terrestrial to marine environments. *Applied Geochemistry*, **19**, 1453–1469, https://doi.org/10.1016/j.apgeochem.2004.01.026

Ohta, A., Imai, N., Terashima, S. and Tachibana, Y. 2005. Application of multi-element statistical analysis for regional geochemical mapping in Central Japan. *Applied Geochemistry*, **20**, 1017–1037, https://doi.org/10.1016/j.apgeochem.2004.12.005

Ohta, A., Imai, N. *et al*. 2007. Elemental distribution of coastal sea and stream sediments in the island-arc region of Japan and mass transfer processes from terrestrial to marine environments. *Applied Geochemistry*, **22**, 2872–2891, https://doi.org/10.1016/j.apgeochem.2007.08.001

Ohta, A., Imai, N., Terashima, S., Tachibana, Y., Ikehara, K., Katayama, H. and Noda, A. 2010. Factors controlling regional spatial distribution of 53 elements in coastal sea sediments in northern Japan: comparison of geochemical data derived from stream and marine sediments. *Applied Geochemistry*, **25**, 357–376, https://doi.org/10.1016/j.apgeochem.2009.12.003

Ohta, A., Imai, N., Terashima, S., Tachibana, Y. and Ikehara, K. 2013. Regional spatial distribution of multiple elements in the surface sediments of the eastern

Tsushima Strait (southwestern Sea of Japan). *Applied Geochemistry*, **37**, 43–56, https://doi.org/10.1016/j.apgeochem.2013.06.010

Ohta, A., Imai, N., Tachibana, Y. and Ikehara, K. 2017*a*. Statistical analysis of the spatial distribution of multi-elements in an island arc region: complicating factors and transfer by water currents. *Water*, **9**, 37, https://doi.org/10.3390/w9010037

Ohta, A., Imai, N., Tachibana, Y., Ikehara, K., Katayama, H. and Nakajima, T. 2017*b*. Influence of different sedimentary environments on multi-elemental marine geochemical maps of the Pacific Ocean and Sea of Japan, Tohoku region. *Bulletin of the Geological Survey of Japan*, **68**, 87–110, https://doi.org/10.9795/bullgsj.68.87

Park, S.C. and Yoo, D.G. 1988. Depositional history of quaternary sediments on the continental-shelf off the southeastern coast of Korea (Korea Strait). *Marine Geology*, **79**, 65–75, https://doi.org/10.1016/0025-3227(88)90157-0

Pawlowsky-Glahn, V. and Egozcue, J.J. 2006. Compositional data and their analysis: an introduction. *Geological Society, London, Special Publications*, **264**, 1–10, https://doi.org/10.1144/GSL.SP.2006.264.01.01

Reimann, C. 2005. Geochemical mapping: technique or art? *Geochemistry: Exploration, Environment, Analysis*, **5**, 359–370, https://doi.org/10.1144/1467-7873/03-051

Reimann, C., Birke, M., Demetriades, A., Filzmoser, P. and O'Connor, P. 2014*a*. *Chemistry of Europe's Agricultural Soils, Part A: Methodology and Interpretation of the GEMAS Data Set*. Geologisches Jahrbuch Reihe B, **102B**.

Reimann, C., Birke, M., Demetriades, A., Filzmoser, P. and O'Connor, P. 2014*b*. *Chemistry of Europe's Agricultural Soils, Part B: General Background Information and Further Analysis of the GEMAS Data Set*. Geologisches Jahrbuch Reihe B, **103B**.

Richardson, J.T. 2011. Eta squared and partial eta squared as measures of effect size in educational research. *Educational Research Review*, **6**, 135–147, https://doi.org/10.1016/j.edurev.2010.12.001

Roser, B.P. 2000. Whole-rock geochemical studies of clastic sedimentary suites. *Memoirs of the Geological Society of Japan*, **57**, 73–89.

Sakamoto, H. 1985. The distribution of mercury, arsenic, and antimony in sediments of Kagoshima Bay. *Bulletin of the Chemical Society of Japan*, **58**, 580–587, https://doi.org/10.1246/bcsj.58.580

Salminen, R., Batista, M.J. *et al.* 2005. *Geochemical Atlas of Europe. Part 1 – Background Information, Methodology and Maps*. Geological Survey of Finland, Espoo, Finland.

Shapiro, S.S. and Wilk, M.B. 1965. An analysis of variance test for normality (complete samples). *Biometrika*, **52**, 591–611, https://doi.org/10.2307/2333709

Shaw, T.J., Gieskes, J.M. and Jahnke, R.A. 1990. Early diagenesis in differing depositional environments: the response of transition metals in pore water. *Geochimica et Cosmochimica Acta*, **54**, 1233–1246, https://doi.org/10.1016/0016-7037(90)90149-F

Smith, D.B., Cannon, W.F., Woodruff, L.G., Solano, F. and Ellefsen, K.J. 2014. *Geochemical and Mineralogical Maps for Soils of the Conterminous United States*. United States Geological Survey Open-File Report, **2014-1082**, http://pubs.usgs.gov/of/2014/1082/pdf/ofr2014-1082.pdf

Sudo, S., Watanabe, Y. and Kobayashi, K. 2003. *Mineral Resources Map of Kyushu, 1:500 000*. Geological Survey of Japan, AIST, Tsukuba, Japan.

Taylor, S.R. and McLennan, S.M. 1995. The geochemical evolution of the continental crust. *Reviews of Geophysics*, **33**, 241–265, https://doi.org/10.1029/95RG00262

Templ, M., Filzmoser, P. and Reimann, C. 2008. Cluster analysis applied to regional geochemical data: problems and possibilities. *Applied Geochemistry*, **23**, 2198–2213, https://doi.org/10.1016/j.apgeochem.2008.03.004

Ward, J.H., Jr. 1963. Hierarchical grouping to optimize an objective function. *Journal of the American Statistical Association*, **58**, 236–244, https://doi.org/10.1080/01621459.1963.10500845

Webb, J.S., Thornton, I., Thompson, M., Howarth, R.J. and Lowenstein, P.L. 1978. *The Wolfson Geochemical Atlas of England and Wales*. Clarendon Press, Oxford.

Xie, X.J. and Chen, H.X. 2001. Global geochemical mapping and its implementation in the Asia-Pacific region. *Applied Geochemistry*, **16**, 1309–1321, https://doi.org/10.1016/S0883-2927(01)00051-8

Xu, J.P., Swarzenski, P.W., Noble, M. and Li, A.C. 2010. Event-driven sediment flux in Hueneme and Mugu submarine canyons, southern California. *Marine Geology*, **269**, 74–88, https://doi.org/10.1016/j.margeo.2009.12.007

Yang, S.Y., Jung, H.S., Lim, D.I. and Li, C.X. 2003. A review on the provenance discrimination of sediments in the Yellow Sea. *Earth-Science Reviews*, **63**, 93–120, https://doi.org/10.1016/s0012-8252(03)00033-3

Habitat mapping for human well-being: a tool for reducing risk in disaster-prone coastal environments and human communities

Yuri Oki[1]*, Hiroshi Kitazato[2], Toyonobu Fujii[3] and Soichiro Yasukawa[4]

[1]Center of Marine Research and Operations, Tokyo University of Marine Science and Technology, Tokyo 108-8477, Japan

[2]School of Marine Resources and Environment, Tokyo University of Marine Science and Technology, Tokyo 108-8477, Japan

[3]Graduate School of Agricultural Science, Tohoku University, Sendai 980-8572, Japan

[4]Natural Sciences Sector, UNESCO, Paris 75007, France

YO, 0000-0002-4868-7978; HK, 0000-0003-4990-3908

*Correspondence: oki.y.ae@m.titech.ac.jp

Abstract: Coastal ecosystems consist of diverse habitats, such as reed beds, salt marshes, mangrove swamps, tidal flats, river deltas, seagrass fields, coral reefs, sandy/rocky-shore beaches and other habitats that harbour biodiversity. The Great East Japan Earthquake of March 2011 caused severe damage to one-third of the fishing communities along the Pacific Ocean of NE Japan. Coastal species, such as seagrasses, function as nursery areas for commercially important species. Coastal ecosystems provide natural infrastructure for the prevention and reduction of hazardous events, a process known as ecosystem-based disaster risk reduction (Eco-DRR). The preparation of topographic and thematic maps of coastal marine environments is essential to establish and visualize the concept of Eco-DRR. Experience gained following the Japanese earthquake, as well as examples from Indonesia and Thailand in the wake of 2004 Indian Ocean tsunami, showed that Eco-DRR is an affordable and sustainable approach. Dissemination of habitat maps should be further promoted as a way to 'Build Back Better'. To scale up and promote Eco-DRR, scientists must work in a transdisciplinary manner and engage with society by understanding the roles of ecosystems by monitoring and analysing, providing solutions and raising the awareness of community and policy makers, enabling them to better implement Eco-DRR.

Human populations and activities are often concentrated along coasts. As a result, coastal ecosystems are altered, making coastal areas prone to natural hazards and risks. According to Adger *et al.* (2005), 23% of the world's population live within 100 km of a coast, and this percentage is expected to rise to 50% by 2030. Consequently, half of the world's population will be exposed to coastal hazards, such as flooding, tsunamis, hurricanes/typhoons and other marine-related disasters (Mora *et al.* 2013). Resilience can erode over time, driven by natural and human-induced threats, such as environmental change and human actions, making coastal areas more vulnerable to hazards and disasters (Adger *et al.* 2005). Approximately 10% of the world's population lives in vulnerable coastal lowlands, located more than 10 m below sea-level (The United Nations 2017). For example, 25% of the citizens of SE Asian countries (e.g. Vietnam, Thailand and Indonesia) are living in these lowlands for at least some period of their lives (Ikeda 2016).

The problem

This article serves as a vision paper to show the future direction of geo-referenced marine habitat maps that can be utilized by scientists and public and non-academic communities as blueprints to wisely use coastal ecosystems to secure local stakeholders' livelihoods and human well-being. The idea of human well-being is derived from The Declaration of the 9th World Science Forum 2019 – science, ethics, and responsibility. It emphasizes scientists' contribution in applying science for the interest of humanity, for well-being and human rights (accessed 29 January 2021 from https://worldscienceforum.org/contents/declaration-of-world-science-forum-2019-110073). Seeing the local seascape through habitat maps helps us to visualize not only the coastal ecosystems and their biodiversity but also human activities, lifestyles and livelihoods in the coastal realm

Here, we outline a plan for the use of visualization tools to facilitate communication and

From: Asch, K., Kitazato, H. and Vallius, H. (eds) 2022. *From Continental Shelf to Slope: Mapping the Oceanic Realm.* Geological Society, London, Special Publications, **505**, 271–282.
First published online December 29, 2021, https://doi.org/10.1144/SP505-2021-26

cooperation between scientists and the people living in coastal areas, particularly those engaged in fishing. These populations are very vulnerable to natural disasters associated with the sea. Working as equal partners in collaboration with scientists will enable coastal communities to access scientific knowledge, prepare better for catastrophic events and rebuild their lives and their environments in the aftermath of such disasters.

Through the temporal analysis of spatial usage in the coastal realm, visualized through different layers in the habitat map database, we can understand how human activities and livelihoods have changed over time, as well as natural disturbances and changes to coastal ecosystems.

Nature of coastal habitats

Despite the hazardous risks posed to coastal communities, natural ecosystems bring benefits such as food security, commercial fishery and the protection of human livelihoods. Coastal marine ecosystems consist of diverse habitats such as salt marshes, mangrove swamps, tidal flats, river deltas, seagrasses, seaweed grounds, coral reefs, sandy- and rocky-shore beaches and other habitats that harbour biodiversity and attract large human populations. These complex coastal ecosystems are sustained by linkages between land and sea. For example, basic landscape components, such as river deltaic sediments, are transported through on-land river systems. Most sediment grains come from land areas (Milliman and Meade 1983) and are re-distributed and deposited after lateral transportation by coastal currents or tidal waves. Thus complex coastal habitats are created as a result of various interactions between land and sea.

Coastal ecosystems provide a wide range of ecosystem services and processes that are important for natural environments, fisheries and human livelihoods (United Nations 2020). The protection of coastal ecosystems supports a material cycle that is fundamental for sustaining human livelihoods in coastal communities prone to disasters. Additionally, biodiverse coastal species, such as seagrasses, function as nursery areas for commercially important seafood species such as fishes, clams and shrimps (Gillanders 2006; Kitazato *et al.* 2020). Coastal ecosystems provide natural infrastructure for both the prevention and reduction of hazardous events; this is known as ecosystem-based disaster risk reduction (Eco-DRR) (United Nations 2020). Eco-DRR is not limited to coastal environments. It can extend to wetlands and forests that function as natural infrastructures that reduce physical exposure to natural hazards, build the resilience of communities and sustain livelihoods. These ecosystems provide essential natural resources that provide food, water and other necessities for human sustenance.

Preservation of coastal ecosystems as a scheme for disaster risk reduction

Coastal ecosystems occupy particular landscapes or underwater seascapes, each with specific topographical characteristics. Precise coastal mapping is important for understanding the kinds of landscapes or seascapes that exist in coastal areas. This vision paper explains how coastal environments and ecosystems should be conserved, not only for maintaining biodiversity but also for Eco-DRR. These activities should be led by local citizens who live on the coast and receive abundant benefits from the coastal ocean. Citizen science should enhance Eco-DRR in collaboration with scientists. Habitat mapping techniques can play an important role as catalysers for empowering local stakeholders.

International initiatives

The UN proclamation on the Decade of Ocean Science for Sustainable Development calls for marine science communities to support countries' actions to achieve the 2030 Sustainable Development Goals agenda 14, which calls for the sustainability of life below water through conservation and for the sustainable use of ocean, sea and marine resources.

Ocean science is an interdisciplinary study that provides data on the global marine environment, including ecosystem dynamics, marine organisms, ocean currents, seafloor geology and fluxes of chemical substances within the ocean and across land and sea boundaries. Ocean science supports management activities and conservation of coastal communities through the prediction of ocean hazards, thereby preventing and mitigating disaster risks (UNESCO 2019).

The Global Ocean Science Report (2020, p. 226) recognizes the importance of indigenous and local knowledge for the successful coordination of ocean science activities, such as joint data-gathering programmes and the co-production of knowledge through science by academia and the private sector (Global Ocean Science Report 2020). To realize the vision of the UN Ocean Decade and practise activities that support the slogan 'the science we need for the ocean we want', the Global Ocean Science Report (2020) aims to strengthen ocean science capacity development under the governing principle of 'leaving no one behind'. This idea supports the principal that all countries, genders, age groups and local and indigenous knowledge have equal opportunities. The best practices of ocean science that follow community-approved guidelines should be adopted when endorsing ocean science capacity development at the national and regional levels (Global Ocean Science Report 2020, p. 43).

Ecosystem-based management (EBM) is an approach that holistically captures interactions in marine ecosystems, including the activities of species, ecosystem services and human interactions (McLeod *et al.* 2005; Katsanevakis *et al.* 2011). By applying EBM, we can monitor and help maintain healthy and sustainable marine ecosystems, aligning with the targets of the Decade of Ocean Science for Sustainable Development. Using geographic information systems (GIS), scientists can integrate a wide range of information on marine environments to help visualize marine ecosystems. Relevant information includes satellite images of land cover/use, surface water analysis, coastal/pelagic and benthic analysis, seafloor topography, marine zoning areas, satellite imagery of aquaculture rafts and much more. All of these data can be organized and synthesized for spatial and temporal analyses. Furthermore, GIS allows local/sublocal stakeholders, such as fishermen cooperatives, to use grid analysis to observe diverse marine ecosystem data on a thematic scale.

The European Union initiative on the *European Atlas of the Seas*, supported by a network of EU organizations, the European Marine Observation and Data Network (EMODnet), is a useful tool for GIS data organization and synthesis at the regional level (Vallius *et al.* 2020). This free-of-charge internet platform allows everyone, including scientists, policy makers and the public, to access and use European marine data grouped into to the following themes: bathymetry, geology, seabed habitats, chemistry, biology, physics and human activity (Vallius *et al.* 2020).

Accessing these processed data and layering the data for different purposes, such as spatial planning and coastal management decision-making, opens future opportunities for innovative ways to utilize a particular marine environment and its resources, allows data/knowledge management of the marine environment and creates a common platform for EU marine data to be open and accessible to all users worldwide (Vallius *et al.* 2020). With the launch of this EU initiative (and its easy-to-use interface), people can generate customized packages of maps, data organization and visualization that are open to society and public stakeholders.

The *European Atlas of the Seas* is a showcase for future directions in the utilization of marine GIS data for real-life marine- and ocean-related global agendas. There are examples of EMODnet cases available on EMODnet's central portal webpage (https://www.emodnet.eu/en/use-cases) and we are excited by the prospect of further examples being added as the portal receives more attention from its users. Building on the EMODnet initiative, we pose the following question regarding the on-ground application (or implementation) of EBM as a marine information database to meet societal needs: how should we utilize a marine GIS database, such as the *European Atlas of the Seas*, for human well-being and a better and more sustainable society? Given how a marine GIS database can combine a wide array of data ranging from biological and ecological to human oriented (e.g. the density of aquaculture facilities and fishery production capabilities), how can local stakeholders (e.g. fishermen managing aquaculture farms) benefit from it, especially when the marine environment is prone to natural disasters such as tsunamis? The following parts of this article explain how compiled habitat maps of coastal ecosystems in the earthquake- and tsunami-affected town of Onagawa in Miyagi Prefecture, Japan, have the potential to identify the vulnerabilities of marine species by gathering multiple thematic maps and organizing them into a single unified map database for effective coastal management

Habitat mapping and the Great East Japan earthquake

To achieve the objectives of the previously mentioned 2030 United Nations Sustainable Development Goalss agenda 14 (i.e. sustainability of life below water through conservation and sustainable use of ocean, sea and marine resources) and pursue actions under the Decade of Ocean Science for Sustainable Development by 2030, habitat mapping has the potential to become a showcase of good practice at the smallest administrative (i.e. fishing-village) level. Habitat maps include information ranging from the life-history traits of species to environmental properties. Physical disturbance to the habitats of species can be used to predict species traits and to capture environmental gradients, especially those of benthic communities, which are spatially and temporally stable compared with those of the pelagic realm (Kostylev and Hannah 2007).

Visualization of biogeochemical and physical sediment properties, and benthic and pelagic biotic components, using a habitat database helps to establish a common language between a wide array of ocean beneficiaries and/or stakeholders. It can also help to predict spatial and temporal patterns and traits of species in a fixed community. The database serves as a promising tool for visualizing past-to-present data, planning/managing human activities and estimating future fishery productivity and the extent to which the environment can hold and protect ecosystems from degradation owing to climate change or from human activities. Once this database is open to the public, or to any fishing industry-related stakeholding organization, its accessibility allows for third-party monitoring of coastal habitats. If the database is made available to local citizens benefitting from marine ecosystem services because they live along a coastline, this brings local citizens and marine researchers to the same discussion table, enabling them to jointly participate in designing the sustainable use of marine resources and ecosystem protection plans.

On 11 March 2011, mega earthquakes and tsunamis hit the coastal regions of NE Japan, known as the Tohoku region, and caused serious damage to areas facing the Pacific Ocean. The piling of rubble, loss of seaweed beds and tidal flats and the accumulation of sand and mud on reefs resulted in drastic changes to marine ecosystems, coastal aquaculture farms and fishing grounds.

The Tohoku Ecosystem-Associated Marine Sciences (TEAMS) project, fully supported by Japan's Ministry of Education, Culture, Sports, Science and Technology, was launched in 2011 to monitor, research and develop the Pacific coastal areas of the Tohoku region affected by the earthquake and tsunami and support the restoration and reconstruction of the local fishing industry through science and technology (Kijima *et al.* 2018). The project lasted 10 years. The first 5 years were dedicated to monitoring the recovery process before transferring knowledge to fishermen and citizens for the restoration of fishery and social systems during the second 5 years. Multiple baseline studies were conducted by TEAMS researchers, as shown in Figure 1, and scientific data, such as precise topographic mapping of coastal regions and the distribution of benthic and planktic organisms, were collected (Kasaya *et al.* 2018).

The TEAMS project has contributed to the restoration and reconstruction of the local fishing industry by developing new science technologies and applying them to the field. The scope of TEAMS's contribution to society extends further by supporting the livelihoods of local people living along the coastline. The need to secure livelihoods is the fundamental basis of all actions relating to coastal ecosystems. The question is, how can scientific knowledge contribute to this objective?

From the fieldwork interviews conducted by Oki and Kitazato (2019), it is clear that coastal livelihoods and their associated human activities depend on whether local people can make the decisions necessary to protect and sustain their day-to-day activities, such as aquaculture farming, or whether they have access to the nearby shore and benefit from ecosystem services. Creating a marine GIS database is not only about visualizing diverse kinds of data; it also serves to illustrate the potential options that citizens can choose from in order to protect and sustain their common marine resources. Being able to visualize past, present and possible future scenarios for the marine commons means that the local beneficiaries are better informed about the state of their ocean. This enables local people to make better decisions when confronted with big life questions, such as how they envision their future lives in coastal settings, what kinds of lifestyles to adopt and how to sustain their livelihoods that are dependent on the coastal marine environment.

One of the main achievements of the TEAMS project is the spatio-temporal dynamics of Onagawa Bay's marine ecosystem, compiled as a comprehensive habitat map database (Fujii *et al.* 2019, 2021), as

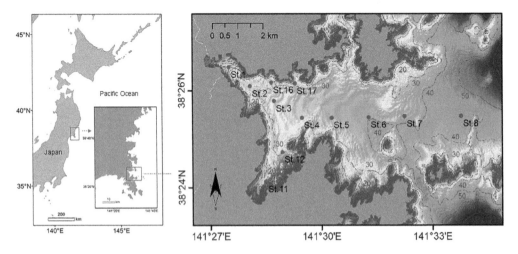

Fig. 1. Map of Onagawa Bay and the baseline study at station (St.) points. The colour contours depict the ocean depth.

summarized in Figs. 1 and 2. This database covers only the coastal areas of Onagawa Bay, which is an extremely local sample area and a microscale that may appear contradictory to the definition and scope of a habitat map. However, scientists of the TEAMS project consider this map to be a good example of a habitat map, as it contains attributes and functions (e.g. a topographical map as its baseline, while including layers of thematic maps) that are typical of habitat maps.

When creating best practices to achieve sustainable development of marine resources, the better defined the best practices are, the more they are focused on the entities utilizing the resources. In this way, local citizens living along the coastline have more chance of benefitting from marine ecosystem services. Habitat mapping at the microscale, therefore, is an effective method for approaching local citizens who are direct beneficiaries of the marine commons.

Ecosystem-based disaster risk reduction

The concept of Eco-DRR developed after Sumatra's large earthquake and tsunami took place in 2004. At that time, mangrove swamps in Sumatra played a role as green walls against tsunamis. Coral reefs and wide tidal flats fulfil the corresponding role of blue walls, reducing the impact of tsunamis. Mangroves, and sandy or rocky shorelines, function as natural barrier from storms, waves and surges. These 'blue and green walls', together with artificial infrastructures, protect coastal livelihoods and pristine coastal ecosystems.

Coastal ecosystems, such as seagrasses, corals and mangroves, function as nature-based protective solutions against extreme weather events (Guannel et al. 2016), as well as being critical habitats. They help to protect coastal shorelines, and their management is an essential tool for safeguarding habitats (Spalding et al. 2014). In this current paper, we introduce examples of nature-based solutions to disaster risk management from countries that experienced the 2004 Indian Ocean tsunami and the 2011 Great East Japan Tsunami.

The essence of Eco-DRR is to bolster resilience and enhance natural disaster preparedness. Baseline maps of coastal ecosystems can provide fundamental information about the state of the natural environment before the disaster, greatly increasing the effectiveness of Eco-DRR in disaster risk preparation. A baseline map of a bay allows us to refer to the natural state of the ecosystem at the microlevel and provides a strong foundation on which we can base our responses in post-disaster times.

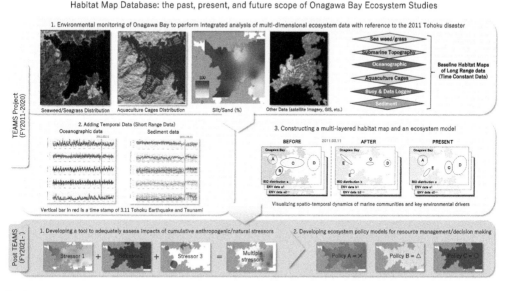

Fig. 2. Habitat map database of Onagawa Bay. During the first 5 years of the TEAMS project, researchers collected both long- and short-range data for generating thematic maps, such as seaweed/seagrass distribution, aquaculture cages distribution, silt/sand distribution and other data including satellite imagery. Based on time-constant, long-range data, each thematic map can be broken down further into short-range data, constructing a multi-layered habitat map of Onagawa Bay. The scope of the post-TEAMS phase is rooted in developing an assessment tool for measuring natural stressors and reflecting findings when designing public policy models for resource management and policy decision-making.

Generating a habitat map database, therefore, is a measure to prepare for an upcoming disaster and build back better the damaged ecosystems and human activities. The concept of build back better (BBB), or the idea of building community resilience is described in an UN project report written in response to the 2004 Indian Ocean Tsunami (Manu *et al.* 2010).

Good practices of eco-DRR

The case of Thailand: multi-stakeholder coral reef reconstruction

After the 2004 Indian Ocean tsunami, the impact of the tsunami on coral reefs was investigated by several universities and institutions in Thailand. The overall results showed that the damage to corals was less than expected, although damage was found as deep as 27 m. Most coral damage occurred between 10 and 20 m, where the seafloor was steep (Pataporn *et al.* 2017).

Intensive surveys showed that the massive coral form was the most susceptible to the tsunami at almost every study site. Several measures were established to reduce the disaster and restore damaged areas, particularly coral reef areas (Pataporn *et al.* 2017). These included the clean-up of debris, coral rehabilitation and coral health monitoring. In addition, the management of multi-use areas, such as zoning and carrying capacity analysis, was implemented, and collaborations between domestic and international organizations for information and technology exchange were established (Pataporn *et al.* 2017).

The case of Indonesia: planting of mangroves in aquaculture ponds as an DRR measure

The Green Coast project in Indonesia was developed as a response to the 2004 tsunami disaster. Wetlands International, with other international organizations, implemented a programme for coastal ecosystems rehabilitation (e.g. mangroves, beach forests, lagoons and coral reefs) combined with alternative livelihood development in tsunami-affected areas (Wibisono and Sualia 2008). Among the activities facilitated by the Wetlands International project, a silvofishery system (i.e. planting mangroves in aquaculture ponds) was applied in the tsunami-affected pond areas. Field observations confirmed that silvofishery was capable of improving environmental quality and attracting wild shrimps to enter the ponds (Wibisono 2017). By using traditional devices installed at water gates, farmers received additional income from their daily catch. From the landscape perspective, silvofishery combined with sandy beach rehabilitation is designed to establish a green belt as part of Eco-DRR.

The case of Japan: setting back the seawall in Okirai Bay, Iwate prefecture

This case study, which involved setting back a seawall to protect the coastal ecosystem, was realized when scientists, local fishermen and citizens and municipal bureaucrats acted together in the decision-making process. After the 2011 tsunami, physical disturbances in Okirai Bay's coastal ecosystem created a sandy habitat along the new shoreline, which became home to a wide range of species. A group of researchers from Kitasato University monitored these changes and provided science-based evidence of the coastal ecosystem biodiversity found in the newly created habitat (Asahida 2020). Noting that the town was designing and implementing a seawall in the location of the newly created habitat, local fishermen, with other stakeholding citizens and researchers providing scientific evidence about the coastal ecosystem, lobbied municipal policy makers to revise the seawall construction design, setting the wall back 200 m inland (see the aerial image in Fig. 3) to protect the new, sandy habitat.

The success of repositioning the seawall resulted from citizen participation in the decision-making process. Local citizens acquired science-based ecosystem information from researchers and local fishermen, including findings on new species in the newly created habitat. This served as a baseline study of the new ecosystem created in response to the natural disaster. This is a good example of how citizen participation led to a chain of actions, all of which contributed to the 'building back better' of what had been damaged and lost. First, the baseline study conducted by researchers after the disaster provided scientific evidence that was shared. Second, the knowledge- and science-based information regarding the new habitat was shared among the local public, as well as with policy makers at the municipality level. Third, the local people's decision and their initiative in submitting a petition to the local government to reposition the seawall 200 metres inland was endorsed by the local elite (i.e. community leader of the fishing village).

Future direction

Paradigm shift from 'science for science' to 'science for society'. The three components that led to the successful example of Okirai Bay could not have been realized without engagement from multi-stakeholder parties in deciding what to do with the

Recovery from Tsunami by Great East Japan Earthquake

Different stakeholder engagement on science for DRR

Scientists
Monitoring/Studying
<Scientific Knowledge>

Fishery industry
Implementing fishery
<Local Knowledge>

<Knowledge sharing/Discussion>

Municipalities
Rehabilitating Infrastructure

With the discussion of stakeholders:

- The techniques to recover fishery (e.g. oyster farm) has been identified and implemented (in Miyagi Prefecture, Togura in Shizugawa Bay)

- The location of the sea wall was identified to best protect the marine ecosystem and fishery industry (in Iwate Prefecture, Okirai Bay)

Example of the Sea-wall in Okirai Bay

- Some researchers had been monitoring coastal ecosystem of Okirai Bay before the Great East Japan Earthquake occurred

- They found that the sea-wall that the municipality was planning to 'BBB' after the earthquake, might destroy the coastal ecosystem on the course of recovery

- They shared this information with the local fishery industry and proposed to set the sea-wall back inside from coast line

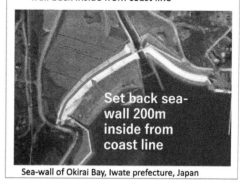

Set back sea-wall 200m inside from coast line

Sea-wall of Okirai Bay, Iwate prefecture, Japan

Fig. 3. Eco-DRR and multi-stakeholder engagement.

newly created habitat after the disaster. Setting back the seawall in Okirai Bay is a showcase of how scientific research and science-based evidence were utilized to help achieve a better coastal livelihood through stakeholder participation in the public policy decision-making process – in this case, protecting the shore and the coastal ecotone.

Restoring the coastal area better than it was before is well aligned with a disaster risk reduction scheme because natural disturbances are cyclical events, and history tells us that human lives and livelihoods are constantly at risk. If humans possess knowledge of the science behind the dynamics of coastal ecosystems, they can better know how to respond in the wake of disasters. Once we know how to respond and build the coastal livelihood better after a disaster, we can then backcast and prepare what is needed between the cycles of natural disturbances.

Science and science-based knowledge should find their roles and functions in making society and livelihoods better so that we can all benefit from human well-being. Not only does science provide methods for building better livelihoods through technological innovation and growth, it also provides the basis – the platform – when thinking/re-thinking

about life security. The example of Okirai Bay showed us the humanitarian dimension of protecting the newly created microhabitat and its coastal biodiversity, balanced with the engineering solution of setting the seawall inland from its originally planned construction site. Citizens' participation in science and multi-stakeholder participation and cooperation in policy decision-making processes are key components in realizing sustainable society and human well-being. As seen from the Okirai Bay case, mapping habitats and acquiring scientific information that explains how regional coastal ecosystems are sustained pre- and post-disaster can be used as a strategic map (or strategic database) by the local public and policy decision-makers to chart the wise usage of marine ecosystems and their services. This strategic database could benefit local stakeholders – the fishermen and the citizens who use the coastal ecosystem for sustenance – at the local level, as it visualizes and preserves a record of the micro-seascape in the past and during the present.

Scientists have worked to address scientific questions prompted by their curiosity and to explain outcomes to their affiliated scientific communities. Until the twentieth century, scientists lived in the world of 'science for science' and thus did not need to think about the wider society and human livelihoods, even though they were living as members of that wider society. In 1999, the International Council of Science (ICSU), with the United Nations Educational, Scientific and Cultural Organization, held the World Conference on Science for the Twenty-first Century: A New Commitment, in Budapest, Hungary. That conference discussed various subjects in connection with how scientists play a role in and for global communities. After the conference, ICSU released the Declaration on Science and the use of Scientific Knowledge (ICSU 1999).

In the Declaration, ICSU stresses that scientists should consider where the natural sciences stand today and where they are heading, what their social impacts have been and what society expects from them. We should assert the following four points: (1) science for knowledge, knowledge for progress; (2) science for peace; (3) science for development; and (4) science in society and science for society.

The first and third items are ongoing ideas. However, the second and fourth are rather difficult to carry out. 'Science for peace' sounds respectable but is difficult to act upon and achieve, as we face the dual-use dilemma. Some sciences pursue pure science, but unfortunately, they can be used as weapons of war. 'Science in society and science for society' is even more difficult to make happen in practice. Applied science is easy to adapt for social systems. However, for many disciplines of the natural sciences, such as mathematics, physics, chemistry, biology and geological sciences, it is difficult to apply their results to society. How should we transfer our scientific results to citizens and local governments and trickle them down to policymaking? This is a challenging topic for all scientists.

From interdisciplinarity to transdisciplinarity. Let us think about scientific activities step by step, as shown in Figure 4. This figure is a model showing the steps by which scientific knowledge and/or technologies are transferred from scientific communities to the public. The model flows as follows. First, scientists collect observations, measurements and data. Second, scientists synthesize and generate models based on these data. Third, implications of the second step are processed into practical information of societal value. Fourth, the information and its implications are delivered to society. Fifth, societal stakeholders take action based on scientific evidence and build better societies. These steps illustrate the transition from interdisciplinary to transdisciplinary science and technology to citizen sciences.

Scientific results should be presented to both scientific and public communities. Simplified research results are easier to transfer from scientists to citizens using diagrams, cartoons, PowerPoint presentations and many other eye-catching media. These visualization techniques are common procedures for presenting scientific results to the scientific community and are easily disseminated among circles of scientific communities, as they involve commonly used words and expressions. However, the languages used by scientists and citizens differ. We, as scientists, should explain the results directly to local stakeholders using simple and easy-to-understand language.

The word 'interdisciplinarity' refers to the integration of different scientific fields. The TEAMS project has worked with liberal arts scholars to properly contextualize societal problems embedded in natural disturbances, such as earthquakes and tsunamis, and through multi-stakeholder dialogues and workshops that include community stakeholders, reframing the problem as a common global agenda.

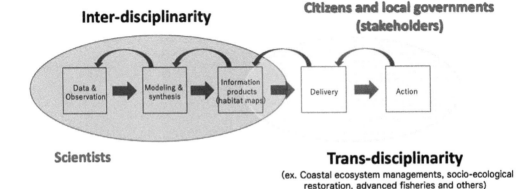

Fig. 4. Interdisciplinary to transdisciplinary: the social application of scientific results. Pathways moving from 'science for science' to 'science for society'.

Applying the results of the TEAMS project has gradually extended its scope to public/local citizens, who are the survivors of the earthquake and tsunami. Besides the local public, fishery and fishery-related industry workers and local policy makers are joining the circle of this transdisciplinary effort.

Sustainable marine commons and the environment. Transdisciplinarity is an approach in which scientific results are applied to society, encompassing both the public and non-academic populations. The translation of science, the scientific way of thinking, the understanding of scientific concepts and the ability to transcribe scientific language into everyday language are required when engaging with the public. Maps, diagrams and photos/videos are the most intuitive methods of conveying new information and ideas to those who neither have the background nor the prior training, especially regarding scientific topics.

Habitat maps can fill this gap between scientists and the wider society. They can be used to collect, organize and present data to the public; the public can then build on these habitat maps by adding information from the past and present. In order not to confuse scientific data with data collected by the public, different layers corresponding to different data source can be added to the base map. This map can be used by local stakeholders who wish to protect and sustain marine resources. The EMODnet's *European Atlas of the Sea* is a good example of a communication tool that has high potential for disseminating large amounts of thematically categorized data and making it readily accessible for academic and non-academic audiences.

The Great East Japan Earthquake of March 2011 caused severe damage to fishing communities along the Tohoku coastal areas, both land and sea. Along the coast of Japan, we enjoy the benefits of ecosystem services in our daily lives, whether we live close to the coastal area or not. The ocean brings both benefits and disasters to our livelihoods. However, we are often unaware of the habitats and ecosystems of the coastal land/sea transition (ecotone), the ocean and the seafloor.

Needless to say, coastal fisheries depend heavily on marine and freshwater resources. At the same time, coastal marine ecosystems and environments provide important habitats for marine species. Although researchers continuously monitored the environment and marine ecosystem after the Great Earthquake, data were not visualized and communicated to local fishermen and residents, who are the main stakeholders of marine commons. The local government and other public institutions have failed to provide an opportunity for discourse and dialogue about the changes in the sea caused by the earthquake and tsunami. Discussing and thinking about the two contrasting sides of the same coin of Earth's blessings – the benefits we receive to sustain human well-being and the dynamics which at times inflict disasters on human lives – should be considered actions that enhance understanding of the ocean and its ecosystems, hence raising marine literacy.

The Global Ocean Science Report 2020 reminds us that the ocean has become a new frontier in the global challenge against climate change, as it absorbs 93% of the heat from human-generated carbon dioxide emissions (Global Ocean Science Report 2020–Charting Capacity for Ocean Sustainability 2020, p. 189). Enhancing stakeholder understanding of the ocean would help nurture marine stewardship and provide an additional impetus for sustainable coastal management. Acknowledging the indigenous/local knowledge acquired by local fishermen/citizens should be treated as equally significant data/information compared with scientific data when dealing with climate change on a global scale. In doing so, partnerships across disciplines and all stakeholder groups should be established and supported in order to sustain our common ocean.

Concluding remarks

A habitat database has many functions and covers a wide range of audiences. For policy makers, it can serve as a management toolkit, allowing them to base their assumptions (and decision-making) on scientific facts regarding coastal ecosystems and their richness in biodiversity. Similarly, local citizens can raise their marine literacy by utilizing past/present maps to forecast the future and then backcast based on past/present knowledge and on how they wish to foster, enhance and manage their surrounding marine ecosystem. Thus a habitat database can be used by all those with a stake in coastal marine commons, as well as for designing the future at the community level. In the same way that we humans have medical records of our past/present illnesses, prescriptions and descriptions of lifestyle habits, documented by our tending physicians, the coastal habitat database serves as a record of ecosystems health. By understanding the past medical records of our coastal ecosystem, we are more informed of current and future risks and can change our attitude from passive to proactively taking preventive measures before the next disaster strikes. After the events of 11 March 2011, hard infrastructures were built back better and lifelines were reconstructed to protect against future earthquakes and disasters. However, we must question whether the environments surrounding local citizens have been reconstructed so that they can make better choices and improve their well-being so that it is better than before.

Eco-DRR is not limited to coastal environments. It is also applicable to wetlands and forests that function as natural infrastructures that reduce physical exposure to natural hazards, build communities' resilience and sustain livelihoods, and as ecosystems that equip us with essential natural resources providing food, water and other necessities for human sustenance. To extend local citizens' capabilities in Eco-DRR and to enhance their well-being, scientists must exert due diligence in terms of information disclosure and public outreach by advocating science-based facts. This is so that when local citizens face making public choices in their communities, they will have a full set of options to choose from, supported by scientific evidence. Having not just one but a set of choices could expand the power of resilience, bolster the well-being of local people, and reduce disaster risks, thereby making communities better.

The examples given and the lessons learned from them speak to a wider audience of people in the world today that rely on natural ecosystems in inland areas away from the sea. For example, the logic behind the case of Okirai Bay and the construction of the seawall can be extended to the issue of construction or strengthening of river embankment systems in towns and cities in response to the flooding of rivers during heavy rainfall and storm disasters. The recent landslip disaster in the steep mountainside city of Atami, Japan, is thought to have been triggered by an artificial multi-tiered embankment constructed in the mountains above the city (International Science Council 2021). Similar landslip/landslide disasters in developed (and developing) countries could result from population growth and urbanization that cause changes to land uses. Earthquakes, heavy rainfalls and storm-related disasters in inland areas will have similar effects on human lives and welfare so long as humans remain as the main beneficiaries of ecosystem services.

Natural hazards like landslides owing to heavy rainfall do not create disaster in isolation. Instead, some kinds of human activities, such as unplanned construction works in upstream areas, alter effects happening downstream, where peoples' lives are lost by sudden floods and debris flows. Ironically, it is therefore the development activities of our own people preceding the onset of natural hazards that create negative consequences for us. In the longer run, these activities exacerbate the impact of disasters on human lives and well-being. To mitigate these negative consequences, the local beneficiaries of coastal ecosystems can utilize the database in order to monitor the effects of human activities on natural phenomena.

Mapping small-scale coastal habitats, and compiling the information with spatio/temporal thematic maps into one comprehensive database that is easily accessible free of charge through a portal site is the exit strategy for the future of EBM. A habitat map incorporating abundant data on coastal ecosystems thereby functions as a strategic tool for making better and scientifically informed decisions. The choices that local citizens make strongly influence the future of disaster-stricken communities. Knowing what coastal ecosystems were like in pre-disaster times would help to facilitate science-based, informed outcomes.

Scientists should work towards collecting, categorizing, synthesizing, visualizing and disseminating data and its implications for communities. The role of science and scientists in developing Eco-DRR plans and building back communities better than before a disaster is to support citizens' decision-making by assembling science-based information, using tools such as a habitat map database and informing the local community about the risks in advance so that people can build on their ownership of the choices they are making and their consequences. All of these actions will contribute to bettering the livelihoods and well-being of local human populations.

Acknowledgements This vision paper is a fundamental contribution by both HK and YO as members of the Tohoku Ecosystem-Associated Marine Sciences Project (TEAMS). TEAMS is a research programme created to establish a centre for research on and development of marine ecosystems; it was subsidised by the Ministry of Education, Culture, Sports, Science and Technology (MEXT) from 2012 to 2021. Both YO and HK are tasked with transferring research results to multiple stakeholders, such as local fishermen communities, citizens and governments, regarding coastal ecosystem disturbances – monitoring recovery processes and restoration from the big earthquake and tsunami on 11 March 2011. Tentative reports were given to scientists and citizens at several conferences, such as European Geosciences Union (EGU), Geological Society of America (GSA), Resources for Future Generations (RFG) and other international meetings. We are very grateful to have had the chance to occasionally give presentations and workshops to public citizens. We thank TEAMS's executive scientists and MEXT officials for permitting us the opportunity to conduct research and outreach activities. Last but not least, we would like to extend our acknowledgements to both reviewers who made valuable comments in improving our manuscript.

Author contributions YO: conceptualization (lead), writing – original draft (lead), writing – review & editing (lead); **HK**: conceptualization (equal), writing – original draft (equal), writing – review & editing (supporting); **TF**: visualization (equal); **SY**: conceptualization (supporting).

Funding This work was supported by MEXT Tohoku Ecosystem-Associated Marine Sciences Project Grant Number JPMXD1111105258.

Data availability Data sharing is not applicable to this article as no datasets were generated or analysed during the current study.

References

Adger, W.N., Hughes, T.P., Folke, C., Carpenter, S.R. and Rockström, J. 2005. Social-ecological resilience to coastal disasters. *Science (New York)*, **309**, 1036–1039, https://doi.org/10.1126/science.11121 22, https://doi.org/10.1126/science.1112122

Asahida, S. 2020. Report of Tohoku Ecosystem-associated Marine Sciences FY2011–2020: Larvae Appearance in the Seagrass Field in Okirai Bay [in Japanese].

Fujii, T., Kaneko, K. *et al.* 2019. Spatio-temporal dynamics of benthic macrofaunal communities in relation to the recovery of coastal aquaculture operations following the 2011 Great East Japan Earthquake and tsunami. *Frontiers in Marine Science*, **5**, 1–13, https://doi.org/10.3389/fmars.2018.00535, https://doi.org/10.3389/fmars.2018.00535

Fujii, T., Kaneko, K., Nakamura, Y., Murata, H., Kuraishi, M. and Kijima, A. 2021. Assessment of coastal anthropo-ecological system dynamics in response to a tsunami catastrophe of an unprecedented magnitude encountered in Japan. *Science of the Total Environment*, **783**, https://doi.org/10.1016/j.scitotenv.2021.146998

Gillanders, B.M. 2006. Seagrasses, fish, and fisheries. *In*: Larkum, A.W.D., Orth, R.J. and Duarte, C.M. (eds) *Seagrasses: Biology, Ecology and Conservation*. 503–536, https://doi.org/10.1007/978-1-4020-2983-7_21

Global Ocean Science Report 2020–Charting Capacity for Ocean Sustainability. 2020. IOC-UNESCO, https://unesdoc.unesco.org/ark:/48223/pf0000375147

Guannel, G., Arkema, K., Ruggiero, P. and Verutes, G. 2016. The power of three: coral reefs, seagrasses and mangroves protect coastal regions and increase their resilience. *PLoS ONE*, **11**, https://doi.org/10.1371/journal.pone.0158094, https://doi.org/10.1371/journal.pone.0158094

International Science Council GeoUnions Standing Committee on Disaster Risk Reduction. 2021. How can we communicate anthropogenic factors that contribute to natural disasters? – Transdisciplinary approaches enhance cooperation among different stakeholders. http://www.iscgdrr.com/plus/view.php?aid=53 (accessed 25 Nov. 2021)

ICSU. 1999. Declaration on Science and the Use of Scientific Knowledge, http://www.unesco.org/science/wcs/eng/declaration_e.htm

Ikeda, Y. 2016. Human being and natural hazards. *Science (Kagaku)*, **86**, 1061–1064.

Kasaya, T., Yamaguchi, M. *et al.* 2018. No current condition of artificial reefs deduced by acoustic data and ROV dives off Otsuchi bay. *Nippon Suisan Gakkaishi*, **84**, 893–896, https://doi.org/10.2331/suisan.WA2566-4

Katsanevakis, S., Stelzenmüller, V. *et al.* 2011. Ecosystem-based marine spatial management: review of concepts, policies, tools, and critical issues. *Ocean and Coastal Management*, **54**, 807–820, https://doi.org/10.1016/j.ocecoaman.2011.09.002

Kijima, A., Kogure, K., Kitazato, H. and Fujikura, K. 2018. Reconstruction and restoration after the Great East Japan Earthquake and Tsunami; Tohoku Ecosystem-associated Marine Sciences (TEAMS) project activities. *In*: Santiago-Fanño, V., Sato, S., Maki, N. and Iuchi, K. (eds) *Advances in Natural and Technological Hazards Research*. 279–290.

Kitazato, H., Oki, Y. and Yasukawa, S. 2020. Habitat map plays an active role for coastal eco-DRR by multi-stakeholders, EGU General Assembly 2020, 4–8 May 2020, EGU2020-3815, https://doi.org/10.5194/egusphere-egu2020-3815

Kostylev, V.E. and Hannah, C.G. 2007. Process-driven characterization and mapping of seabed habitats. *In*: Todd, B.J. and Greene, H.G. (eds) *Mapping the Seafloor for Habitat Characterization*. Geological Association of Canada.

Manu, G., German, V., Suman, N., Abhilash, P., Kuberan, R., Kristi, H. and Rajardo, S. 2010. Building back better for next time – experiences and lessons learnt from the project Building Resilience to Tsunamis in the Indian Ocean, https://www.preventionweb.net/publication/building-back-better-next-time

McLeod, K.L., Lubchenco, J., Palumbi, S.R. and Rosenberg, A.A. 2005. Scientific Consensus Statement on Marine Ecosystem-Based Management prepared by scientists and policy experts to provide information about coasts and oceans to U.S. policy-makers published by the Communication Partnership for Science and the Sea, https://marineplanning.org/wp-content/uploads/2015/07/Consensusstatement.pdf

Milliman, J.D. and Meade, R.H. 1983. World-wide delivery of river sediment to the oceans. *The Journal of Geology*, **91**, 1–21, https://doi.org/10.1086/628741

Mora, C., Wei, C.L. *et al.* 2013. Biotic and human vulnerability to projected changes in ocean biogeochemistry over the 21st century. *PLoS Biology*, **11**, https://doi.org/10.1371/journal.pbio.1001682

Oki, Y. and Kitazato, H. 2019. Towards sustainable fishery: building back better fishing communities after the Great East Japan Earthquake 2011. *In*: Martins, A.N., Hobeica, L., Hobeica, A., Santos, P.P., Eltinay, N. and Mendes, J.M. (eds) *8th International Conference on Building Resilience Risk and Resilience in Practice: Vulnerabilities, Displaced People, Local Communities and Heritages*. 8th ICBR Lisbon Book of Papers, 272–281.

Pataporn, K., Chavanich, S.A., Viyakarn, V., Sojisuporn, P., Siripong, A. and Menasveta, P. 2017. Sumatra earthquake and tsunamis in Thailand: scientist activities toward ecological disaster risk reduction, Conference Presentation at the World Bosai Forum 2019, Sendai, Japan.

Spalding, M.D., Ruffo, S., Lacambra, C., Meliane, I., Hale, L.Z., Shepard, C.C. and Beck, M.W. 2014. The role of ecosystems in coastal protection: adapting to climate change and coastal hazards. *Ocean and Coastal Management*, **90**, 50–57, https://doi.org/10.1016/j.ocecoaman.2013.09.007

UNESCO. United Nations Decade of Ocean Science for Sustainable Development. 2019. *UNESCO*. https://www.oceandecade.org/about

United Nations. The Ocean Conference Fact Sheet: People and Oceans. 2017. https://www.un.org/sustainablede velopment/wp-content/uploads/2017/05/Ocean-fact-sheet-package.pdf

United Nations. 2020. Ecosystem-based disaster risk reduc-tion: implementing nature-based solutions for resilience, https://www.undrr.org/publication/ecosystem-based-disaster-risk-reduction-implementing-nature-based-sol utions-0

Vallius, H.T.V., Kotilainen, A.T., Asch, K.C., Fiorentino, A., Judge, M., Stewart, H.A. and Pjetursson, B. 2020. Discovering Europe's seabed geology: the EMODnet concept of uniform collection and harmonization of marine data. *Geological Society, London, Special Publi-cations*, **505**, https://doi.org/10.1144/sp505-2019-208

Wibisono, I.T.C. 2017. Experience and lessons from Green Coast Project implementation in Indonesia. *World Bosai Forum 2017, Sendai, Japan.*

Wibisono, I.T.C. and Sualia, I. 2008. Final report: an assessment of lessons learnt from the 'Green Coast Pro-ject' in Nanggroe Aceh Darussalam (NAD) Province and Nias Island, Indonesia, Period 2005–2008, https://www.wetlands.org/publications/an-assessment-of-les sons-learnt-from-the-green-coast-project/

Index

Page numbers in *italics* refer to Figures. Page numbers in **bold** refer to Tables.